Beguiled by Birds

A Note on the Author

Ian Wallace is one of the best known and most loved of all British birdwatchers. A writer, artist, and above all purposeful observer of birds, his birdwatching career spans a half-century of radical changes in the hobby, and he has been a key figure at many of birdwatching's defining moments and events. He is a former editor of *British Birds* magazine, a former Chairman of the British Birds Rarities Committee (1972–76), and a founder contributor to the mammoth and groundbreaking *Birds of the Western Palearctic*, as well as author and illustrator of several other books, and many papers and articles. Ian's tremendous experience, both in Britain and on numerous expeditions throughout the world, coupled with his dazzling descriptive ability (in both words and pictures), have earned him the affection of several generations of British birdwatchers.

Now confined to a retirement patch in East Staffordshire, Ian currently spends most of his birdwatching time on conservation surveys. Even so, his need for ornithological adventure still takes him to places as far apart as Donegal and Oman, and his affection for the elders of field ornithology is regularly expressed in articles and the odd rebuke to those who pay them no heed.

Beguiled by Birds

*Ian Wallace on
British Birdwatching*

Ian Wallace

CHRISTOPHER HELM
LONDON

Because of
Dad, Ken, Dougal, James, Guy, Phil and Stanley
who inspired and helped me
and for
Bill, Joe and Peter, Karin, Bob and Pete, Andrew and Irene, Anthony and Dave
who kept good company with me and the birds

HELM
Bloomsbury Publishing PLC
50 Bedford Square, London, WC1B 3DP, UK
29 Earlsfort Terrace, Dublin 2, Ireland

BLOOMSBURY, HELM and the Helm logo are trademarks of Bloomsbury Publishing Plc

First published in the United Kingdom 2004

A catalogue record for this book is available from the British Library

ISBN: 978-1-3994-2101-0

2 4 6 8 10 9 7 5 3 1

Produced by The Hanway Press Ltd
Designed by Fluke Art, Cornwall

Printed and bound by CPI Group (UK) Ltd, Croydon, CR0 4YY

Bloomsbury Publishing Plc makes every effort to ensure that the papers used in the manufacture of
our books are natural, recyclable products made from wood grown in well-managed forests. Our
manufacturing processes conform to the environmental regulations of the country of origin.

To find out more about our authors and books visit www.bloomsbury.com and sign up for our
newsletters

Contents

Prologue

The author's viewpoint and bias – the book's conception and span

Please read my foreword. It is longer than usual but I am anxious to declare early my ornithological bias and to indicate fully this book's germ and purpose.

For me, half Lowland, half Highland Scot sporting some heady drops of illegitimate Stuart blood, Parry's tune to Blake's Jerusalem is altogether too English-triumphant but the verses do supply three apt titles for my main stations with birds and the human activities that they provoke. My 'bow of burning gold', the assembly of early knowledge and critical observation, was given me by one gamekeeper and three schoolmasters. From the first, I acquired an acute sense of birds' seasonal statuses and habitats, in the Dalhousie estate south of Edinburgh, and nature's best instrument for the fine line painting, the pin feather of a Woodcock. Had I been older than 12, I would have envied his evident contentment with his place between man and animals, especially that tribe known as game. My other early mentors gave me entry to the wider theatre of informed and purposeful birdwatching. On a particularly magic day in May 1947, I opened the cupboard of the Loretto School Ornithological Society and found there not just the five volumes of the old Witherby *Handbook* but copies of a green-jacketed, grouse-badged journal named *British Birds*. Compared even to Coward's friendly three volumes *The Birds of the British Isles and their Eggs*, a veritable cornucopia of avian lore opened. By chance of being boarded away from the London Blitz, I was launched into a broad enjoyment of birds, and most of the hormonal torment of my teen years was sublimated by a daily preoccupation with ornithological acquisition.

To go with my bow, I soon had an intellectual quiver full of 'arrows of desire'. Targets and questions multiplied apace. Would I ever meet the great observers named in the *Handbook* and the journal? Where were the rarity hot spots such as Hastings and Cley? How could we have steamed so often to Shetland and never stopped off at Fair Isle? Could I add new information to the national store? The wish to participate in and contribute to a hobby cum amateur science was huge. By my 17th year in 1950, Dad had conceded that his uphill charges would be slowed to allow me time for the birds along the paths; I had found (and had accepted by Bernard Tucker) my first rarity, a Lesser Yellowlegs, and had even seen such trophy enshrined in print in BB. As a damper to any tendency towards over-enthusiasm with rarities, a long letter from Duncan Wood, then the journal's assistant editor, worked well. Calmer pursuits like a search for Sand Martin colonies and the observation of breeding Dunlin followed. It seemed that anywhere I aimed, the target would be hit. So passed two decades of birdwatching adventure and increasing involvement with the establishment of British field ornithology, presented most cogently by first the London Natural History Society and the Cambridge Bird Club, both in their heydays, and then the St. Agnes Bird Observatory, which after half a century of a seeming Fair Isle monopoly of rare birds showed again the early promise of the Scilly archipelago.

Scotland's second-ever Yellowshank or Lesser Yellowlegs, Aberlady Bay, East Lothian, 13th May 1950. Half a century ago, three schoolboys sent a rarity claim to the one monthly journal and its Editor was the nation's sole judge of it.

In the late 1960s, when I was in Africa adding to my earlier brush with the Mau Mau the pleasure of selling cold lager to hot Nigerians, a new ecumenism began to show in the ranks of birdwatchers and with it new creeds. I had recognised that my hobby's freedoms would be reduced as early as the mid-1950s but I was so confident of my rates of hunting and gathering and skill in identification that I was blind to the concomitant risks – inevitable error and loss of face. In October 1971, I mistook a Bonelli's Warbler for a Booted and all at once my personal vehicle was a 'chariot of fire'. After struggling with the establishment, particularly the two national rarity review bodies, for another eight years, I found the recreation in my hobby threatened by a few peers who decided that my visual acuity might well be assisted by invention. There was no proof of such duplicity but in the swelling ranks of ornithological chatterers, some mud stuck and has indeed caked. It was time to step aside and accept that the zealots had won.

My escape was not, however, to the relative wilderness of my youth but to other large tasks that needed a worker. From Stanley Cramp came the honour of being the sole writer/artist for *The Birds of the Western Palearctic* and from Guy Mountfort and Phil Hollom issued the chance to bring Britain's favourite field guide up to date. In 1993, I revised the entire text of 'Peterson', as *A Field Guide to the Birds of Britain and Europe* is known. Rather tellingly, the only person who was upset by my update of simpler identification was Roger Peterson. Against his original aim I had made too many treatments over-complicated. Last but by no means least, I was asked by Chris Dawn to put some soul into *Birdwatching*, an Emap magazine and the first of its genre to attract and sustain mass circulation. Thus although I lost most of my shout on identification and rarity issues, my voice spoke anew in a monthly column wherein I could still be passionate about the fabulous beings that are birds. Happily, too, the last of eight career moves on three continents brought me a close working relationship with the Royal Society for the Protection of Birds. My enthusiasm for birds as beings and my support for their right to exist alongside us are thus undiminished. Both feelings should show in this commentary on British birdwatching but to ensure that my now-declared

"Its breakdown of class barriers in a still far from class-free society".

The fourth choice of James Ferguson-Lees, when asked for the prides of British birding (2000).

On 22nd August 1971, this Bonelli's Warbler posed no identification trauma. Only 43 days later, I mistook my next for a Booted Warbler. Never mind the fact that both have since been split into two species, the real lesson was that ornithological pride is not exempt from falls!

bias is as much corrected as possible, I have undertaken a wide canvass of other experience and opinion. This is detailed in several chapters.

Faced with a daunting prospect by Richard Porter, who proposed my authorship to Andy Richford, it took me nearly a year to come up with a cohesive plan of the book's content and to take soundings from other birdwatchers on its merit. Of all the advices received, the one that helped most to decide on the book's heart came from David Cromack, the caring editor of Birdwatching. "I hope", he said to Andy, "he'll write it in a way that will allow younger birdwatchers to reprise the early adventures of the hobby and to understand how they can also contribute to the science of ornithology." Instantly I knew that I must lay down for others the magic carpet of field ornithology that I had been taught so well by people like Ken Williamson, Dougal Andrew and Phil Hollom. If I could find (to use a business term) the critical path of purposeful birdwatching, its endless rewards to mind and soul would perhaps be easier to demonstrate.

The next big question was to decide when that path was first signposted. Initially, Richard Porter and Andy Richford had thought that the book should cover the last 50 or 75 years. There seemed to be a particularly important period of study development in the 1920s, when Max Nicholson led the small caucus based at Oxford that issued instructions for more precise field ornithology. To start there and then, however, would be to ignore the amazing stocktaking of Britain's nature in the Victorian era. For birds, I knew that this had even involved the nation's lighthouse keepers, then constituting legions of sharp eyes. Chris Mead gave me a further clue to their efforts, "Once at a BOU conference, Bill Bourne filled in for a missing speaker with an amazing discourse on the original lighthouse migration scheme. Try him." I did. Bill obliged by showing me that marshalled, co-operative studies were indeed over 130 years old and by answering a score of other questions by return. Then out of the blue came a note from David Ballance, a friend from the Cambridge Bird Club in its vintage period, telling me that he had just finished a county by county bibliography of all local ornithological publications. This proved that the tradition of county avifaunas flowered in the same period, co-incident with the peak of the then widespread rural interest in finding and shooting unusual birds for gainful sale to collectors. Suddenly the relevance of the astonishing displays of mounted specimens that had caught my attention, for example at Audley End in north Essex, was obvious. I had at last a cohesive sense of the beginnings of an accurate written ornithology and a vision of the first ranks of keen-eyed hunters and watchers.

So this book deals mainly with the last 125 years of the interest in birds expressed by British people and some of their contingent behaviour. The national limitation is more than implicit. There were other acute progenitors of birdwatching and ornithology in Europe and North America but, even though prodded hard by Lars Svensson, I have eschewed such width of canvas. Some foreign credits are given, for they are surely due, and I have included a snapshot of the current images abroad of British birders (as birdwatchers are now sadly increasingly renamed). Finally, if my book disappoints, do not delay. Ask Bill Bourne to do a full history, for surely he knows it.

The ascent of British ornithology in the first two millennia AD.

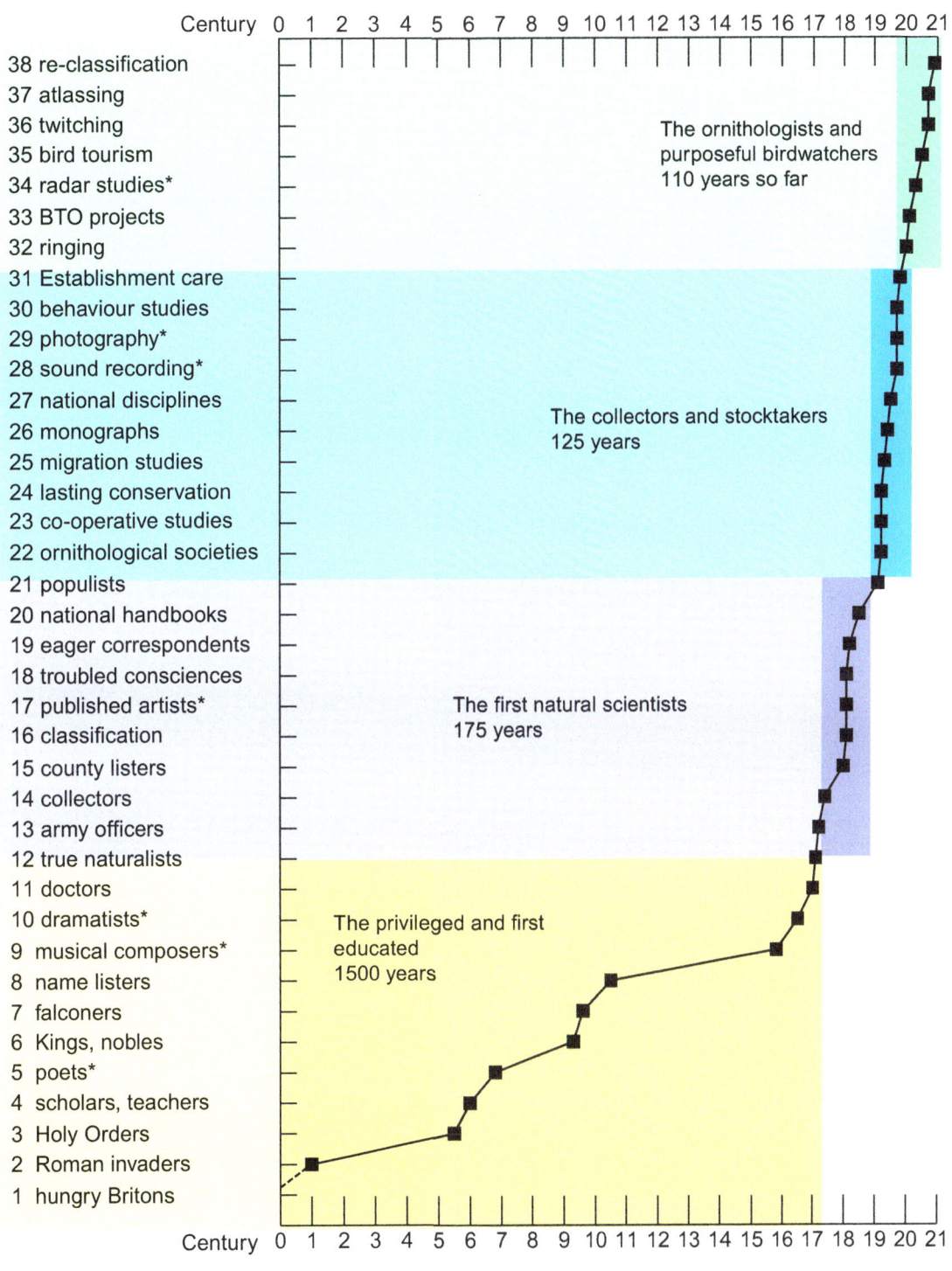

Century 0 1 2 3 4 5 6 7 8 9 10 11 12 13 14 15 16 17 18 19 20 21

38 re-classification
37 atlassing
36 twitching
35 bird tourism
34 radar studies*
33 BTO projects
32 ringing

The ornithologists and purposeful birdwatchers 110 years so far

31 Establishment care
30 behaviour studies
29 photography*
28 sound recording*
27 national disciplines
26 monographs
25 migration studies
24 lasting conservation
23 co-operative studies
22 ornithological societies

The collectors and stocktakers 125 years

21 populists
20 national handbooks
19 eager correspondents
18 troubled consciences
17 published artists*
16 classification
15 county listers
14 collectors
13 army officers

The first natural scientists 175 years

12 true naturalists
11 doctors
10 dramatists*
9 musical composers*
8 name listers
7 falconers
6 Kings, nobles
5 poets*
4 scholars, teachers
3 Holy Orders
2 Roman invaders
1 hungry Britons

The privileged and first educated 1500 years

Century 0 1 2 3 4 5 6 7 8 9 10 11 12 13 14 15 16 17 18 19 20 21

Thirty-eight stations of people, employments and study types are shown. After 1500 years when bird learning was restricted mainly to the privileged and educated, ornithology was developed by the early natural scientists of the 17th and 18th centuries and the collectors and stock-takers of the 19th century. As a science shared by both professional ornithologist and amateur but purposeful birdwatcher, it has prospered most during the last 110 years. Mainly from Fisher (1966), Tate (1986) and Mearns & Mearns (1998). *Themes/classes not covered in book.

Acknowledgements

The origin of this book has already been described but I have again to thank its instigator Richard Porter for his bountiful encouragement and help, Andy Richford for his acceptance of the concept, and Nigel Redman and Marianne Taylor for assisting at the complicated birth. I am also grateful to Richard, Tony Marr, Guy Kirwan, Mark Cocker and Simon Aspinall for reading the manuscripts, making constructive comments upon it, and stating that it merited publication.

To reduce the personal bias in my commentary, I sent two questionnaires to British birdwatchers and ornithologists. The shorter went to 'elders of the Kirk', mostly in their 80s and namely the late Bert Axell, Derek Barber, Baron of Tewkesbury, Peter Cunningham, Richard Fitter, Derek Goodwin, the late James Hancock, Eric Hardy, Phil Hollom, the late Max Nicholson, Tony Norris, the late Geoff Pyman, Keith Shackleton, Eric Simms and David Snow. All fourteen responded, all but one completing written responses which contained enough detail to allow cross-analysis of most answers.

The second, longer questionnaire went to 'members of the Establishment' and 'fact-contributing observers', the latter cell mostly recommended to me by Mike Rogers. Altogether 59 people in their early 70s to mid-20s were approached but response fell (alarmingly) in line with their decreasing age. Five conversations and 34 written responses ensued. Thirty-one of these gave sufficient answers for cross-analysis, coming from Dougal Andrew, Mark Avery, Colin Bibby, Bill Bourne, Peter Combridge, James Ferguson-Lees, Steve Gantlett, David Glue, Mark Golley, Rupert Higgins, John Holloway, Rob Hume, Janet Kear, John Lawton, Tony Marr, Frank Moffatt, Anthony McGeehan, Ian Newton, Norman Orr, David Parkin, Richard Patient, Chris Perrins, Jane Reid, Mike Rogers, Tim Sharrock, Moss Taylor, Keith Vinicombe, Alan Vittery, Stephen Votier, Eddie Wiseman and Jon White.

Hence a total of 44 responses that formed the basis of my descriptions of modern mature birdwatchers. I should have dearly liked more comment from youthful students but their pens hardly moved. The written responses varied from full histories to brief notes and led in 15 cases to further informing correspondence, of which that from Bill Bourne was a constant exhilaration. As already acknowledged, he would have been a better choice than I to write a proper history. All those that I bothered have been thanked but to ensure that none of their fascinating content is lost, all the original papers will be deposited at the EGI Library in Oxford.

A third questionnaire was sent to 12 'European correspondents' of leading journals. Arnoud van den Berg (Holland), Lasse Laine (Finland) and Lars Svensson (Sweden) responded and I am particularly grateful to them. Other European comment and information came from 17 noted conversations at bird fair stands and in conference lobbies, the utterers being named in the next list.

Expert help and criticism on chapter themes came from 73 other birdwatchers, persons and bodies serving their needs. My thanks are due to: African Bird Club,

"Thanks for your nice picture and note of 18 December – bribery will get you nowhere; letters get answered or not according to other considerations, such as the outbreak of disorder in Scotland after the Shortest Day. The birds do look quite like young Longtails – I was caught on the hop by them north of the Falklands, and thought they must be some sort of wader."
Bill Bourne, *in litt.* (2001)

Tim Appleton, Myles Archibald, David Ballance, Franz Barlein, BBC, Birdquest, The Birdwatchers Yearbook, Duncan Brooks, Nick Carter, Jacquie Clark, Mark Cocker, Peter Colston, David Cromack, Ian Dawson, Dutch Birding Association, Steve Dudley, Paul Dukes, Eilat Birding Centre, Lee Evans, Dick Filby, Sue Fleming, Paul Flint, Finnature, Rolfe Green, Alan Greensmith, Robert Gillmor, Gullivers Natural History Holidays, David Holman, In Focus, Chris Harbard, Francis Hicks, Peter Howlett, Ron Johns, Chris Johnson, Algirdas Knystautus, Andrew Lassey, Richard Lear (per *Birdwatching*), London Camera Exchange, Leica, Limosa, Tim and Irene Loseby, Duncan Macdonald, Douglas Maxwell, the late Chris Mead, Chris Mylne, Ian Nisbet, Opticron, Oriental Bird Club, Ornitholidays, Ornithological Society of the Middle East, Doug Page, John Parslow, Peter Rathbone, John Pemberton, Robert Prys-Jones, A Rocha, Ken Shaw, Bob Scott, Sunbird, Tony Swann, Swarovski, Don Taylor, Moss Taylor, John Websper, Chris Wernham, Chris Whittles, The Wildlife Trusts, David Wilson, Paul Willoughby, Stuart Winter, Duncan Wood, John Wyatt, The Wildfowl and Wetlands Trust, Carl Zeiss.

For photographs and permission or help to reproduce them, I am indebted to: Dr H.D. Altmann, Dougal Andrew, Janet Coram, the late Sir Josias (Joe) Cunningham, Jr (by courtesy of Lady Anne Cunningham), David Clugston, Graeme Easy, Bob Emmett, the late Eric Hosking (per David Hosking), John Isaacson, the late John Izzard, Jordan Tourist Authority, Colin Kirtland, Peter Naylor, Brian Newport, Anthony McGeehan, Mike Pearson, John Scovell, Peter Tate, the late Jackson Wallace and David Wilson.

To the minimum of 126 people or bodies that contributed to my book, I send also my appreciation of their delight in birds. I just hope that they find some of it translated herein. Jane Reid will be disappointed; I could not find the right place for the righteous wrath that she poured out over the excessively masculine ethos of modern birding. I am also aware that I have dealt rather meanly with the Irish dimensions; I fear that the recent divide of our once unified ornithology made me shy.

As for the ancients and my departed friends and mentors, hopefully enjoying Elysian fields still full of larks, linties and the odd 'beezer' of a rarity, I can only pray that they are not disturbed by any miscue in my perceptions of them.

Last but no means least, I have to thank Webb Ivory Ltd for allowing me post-retirement use of its office facilities and I cannot praise enough my former secretary Ann Shilton who typed on and on for two years. Without her and Julie Dando who somehow got it all onto the printed page, there would have been nothing.

"Why are there so few women birders? The sex ratio was about 50:50 until I was about 13; then all the other lasses mysteriously vanished by the wayside. Is it because the world of birding quickly becomes so masculine and competitive? But, is it so male-dominated because it's so competitive, or so competitive because it's so male-dominated? I have to justify my presence all the time, and feel that I am often not taken seriously just because I'm female." Jane Reid, *in litt.* (1999)

In horse-riding, the reserve obtains and the competition is only a little less evident. Can we look forward to illuminating studies by a psychologist?

Prehistoric Ramblings

Middenbones and fossils – James Fisher and his ornithological histories –
early records and perceptions – William Turner's initiations of
field observation and fable review

My initial enjoyment of natural history and particularly birdwatching is not accurately dated. The family apocrypha alias photo album pins it to 'before the war' and seemingly 1936. Certainly I do have vivid memories from that time of Shetland summers and creatures as striking as Killer Whale and Shag. With rather less obvious beings like Pink-footed Geese and Bramblings, my first full trysts came from the grounds of my first boarding school, Dalhousie Castle, nestled by the South Esk eight miles south of Edinburgh. The year has to be 1944, during which I became the proud possessor of ex-German Navy 6 x 30 binoculars, but sadly my early diaries were lost in a house move decades ago. Such are the imprecisions of history when its future imperatives are unknown. So it is with the earliest accounts of cognitive reactions between Briton and British bird.

Unlike southern Europeans, we have no vivid cave drawings to demonstrate our close familiarity and aesthetic appreciation of animals as early as 13,000 BC. Unlike the Egyptians, we have no coloured frescos of Lesser White-fronted and Red-breasted Geese to prove our accurate perception of wintering birds in 3,000 BC. Between those periods, our ancestors were experiencing the stress of their final separation from other Stone Age relatives in Europe. Their employments did not include the niceties of recreational art and altruistic science. We have had to dig and scrape for evidence of our ornithological prehistory.

Proof exists of a diversity of man and bird contacts as far back as 250 BC. In the lake dwellings of Ireland and south-west England, the Romano-British settlements of England and the Norse-Viking stone warrens of the Western and Northern Isles, man caught, and also domesticated, birds for food. For the first millennium AD, the evidence of archeological excavation allows at least 63 wild species and 5 domestic strains of species to be identified. Of the latter, two had come, like a Syrian man recently disinterred at Bath, from the farthest eastern sector of the Roman world; they were Hen (or Jungle Fowl) and Pheasant. It comes as no surprise that in addition to these two highly edible species, at least 28 of the wild birds would even today be regarded as pot-worthy. What is more intriguing is the early evidence of bird pets providing relaxing companionship for man. The Romans tamed Ravens, even teaching them to talk. To this first store of certain bird identifications and the early lore that the widespread hunting and eating of birds would have created, 19 more species can be added from the etymology of ancient names. Within the total of 82 wild birds, there is a marked bias towards aquatic species. Among those eaten, their remains then thrown on to middens, grebes, pelecanids, wildfowl and waders, raptors, game birds, rails, gulls, auks and crows all feature in relatively high incidences. Clearly our pre-Norman ancestors had no trouble in discerning the consumer values of birds as food stuffs.

In 1945, 5,000 years after its recognition in Egypt, the Lesser White-fronted Goose (small, centre left) was found to occur in the flock of Russian White-fronted Geese that wintered on the New Grounds, Gloucestershire. No wonder Sir Peter Scott made his last home there.

Ravens were pushed north and west by persecution in the 19th century. From their peripheral strongholds, they are slowly re-occupying lost grounds as far inland as the West Midlands.

There are hints in the Gaelic nomenclature and Celtic lore of wider and deeper acquaintances with birds as far back as the ninth century. This is most notable in swan mythology, which indicates that the great birds with bugling voices could evoke emotional responses as well as smacked lips. However, the only other certain evidence for a greater diversity of possible contacts comes from even earlier natural deposits. Although difficult to discern as fossils, no fewer than 33 passerine species and five more non-passerines have been identified for an earlier interglacial period of about 150,000 BC. Those identified from Devon included the Wren, which had already completed its still-astonishing eastward or westward dispersal from the North American home of its family. Another count from the entire fossil record for the Pleistocene period, ending around about 10,000 BC, lists 142 species. Put together, the palaentological, archaeological, early literary and linguistic evidence suggests that a thousand years ago, our forebears could have had already encountered at least 40% of the birds regularly occurring in Britain in the current inter-glacial. The fact that they had also begun the adaptation of at least five species to permanent domesticity shows also that an early form of economic ornithology had taken root.

James Fisher opined that the Wren spread not east but west from the Nearctic in Ice Age times. This is an adult of the Hebridean subspecies, remarkably like the form of the Eastern USA. Where the Wren faces the Atlantic in Britain and Ireland, it has radiated into five maritime forms.

The greatest of our few recent ornithological historians was James Fisher, who wrote with an awesome width of research and perspective. I met him only once, in 1956. Newly returned from a long vacation in Canada, I was introduced to him and Roger Peterson when they gave their lecture on 'Wild America' at the Zoological Society. Guy Mountfort presented me to the great men with the words, "This is Ian Wallace, he's just spent two months crossing North America". "Oh", exclaimed Fisher, the august, self-styled marine ornithologist, "how many species did you get?" "260", I ventured. "Not enough", he responded, "I still hold the record". Seriously put down, I do not remember the rest of our conversation but happily James Fisher was already more to my generation of birdwatchers than a competitive lister and academic. His was one of the then very few voices that

The Bass Rock is home to one of Britain's oldest colonies of Gannets. Founded before 1521, it was studied by Bryan Nelson in the 1960s. His 1978 monograph remains the classic work on a fine bird.

spoke vividly of birds on the wireless and black and white television. He and Peter Scott were leading members of what he called "the unofficial BBC Bird Club". Furthermore, few of us had not read his *Watching Birds* (1940), being as it was the only cheaply accessible primer of our hobby cum amateur science. In truth, I learnt far more from his *Birds as Animals* (undated) and the treasure-trove of his *Shell Bird Book* (1966). Nobody seeking the origins of birdwatching should leave them or Peter Tate's equally informative *Birds, Men and Books* (1986) unread.

In Britain, words about birds and images of them began to appear in religious texts and even on coins from about 300 AD. Fisher allowed himself a cheeky but educated guess that the first descriptive, even poetic response to birds came late in the seventh century. Whatever its precise timing, the acute and felicitous description of six species by an Anglo-Saxon scholar, in what Fisher fancied to be the "Bass Rock" passage of *The Seafarer*, is remarkable. Using other classic early texts to prime his historical counts, Fisher assessed the total actually recorded (rather than recognised) varieties of birds in England as only 16 in 700 AD, rising to 114 by 1,460 AD. My more Celtic mind suspects that these may have been under-estimates of our ancient ability to tell apart the different beings of birds. What is more impressive than any debate on list length is the degree of perception shown by the early names. The eighth century use by Anglo-Saxons of 'thisteltuige' for Goldfinch demonstrates how acute early British observation of birds could be. For 'tuige', read 'tweak' and an understanding of why the Goldfinch has a long beak is not far off.

In the third millennium, however, lyrically constructed lists are not the stuff of science. So when can latter-day zealots begin to trust the historical measures of British bird recognition? From the end of the 12th century appears to be the answer, with partial list comparisons even possible with Ireland, and a first semi-rarity, a Golden Oriole, mentioned for Wales. About 200 years later, Geoffrey Chaucer set 43 species into the contexts of his romances; four were new to the written record and Fisher hailed him as an "ornithological hero". It was not until the New Renaissance, however, that a first authoritative statement appeared. This was the *Avium Praecipuarum Historia* (1544), one of the first printed books about birds written by the one-time Dean of Wells, William Turner. He added only 15 new species but more significantly gave proof of the relatively difficult distinctions between kites, harriers, larks and finches. Even more importantly, Turner began the process of retrospective review. At last, what was so often mere repetition of Aristotelian tradition began to yield to actual observation. Turner knew that the Hobby was a summer visitor, where it nested and what, in part, it ate. The storing of scientific reality had begun. Fisher awarded Turner the title of "father of British ornithology" and had he not been a refugee from Church politics, his contribution would have been even greater. Another great scribe for birdwatching and ornithology, Richard Fitter went further and accredited him as the "first European birdwatcher", apart from the less harassed Emperor Frederick II of Hohenstaufen. Even if Turner and his close contemporary John Kay were the first real observer-scribes, their work did not provide any broad overview of our avifauna. Moreover, their Latin utterances went unnoticed by the great mass of ordinary folk. By 1600 AD, Fisher's list of recorded birds contained no more than 150 species.

A family of Hobbies decorated the branches of a poplar at Rolleston Park, Staffordshire, in the late summer of 1996. To the delight of birdwatchers, this superb summer visitor has spread north in the last 30 years. Over 2,000 pairs now breed in England and Wales.

Early Perceptions

Correspondents of the 17th and 18th centuries – John Ray and Francis Willughby's foundation of ornithology – Gilbert White's letters – Thomas Bewick's woodcuts – George Montagu, his extraordinary life and dictionary – Macgillivray and Yarrell – Audubon and Gould – the Irish start – early universes of birdfinding and collecting – some perceptions of island folk – the early proving of identifications – the scarcity value of unusual species – the number of 18th century observers

The first real seed change in the conduct of British bird study came in the 17th century. There emerged a small group of corresponding observers cum explorers. In their work, the first recognisable strands of geographical ranges and seasonal occurrences, hence avian statuses, began to appear. Chief among these pioneers was John Ray, born in 1627 and destined to complete 78 years as a Cambridge don. Spurred on particularly by many bird notes from East Anglia's first great birdwatcher Sir Thomas Browne, Ray undertook an astonishingly wide exploration of Britain and Europe. From 1660 to 1672 he was stimulated and ably assisted by a regular travelling companion, the Warwickshire gentleman Francis Willughby. The main outcome of their association was the appearance in 1676 of the Willughby-Ray *Ornithologia*, now regarded as a fundament of that science in western Europe. It was also the first combination of amateur observation and professional review. Crucially, an English version *Ornithology* was published in 1678. At last, the way to easier learning about birds was open. Ray was interested in all natural history and anatomy and is widely considered to be the "father of modern zoology". Importantly, he made direct acute observations in the field and, with the development of better fowling pieces, was able to collect specimens as well. Given his use of notes, Ray could also be called the "father of field identification". His description of the pelicans presented by the Russian Czar to Charles II in 1662 shows them to have been Dalmatian, not White. Thus as the Restoration took hold, two main tools of modern ornithology were already in use.

Dalmatian Pelican, Hayle Estuary, Cornwall, 6th January 1968. With no wild birds surviving west of Albania, it was declared an escape but the early twitchers went to see it, just in case …

Fisher labelled the 17th century the first of modern bird study. In 1608 the first binoculars appeared but had a magnifying power of only x 3. Nevertheless the habits of skilled observation and wide exploration and a first trustworthy textbook were all put in place. Early attempts at county lists were made for Oxfordshire, Norfolk and Staffordshire, and the prototype of island-going adventure and marine ornithology came in the epic voyage of the Reverend Martin Martin to St. Kilda in 1697. There was, however, still no disciplined framework to the British approach to ornithology or the list of British birds, believed to contain 202 species by 1699. The confusion of vernacular names remained a particular stumbling block. The gift of species-defining and -stabilising names came from Sweden but not until 1758. In that year, Carl von Linne, self-alias Carolus Linnaeus, published in his serial work *Systema Naturae Regnum Animale* scientific names for 215 birds then on, or later to reach, the British list. The names consisted of two Latin or Latinised words, the first noun establishing the genus (or tribe) and the

St. Kilda can be a forbidding place but its seabirds have caught human attention since the late 17th century. Coming into Hirta late on 3rd July 1956, eight seasick observers found their boat surrounded by Leach's and Storm Petrels.

second an adjective defining the species (or tribal member). This system of binomial nomenclature was the key to the secure, cross-border classification of flora and fauna, soon the main goal of most early natural historians. With the later addition of another adjective defining the subspecies (usually a discernible geographic population within the tribal member's total range), the adoption of the Linnaean system has in the last 250 years crossed all national boundaries and reduced the confusion of spontaneous vernacular names.

In 1713 Ray published his second major text on birds, the *Synopsis Methodica Avium*. From this and his earlier foundation statement flowed a stream of at least eleven general descriptive works on British and Irish birds and three more forerunners of county avifaunas, lists for Cumberland, Sussex and Dorset. Of this first flood of ornithological literature, John Latham's *General Synopsis of Birds* (1781-82) and Thomas Pennant's *British Zoology* (1766) were the most instructive in the refining of identities and the sorting of species but none caught the imagination of a wide readership. What did, however, was the delightful correspondence of an English curate. Particularly when exchanged with his first mentor Pennant and eventually published in 1788 under the title of *The Natural History of Selborne*, the letters of Gilbert White still make utterly absorbing reading. Clearly his visual and aural skills allowed him a wonderful spontaneity of observation and in every chapter his acute senses, through which nature posed questions to him and with which his perseverance (in thought or consultation) provided answers, are apparent. Fisher opined that White's appreciation of the beauty of his local countryside and its animals could have been a sublimation of his duty of worship, but whatever his motivation, White is accorded the title of "father of British field natural history". To those under his curacy, he was less august, being described as "a little, thin, upright man with stumpy legs" and "the Hussar parson" due to the horseback seat of his observations in early years.

In his most praised ornithological discovery made without binoculars but using his keen hearing, White ended 50 years of debate on the number of British breeding

"Gilbert White whose book The Natural History of Selborne *made me think about animals for the first time, and Niko Tinbergen whose* Herring Gull's World *made me want to go to University to study zoology!"* David Parkin, *in litt.* (2000)

Two cardinal authors of ornithological narrative acknowledged.

From top to bottom, singing Chiffchaff, Willow Warbler and Wood Warbler. Gilbert White realised that all three sang different songs in the Hampshire countryside.

A pair of Cirl Buntings in their last British refuge, a south Devon field with stubble, tall hedge and sentinel trees. In Montagu's day, the species occupied most of southern England and lowland Wales.

leaf warblers. He was the first firmly to realise that Willow Warbler, Chiffchaff and Wood Warbler were all different species. I thought of his perceptions 230 years later when in May 1998 I stood confused by a chorus of warblers in Kazakhstan and declared my impotence as an aural distinguisher of birds. Happily, Lars Svensson was close by. "Calm down, Ian", he said, "and listen to them. They tell you who they are". Clearly White had needed no such advice and he went on to be among the first to distinguish the Lesser from the Common Whitethroat. From his examples of personal discovery, the chances of other amateur observers to make gains in scientific knowledge became apparent but in his total output, it is the human recreation available in natural history study and particularly birdwatching that shines out.

White gave his readers a heartfelt book that remains a model for ornithological narrative or argument but they were soon to have another equally pleasurable utility, the first fairly trustworthy picture-book of birds. This was *A History of British Birds*, a work of great woodcut art created by Thomas Bewick. Its two parts came out of a Newcastle printing works in 1797 and 1804. A farmer's son, Bewick was self-taught as an artist but within three decades of engraving experience, he lavished seven years of care on cutting the blocks for his charming portraits of birds. These were sufficiently accurate in plumage contours and structural details to allow certain diagnosis even by today's standards. Looking back at his texts, it is clear that Bewick was not an originating genius like White. His images of birds were, however, the first to be widely distributed and seen by ordinary people, while the artistry and settings of those species that he knew best could be outstanding. By the end of the 18th century, the British list had reached 240 and the early students of birds had been fully shown both the scientific promise and aesthetic romance in their chosen class of animals.

In the 19th century, the pursuit of birds for science, recreation and collecting gain prospered enormously in Britain and Ireland. To the well-established general discourses and early local surveys were added other types of bird books. These included the first volumes devoted to particular families or signal species, called monographs, and specialised texts on breeding birds, migrants and eggs. The last was a clear reflection of the growing desire to collect and display natural objects. Fisher reckoned that within the first boom in the ornithological literature, thirty-three books contained important advances in knowledge or presentation. Chief among these was the first *Ornithological Dictionary or Alphabetical Synopsis of British Birds*, written and compiled by George Montagu and published in 1802. Directly inspired by White's field successes, Montagu proved his own skills by spotting that the Cirl Bunting was not an odd Yellowhammer and that the harrier which bears his name was not a form of the Hen Harrier. Montagu has claims to be the first contrary eccentric of British ornithology since William Turner. One of thirteen siblings, he was sent by his father, the first Earl of Manchester, to the Army. He rose through the ranks, then eloped with the niece of a former Prime Minister and finally found himself fighting the rebellious colonists in the American War of Independence. Finding the actuality of battle repugnant, Montagu turned to the solace of nature, even collecting American birds to present them as skins to his mother-in-law "as proof of his regard and memorials of his past adventures". About 1780, he left the Army for the lesser force of the Wiltshire Militia, in which he rose to the rank of half-Colonel, only to be displaced by his fellow officers after a court-martial. In a further bout of restlessness, Montagu then refused the full advantage of inherited income by leaving his wife Ann and co-habiting in a South Downs cottage with his lasting soulmate Elizabeth Dorville who had deserted her merchant husband for him. Settled at last and directly assisted by his new partner who was a skilled illustrator, Montagu devoted his remaining sixteen years to natural history, particularly marine biology and ornithology. Given that he was also involved in costly inheritance litigation and battered by the loss of two sons in active service, his last bout in the elucidation of nature is all the more astonishing. A substantial *Supplement* to his dictionary was issued in 1813. As the Napoleonic campaigns convulsed Europe, he had somehow found the time to hew the third pit-prop to modern ornithology. Two days after the Battle of Waterloo, he died from lockjaw. A rusty nail in the foot had widowed Elizabeth, half a dozen spaniels and many caged birds. His shotgun fired no more.

In 1926 one of the few giants of modern natural history, Max Nicholson, was the first to draw attention to the true importance of George Montagu. He did so again in 1956, when the cottage alias Knowle House received a memorial plaque, but Montagu remains a little-appreciated benefactor of British ornithology. His prose lacks the close, almost affectionate approach of Gilbert White and it is difficult to detect in it the intensity of individuality that the rest of his life demonstrates. Like that of Richard Meinertzhagen in the 20th century, his determination to enquire and evaluate was driven by remarkable personal force. A fuller account of Montagu is given by another harrier expert (Clarke 1996) but I cannot leave him without noting two more of his discoveries. In an amazing product of early avian vagrancy, the first scientific description of the American Bittern came from his procurement of the first to be shot in England in 1804. First identified as a Bittern and then as a Little Bittern, the bird finally met its match in 1813 when Montagu realised that it was an unknown species and named it the Freckled Heron. Uncannily it was not until the next year that the expatriate Scot Alexander Wilson recognised the bird in its home continent and gave it its proper name. Montagu's skin resides at the Natural History Museum in Tring; it is 188 years old. If a bittern is tricky, so is any bat. Yet Montagu also discovered and first described the Lesser Horseshoe Bat, so matching White's recognition of the Noctule. With the ignorance now displayed by today's birders of their ancestors, I can only hope that eventually they will learn to salute the old authorities. Montagu's example was followed by many other serving officers in the next two centuries.

Another hero of bird study in the nineteenth century is Scotland's William MacGillivray. Although a friend and collaborator of John Jacques Audubon, he has, like Montagu, been relatively unsung. Yet his *History of British Birds* (1837-52) had a stronger vein of science than all its contemporary rivals. Stormbound in Stornoway in June 1996, I and the other ornithological itinerants aboard the 'Ocean Bounty' chanced on an impressive retrospective exhibition of MacGillivray's work in the town museum. Totally transfixing among the illustrations was a seemingly

In its home range, the American Bittern is often seen in open country but as a transatlantic vagrant, it usually lurks in a reed marsh like the Bittern. Its dark unbarred flight-feathers are diagnostic.

perfect portrait of a Cape Gannet, complete with long gular stripe and black tail, caught on the Bass Rock in May 1831. An echo of an ancient Gaelic myth or the first record for the Western Palearctic? Unfortunately, the person who gave the bird to MacGillivray had interests in shipping and we shall never know the bird's true provenance. Nevertheless, MacGillivray shared with Wilson the distinction of being the first Scottish ornithologists of merit and in 2001, the Post Office gave the Cape Gannet fresh publicity by putting its image on a stamp!

Appearing first in the same year as MacGillivray's book, another *History of British Birds* by William Yarrell was to rival Montagu's dictionary as the longest-lived standard national reference. The final edition of the dictionary had appeared in 1866 and overall the work was credited with 75 years by British ornithological elbows. Yarrell's scholarly science first appeared in 1837, subsequent editions followed until 1899, and modern ornithologists including the child Fisher were still learning from the last of these up to the 1930s. Among Yarrell's personal observations was the separation in 1824 of the two wild swans, the smaller one being named after Bewick. In its later stages, Yarrell's work attracted competition from the first really magnificently illustrated hand book. This was *The Birds of Great Britain* by John Gould (1862-73).

In their day and still to their current owners, Gould's works were the first European books to approach, even in part match, the majesty of John Audubon's *Birds of America* (1827-38). Both authors presented birds in the large coloured plates available from a fast-developing reproduction technology. Although it portrayed Nearctic species, the first edition of Audubon's work was actually originated in Edinburgh and London. The Scottish engraver responsible for the first templates was William H. Lizars. His descendants still ran a retail business in spectacles and optical equipment when I was a boy in Edinburgh. It had the first shop window into which I poured savage desire for unaffordable field glasses, as binoculars were once called. The production of Audubon's 435 plates and their publication in 87 parts, also bound in four volumes, took eleven costly years. Of 279 subscribers, only 161 stayed loyal to the last issue. In America, the whole work cost $1,000; in Britain, it sold for £182 14 shillings. The figures allow the incidental calculation of an exchange rate of $5.47 to £1 ($1.42 to £1 today) and, given 180 years of inflation, a first transatlantic ornithological product cost of around £1,900. Audubon's approach to bird study included early experiments in behavioural science, as is recorded in his *Ornithological Biography*, which was written with further help from MacGillivray and valued by William Vogt above many a 20th century tome. Audubon began his artistic odyssey as a portrait painter and his creative genius was accompanied by a capacity for self-dramatisation. He would hint that the French Revolution had cost him a throne. His parents were actually a French naval officer and a Creole lass. Never mind, a love-child made a wonderful birdwatcher and artist.

John Gould's life story is scarcely less dramatic than that of Audubon. He rose socially from gardener's son to museum curator and on to gentleman naturalist and publishing impresario. Although poorly educated and no great field man, Gould was a born manager and a workaholic. Attaching such attributes to his

"The varying modes of flight exhibited by our diurnal birds of prey have always been to me a subject of great interest, especially as by means of them I have found myself enabled to distinguish one species from another, to the furthest extent of my power of vision." John James Audubon

The holistic approach to field identification is generated by repeated perceptions. As Bob Scott has long observed, artists have an eye-start in seeing differences in jizz.

passion for birds and mammals, he delivered eleven major works and hundreds of original descriptions. By travel and collection or by import of others' specimens, he secured more and more knowledge of both individual species and whole avifaunas. He and his team of artists, which included his wife Elizabeth and the keen-eyed Joseph Wolf, achieved a new veracity of coloured plumage illustration and set their birds against backgrounds that indicated their niches. Gould's sense of what would interest the British public was acute and much used to secure both wealth and fame. His special display of hummingbirds at the London Zoo in 1851 attracted 75,000 visitors; one was Queen Victoria. Not shy of advantageous field exploration in an age that paid for the novel and curious, Gould hastened to be among the first collectors of Australian wildlife. Through the coincidence of heavy rains and an associated concentration of birds, he once found himself swamped in Budgerigars. So impressed was he by them (and further seduced by the local aborigines' opinion that they had flown in to see him, the bird man), Gould organised some captures. They were brought alive to England. His *Birds of Great Britain* took, like Audubon's masterpiece, eleven years to complete.

Budgerigars at an Australian waterhole. Fifty years ago, they were favourite cage birds. The trade in them was stimulated by the 19th century curiosity in the world's nature.

For Irish birds throughout the 18th and 19th centuries, the ornithological cupboard remained rather bare. It took a refugee from the Belfast linen trade, William Thompson, to write the first worthwhile Irish avifauna. The scientific disciplines applied to the ornithological volumes of *The Natural History of Ireland* (1849-57) shows that he was a true peer of MacGillivray. Sadly Thompson died before the full publication of his work. His early death and the fact that the early Irish recording universe lacked enquiring priests like Gilbert White did not assist the natural description of the Green Isle. For Welsh birds in the same period, there are only two references of any note. The example of Giraldus Cambrensis, who had made observations in Wales and Ireland in the 12th century, had not been followed. Even so, Fisher calculated that the total number of recorded British and Irish birds grew steadily to 214 in the first Linnaean year of 1758, about 240 in 1800 and close to 380 a century later, with the first official list coming from the British Ornithologists' Union in 1883.

One of the troubles with praising famous men is that even in their close choruses, the surrounding ordinary folk are only rarely identifiable. So it is with the early scenes of birdwatching. Seemingly monks and other clerics, doctors, nobles and landed gentry abounded as chief characters, clearly brought to science by enquiring minds acquired from higher education, but who else took early part? To try and answer this question I took out my copies of *A History of British Birds* (1851-57) by the Rev. F. O. Morris, BA. Although a member of the august Ashmolean Society, Morris was not a particularly original observer and his volumes do not rank among the signal works praised by Fisher. His popular style is pleasantly discursive, however, and he tells many tales from which it is possible to add some social compass to the man and bird contacts of the 19th century. By far the most compelling evidence in Morris of a wide native recognition of birds throughout England is the proliferation of popular bird names. As many as ten are listed for common species such as Great Spotted Woodpecker and Yellowhammer. I was particularly fascinated by the alias of Winter Wagtail for Grey Wagtail, a clear

Oxeye
Tomtit
Black Cap
Garden Tit
Sit-ye-down
Great Titmouse
Great Black-headed Tom Tit

Large and bold, the Great Tit attracted a variety of local names but all are now almost forgotten.

sign that the bird's habit of finding a farmland niche in which to over-winter was known over 140 years ago. As I hinted earlier, such acute observation can even be traced in peripheral ancient communities. In Shetland, 61 species have local names, many showing Viking and Norse roots (Venables and Venables 1955). Of these, 49 are breeding birds; within this group, 12 names show perceptions of size, plumage mark and age, 11 have indications of habitat and feeding preferences and ten are oenomatopaeic. Another 11 names display incomplete separations from other species but in most, the acuity of their human namers is obvious. Thus the Lesser Black-backed Gull is also the Herring Maa and Saithe Fool, both old names identifying two commonly eaten fish, and the Peerie Swaabie, truly a smaller version of the Great Black-backed Gull. The Rock Pipit is the Banks Sparrow in summer, the Midden Fool (fowl) in winter and, to some, the Tang (seaweed) Sparrow all year round. The Corn Bunting's seven alternative names include three indicating a close association with dock (a robust weed affording a songpost), three echoing its song and one which translates aptly to 'untidy Skylark'. A charming intimacy with nature in our most northerly crofting and fishing community shines out, as it does in the Faeroes. The Gaelic nomenclature of birds in the Outer Hebrides and Ireland presents the same lessons.

Particularly where the bird of his text was unusual, or its habits still disputed, Morris dropped names freely. Thus for the 60 species in his first volume, a human retinue of at least 225 finders, observers and shooters is discernible. The most obvious class is indeed the landed gentry, with nine nobles and 88 gentlemen with seats attended by at least 21 gamekeepers, four gardeners and two labourers. Next come the clerics, one bishop and 15 of lesser orders, nearly all sporting shotguns as Morris himself did. Then there are five military officers and one ship's captain, clearly set apart from two falconers and six bird trappers. A motley rearguard comprises four boys or youths, three egg-collectors, two each of fishermen and crofters and one each of boatman, boat passenger, walker, pigeon fancier, zoo keeper and lady. Above and around this universe of direct association, Morris paid dues to another 14 nobles (as landowners), seven European and American authorities, 13 British experts, 17 national and 15 regional handbooks, five each of natural history memoirs and journals (including the already popular *Zoologist*), five taxidermists and two local papers. Morris's notations are sufficient to dispel the persistent rumour that most ancient birdwatchers were clergymen, the latter making up only 15% of his direct associations, but it is still difficult to sense from them the everyday character of bird observations.

Turn the years back to White's era, however, and the curtains can be drawn further apart. Unlike Morris, White was extremely reticent with the names of his direct informants, giving only seven in his field reports, but his descriptions of their social types were more telling. The most numerous is again the landed and sporting gentry, with 19 mentions, but significantly next come eight skilled hunters of his own rural community. They include not just the still-extant poacher and wildfowler but also the long-departed specialists in lark netting, bat and Redwing fowling, Wheatear trapping and plump duckling capture. Next come a group of six village neighbours and seven country folk which includes four shepherds. White compliments four people on being excellent observers or intelligent persons and ages another four as young men or boys. Only after these do three clerics (only 5% of all) appear and the tail of his ornithological retinue consist of a Swedish naturalist, a servant, a musical friend who judged that most owls hooted in B flat, a brother in Spain and one little girl. The last along with Morris's lady are the only distaff observers of birds in the two samples. Above and around White's immediate universe of man and bird contacts, he noted three nobles, a colonial governor, three gentlemen and a mariner, eight English experts, and at least 14 national and international reference books. Intriguingly, White also preached the use of the *Naturalist's Journal* to facilitate a discipline of dated observations. It is strange that no equivalent book has survived the 20th century.

Another measure of the behaviour of ornithologists in the 18th and 19th centuries is the span of their references and correspondence. Clearly Gilbert White's library was still dominated by the collaboration of Willughby and Ray, cited 25 times in a total of 66 references. Next came ten mentions of the work of Pennant,

In the 18th century, roosting Redwings were caught by skilled fowlers. Here four autumn migrants (bottom with red flanks) have piled into an East Coast hedge in company with a Ring Ouzel (fourth up on left) and Fieldfares.

The best known of British petrels, the Fulmar, clearly appealed to Gilbert White. The early history of its spread from St. Kilda to almost the entire periphery of Britain and Ireland was told by James Fisher in one of the classic monographs of modern ornithology.

" 'If any of my readers', says Macgillivray, 'should be anxious to know how an author may contrive to talk a great deal about nothing, he may consult the article 'Crested Tit', in an amusing work entitled 'The Feathered Tribes of the British Islands'. I have only hereon to remark that Mr. Macgillivray is very seldom wrong, and this is not one of the few instances in which he is. Mudie (the inventive author) certainly disproved the truth of the proverb, 'ex nihilo nihil fit', for though his stock of knowledge of any bird might be 'nil', that had nothing whatever to do with the 'quantum' he wrote about it; and thus he made his book."

The Rev. F. O. Morris's salute to the Scottish ornithologist William MacGillivray's criticism of vacuous writing, from the former's *A History of British Birds, Vol. 1* (1865).

nine of Linnaeus, and eight of Scopoli. No other book or sage was mentioned more than twice. With the English language literature on birds only a century old, the bachelor cleric of only light parish duties was perforce an energetic and sympathetic correspondent and his loyalty, even in a modern sense confinement, to Selborne made him demanding of others' more travelled eyes. Surely there was a shred of potential envy in this counsel dated 26 May 1770 to his brother in Spain, "Look after the genus of birds called Petrels; they are very peculiar in their way of life, and are in the Atlantic…". With the wits to tell Willow Warbler from Chiffchaff, how he would have enjoyed a seawatch!

A hundred years on, Morris was another energetic correspondent but the literature that could inform him was by then at least twice as varied and substantially more accurate. The authority of Montagu's *Dictionary* had become pre-eminent, being the most frequently quoted systematic and general reference with, for example, 54 quotations in the first 120 species treated by Morris. Next in this sample came 31 mainly nomenclatural references to John Selby, 30 to Pennant and about 20 each to five other major authorities which included John Latham, the first British Museum ornithologist who, in three works published between 1781 and 1828, attempted a world-wide review of birds. One puzzle among Morris's sources is the full identity of an expert called Meyer. He was an illustrator, visiting London Zoo to sketch captive raptors, and clearly had knowledge of subjects as varied as hawk plumages and titmice voices. Indeed he appears to be the first Briton to have written transcriptions of bird calls. Strangely, no plaudit has been offered him by other historians.

A close read of Morris shows that disagreements between authors and betwixt experts were fully overt by the middle of the 19th century. There were signs of some tension in Gilbert White's letters but for the direct debunking of a charlatan author, the opening paragraph of Morris's Crested Tit text set alongside is quite splendid. Writ of dissension reminds me of rarity issues. From the 17th to 19th century, these concerned not just the hapless birds, so quickly corpses, and their identifications but also the scarcity values placed upon them by the collectors' and museum marketplaces. The prices paid for the specimens that proved first and early occurrences were often small fortunes. Britain's first Cream-coloured Courser was an immature bird shot in Kent in 1785. It was presented to Latham who skinned it and placed it in his museum. Along with other exhibits, it was auctioned in 1806, the hammer coming down on a bid of 83 guineas from one Donovan despite interest from no less a person than the Viennese Emperor. Later in its stuffed afterlife, it reached Cromwell Road in London, only to be blitzed in World War II about 160 years after the initial vagrancy of the live bird. Given inflation, the equivalent bid today would be £3,800.

To check the links in the chains of rarity identifications made up to the late 19th century, I turned to Phil Palmer's informative catalogue of first records, entitled *First for Britain and Ireland 1600-1999* (2000). In a sample of 48 such occurrences dated up to 1875, the number of people involved in the initial diagnosis ranged from one to five, averaging nearly three. In 25 of the records, there were further inspections by one to five experts who confirmed 20 original identifications, solved one puzzle

Crested Tits in their original Scottish niche of mature pine forest. About 6,000 birds form a relict population.

and found four mistakes. The final corrections of the last took up to 40 years, a far slower process than the immediate, essentially commercial transfer of the corpses to local or distant taxidermists, natural history (sale) agents or exhibiting authorities, and, thereafter, to the display cases of country seats, great houses and museums. The power of this entrepreneurial process was amazing, its motivation persisting until the 1920s and occasionally even after World War II. The exhibits at an 1819 sale included a stuffed Great Auk and egg which made £16 (£730 today) and the first British-stolen egg of a Leach's Petrel which reached £5 15 shillings (£390 today).

To return once more to Morris and the view that his discourse allows of the ornithological scene of the mid-19th century, I have tried to size the community of observers sufficiently aware of bird identities to spot unusual species. On a sample of 120 species accounts, a mainland universe of between 950 and 1250 perceptive observers is indicated. Amazingly this bracket covers the number of 1,100 trained purposeful observers that emerged after World War II to initiate the most modern phase of British bird study (Wallace 1981). Within the general growth in the study of natural history, the desire in some minds to marshall its ornithological part was made evident in 1858 when the British Ornithologists' Union was founded. So end my brief attempts to describe the earliest scenes of British birdwatching and ornithology. I have tried particularly to add some colour to them but more precise details of the many proponents can be found in the works of Fisher, Tate and Walters (2003).

All wandered desert birds exert a powerful fascination and none more so than Norfolk's fourth Cream-coloured Courser in October 1969. Found by Andrew Lassey near Blakeney on the 18th, it moved to a sandy sugar beet field at Ormesby, East End on the 29th and died there on the 20th November. The farmer Ernest Daniels was a family friend and I was able to inch my car right up to the bird and make these sketches.

Ancient Bird Harvests

*A Pintail in Leadenhall Market – the ancient fenland and its fowlers – duck decoys –
Tudor and later conservation – the end of a wilderness and its former bird diversity –
the fenmen's names for Savi's Warbler – Calais, source of birds for feasting and
falconry – Henry VIII's greatest avian desire – the downland captures of Skylarks and
Wheatears – the London catch – favourite cage birds – market supplies and prices –
an instrument-playing Bittern – the skill and persistence of fowlers – the fowlers'
harvest of new British and Irish species – seabird harvests, particularly at
Flamborough Head – Victorian conservation*

Old enough to remember the Second World War displays of wildfowl in London's
Smithfield and other markets and once trained as fishmonger and poulterer, I am
accustomed to see more birds than domestic fowl as food and objects of commerce.
My first experience of the incomparable Pintail came from a drake hanging above
a Macfisheries slab in Leadenhall Market. So I think it right to include some
observations on the British consumption of wild birds.

The first scene shall be 'the undrained fen', so well evoked by H. C. Darby in
The Birds of Cambridgeshire (Lack 1934). Drawing on accounts from the early
eighth to the early 18th centuries, Darby showed that a "fen of immense size" had
stretched from Grantchester north to the sea, difficult to penetrate, frightening to
live in but seething with birds. According to the monastic scribe Felix whose Life
of St. Guthlac and Legend of Crowland is the earliest reference, the saint was
approached by the wild birds of the wilderness – Raven and Swallow are named –
and he fed them according to their kind. In the 12th century, another monk called
Thomas of Ely saw the congregation of wildfowl as "numberless" and the captures
by lime, net and snare as "by the hundred, and even three hundreds, more or less".
The fowling and egg-harvesting continued, reaching such a scale in the mid-16th
century that Henry VIII sanctioned an Act "agenst the Decstruccyon of Wyldfowle".
The drafting made it clear that the taking of moulting adults and flightless young
had become so excessive "that the brood of wild-fowl is almost thereby wasted
and consumed". For sixteen years, a close season was imposed from 31st May to
31st August, while egg-theft from certain species became punishable by a year in
prison. In 1550, however, Edward VI agreed to the repeal of the first stricture;

In prehistoric times, the English Fens may
well have held as diverse a community of
aquatic birds as the Danube Delta. Fossil
evidence suggests that pelicans fished
alongside wild geese.

The ancient fenland exists only in a few relict localities such as Wicken Fen, Cambridgeshire, but it is still possible to glimpse its wetland magic in the carefully managed reserves of the RSPB and other conservation trusts. Here a Marsh Harrier flushes three Teal.

"the poor people that were wont to live by their skill in taking the said fowl" could not survive without its proceeds of food and profit. So the wildfowl harvests recommenced and the wilderness continued uninterrupted until its first major breach, the draining of the Bedford Level between 1630 and 1653. As the new dyke and resultant wash system were extended, the old method of driving birds into nets gave way to shooting and decoying, the latter concept and practice imported from Holland. In 1678, John Ray referred to the first decoy-ponds as "a new artifice" and only four years later in 1682, he noted the by then "abundance of those admirable pieces of art call'd duckoys", the ponds having acquired the final seduction of live "decoy-ducks". Defoe gave an example of a "duckoy" near Ely that paid a rent of 500 louis (over £1,000) a year to the landowner and maintained "a great number of servants". From it alone, its operators sent to London "three thousand couple (6,000 birds) a week" and from others around Peterborough, there were two despatches weekly by waggons "loaden so heavy" that they needed teams of "ten and twelve horses a piece". The fowling enterprise and commerce had reached "incredible" proportions and both Queen Anne and George II approved new Game Preservation Acts, eventually re-imposing a close season from 1st June to 1st October.

Soon after, however, the agrarian development of the fenland economy took a different course. "A tame sheep was better than a wild duck" was a quaint but economically forcible argument, and when the agricultural productivity of the substrate began to be exploited, the years of the fowlers were numbered. By the 19th century, the southern fen could still present winter wildfowl spectacles but cleared of cover, it was no longer a bird-harvesting ground. In 1851, Whittlesea Mere, the last big open water, was drained. In 1886, only three Cambridgeshire duck decoys could be remembered. Most of the aquatic birds and the fowlers and decoymen had gone, their intimate relationship of over 600 years already ghostly.

As ever, the old records of the actual species taken or encountered in the fen-fowling industry are fragmentary and subject to incomplete identification. Nevertheless, it is certain that the ancient fen – England's greatest wetland – gave sanctuary to a profile of at least 70 aquatic and wasteland species, as diverse as that of the best reserves in Holland today. Many of the birds have been sustained or recovered in the new reserves of the late 20th century but the wonderful substance of the pre-drainage community will not recur. Among the most intriguing of the former, at least seasonal inhabitants were many Dotterels – "great men and Kings are keen in the chase of this bird … wont … to be caught by silly imitation" – Black-tailed Godwits – the young caught for fattening – Ruffs – breeding plentifully, especially around the mound of Ely – Great and Jack Snipes – respectively in autumn and commonly in winter – Black Terns – breeding annually until 1824 and still in large flocks up to 1831 – Spotted and Baillon's Crake – the former common until 1850, still breeding in Wicken Fen around 1860, and the latter elusive but breeding in 1858 – Savi's Warblers – breeding until 1850 – Bearded Reedling – never a tit, hanging on until 1902 – and Hooded Crows – swarms in winter. No need for bird tours to Poland two centuries ago!

Fifty years ago, the Hooded Crow was still an expected sight in late autumn and winter. Nowadays it is less often seen than many so-called rarities. The east coast of England is still the best place for it.

Sadly, apart from the kindness of Guthlac, the British counterpart to St. Francis of Assisi, there is in fen history no record of birdwatching joy, but I take the fenmen's ability to distinguish Savi's from Grasshopper Warbler to be further proof that daily familiarity with an ecosystem conveyed remarkably acute diagnosis of its natural inhabitants. The imperative was not ornithological but the result in one difficult genus was just that. The fenmen had three names for Savi's Warbler: Brown- , Red- and Night-reeler.

Mention of Kings drooling for the taste of Dotterel leads me on to another fascinating scene, the wholesale trade in birds for feasting and falconry between England and Europe in the first half of the sixteenth century. The information on this remarkable business in live birds comes mainly from the voluminous correspondence of Arthur Plantagenet, Viscount Lisle, bastard of Edward IV, who was the Lord Deputy (Governor) of Calais from 1533 to 1540. At that time, the quality of life enjoyed by nobles was much influenced by the barter of favours and goods, and Lisle and his lady, effectively commanding an English colony around Calais, were frequently asked to facilitate these or benefited from them themselves. The transactions were examined in detail by Bill Bourne in the *Archives of Natural History* (1999) and his findings demonstrate truly sophisticated demand and supply. The most numerous delicacy was the Quail, shipped in cages containing up to 540 birds (a good whole summer's count for the entire British Isles in the 20th century!). The other dishworthy species included Grey and Night Heron

Where in the ancient fen Dotterels occurred has been lost. In the 1950s, spring trips were re-discovered on beet fields on the chalky ridge south of Cambridge, particularly near Royston.

chicks brought in for fattening, Dotterels and young gulls, all taken locally in France. More exotic imports comprised Pheasants, Guinea Fowl and Peahen and even 'fast food' in the form of a baked Crane and partridge pies. In total, over 45 species can be distinguished in the Calais trade and five other importations as far north as Northumberland up to about 1620.

The most common raptor imported by falconers appeared to be Goshawk. which was even swapped for English-taken falcons. Henry VIII had a well-stocked mews but still dreamed most of all of owning and flying a Gyr Falcon, while a note of a Boulogne escape of a Saker showed that the demand of falconers was drawing birds from eastern Europe.

A young female Goshawk scares some Wood Pigeons from their roost. This fine raptor's comeback to Britain stems mainly from falconers' lost birds.

In his paper, Bourne also reprised some of the references to the purely English harvest of small birds. These include Pennant's estimates of a five-month catch of 48,000 Skylarks on the Dunstable Downs and an annual crop of 22,000 Wheatears on the South Downs above Eastbourne. In the middle of the 19th century prior to the slow pricking of conservation conscience and early legislation to protect birds, a community of over 200 birdcatchers supplied the London market with singing male cage birds, dumping the killed females as food items. An estimate by H. Mayhew (1861) indicated a total annual take of 250,000 passerines or over 1,000 birds per catcher. In my own youth in the late 1930s and early 1940s, the residue of this trade was still visible in the caged Goldfinches that brightened the then black Edinburgh tenement walls with their sparking colour and sound. As Bourne points out, the full magnitude and diversity of the ancient bird trade has been lost but amongst the commoner species caught for caging like the Goldfinches and Linnets, there were also many Bullfinches, Song Thrushes and Skylarks and even Woodlarks and Nightingales. The skills of the catchers included the ability to change the physiological state of these birds, with early moults induced to produce fresh plumage and full song.

Finally, it should be noted that once the long-distance shipment of frozen meat began (from 1873), the transport of similarly preserved birds soon followed.

Charles Raven, whose schoolboy eyes searched Leadenhall Market at the end of the 19th century, saw an endless stream of birds "from the opening of the wildfowl season when 'flappers' (flightless ducks) appear late in August, down to the Dutch and Danish imports in March, and the arrivals of barrels of frozen birds from Siberia in April and May". Some imports came even from China and Australia, completing journeys as long as the shipments of salted and later frozen fish from Scotland and the North Sea ports. As someone who respects skilled tradesmen, I find it hard to condemn the old commerce in birds. I do find it frustrating, however, how few glimpses there are of the prices paid by merchants and final consumers. A dozen quails cost 2 old pence in 1538 (£25.20 for French-reared birds at Harrod's Food Hall today) and a well-fattened Black-tailed Godwit brought its feeder 17½ old pence profit in 1611 (as expensive then as smoked salmon was before fish-farming). In the 18th century, a dozen Wheatears raised 6 old pence at Eastbourne (about £25 if still eaten today) and a dozen songless female passerines were sold off anywhere at 3 to 4 old pence (about £12 to £17 if still eaten today). Given that the last two classes would provide little more than two nibbles of breast muscle to a hungry person, it is clear that they must have fallen in the once cheap food category formerly inhabited even by such seeming gourmet items as oysters. Going further with some weight calculations, I was astonished to find that the range of 3 to 6

No passionate member of a Fur, Fin and Feather Group would have grasped that the Skylark had been made common by forest clearance and then would be made uncommon by modern agricultural practice. Happily, however, their mass capture for food has long ceased.

The original population of the Bittern was extinguished by 18th century drainage. Its slim modern counterpart has hardly averaged 20 males in the 1990s but the current recreation of fenland reserves may yet secure their strange booming in more places.

The modern series of American Golden Plover records began with this bird on St. Agnes, Scilly from 30th September to 10th October 1962.

old pence per dozen small birds equates closely to the value variation in today's worst to best chicken meat. Recently Bill Bourne has unearthed further poulterers' prices (*British Birds* 96: 339)

More relevant to my general theme, however, is what might be the commercial ornithology of all the hunters. However motivated, the old fen-fowlers, the poachers (already being caught and tried for netting plovers at Landbeach and Cambridge in the 13th century), the fatteners and above all the passerine catchers applied skilled observation and field-craft to their trade. The Bittern was seen as "about as big as a hen, in colour a mixture of yellow and grey, etc, having long legs … It is said to be in the habit of introducing its bill into one of the nearest reeds, and of thundering forth a voice so horrible that those unused to the thing say it is that of an evil spirit, and so loud that two gentlemen assured me it could be heard for three or four miles" (Isaac Casaubon in 1611). Apart from the delightfully but wrongfully attributed ability to play a wind instrument, that is no thin piece of birdwatching and listening. Their pursuits involved real risks – in a 14th century Assize Roll, there is a record of a lad sent out on pattens to collect duck eggs only to lose his footings and be drowned – and the sheer persistence of effort required to lure Skylarks down to clap-nets with the twinkling lights of revolving mirrors, and to set snare after snare for Wheatears under the earth clods of the Downs, would have tested even the most ardent of today's ringers. Having once with eight other helpers taken two hours to secure a Tawny Pipit in a clap-net measuring 8 x 3 foot, my mind boggles at the thousands of man-hours that went into the major captures noted above. That of the Skylarks is particularly impressive, for as the late Chris Mead, one of the Doyens of British ringing, rightly observed, "Nobody catches them any more!" One method was really cruel, the smearing of twigs with a dreadfully sticky white lime. I can just remember one sight of hopelessly fluttering birds, Snow Buntings I suspect, on the South Denes of Great Yarmouth at the outbreak of the Second World War. The area was a long-established trapping site and it gave me the most unpleasant of my early birdwatching memories.

About 15 first British or Irish records of hitherto unrecorded species stem from the activities of those who traded birds. Of these, the most striking concentrations were of Crested Lark, Blyth's Pipit, Common Rosefinch and Rock, Rustic and Little Buntings along the Sussex coast between about 1845 and 1902 and of Richard's Pipit, Black Redstart and Ortolan from the London area between 1812 and 1829. The southern fen delivered Lapland Bunting in 1826 and Shrewsbury provided Short-toed Lark in 1841. The Black Duck, American Wigeon, American Golden Plover and Eskimo Curlew are the only species whose first records are still sustained by imprecise discoveries in shops and markets.

Among the birds shipped across the English Channel over 400 years ago were gulls, but for major harvests of seabirds, two Scottish and one English adventure stand out above all others. These were the cropping of particularly Fulmars and Puffins in St. Kilda, of young Gannets from Sula Sgeir and of auk eggs and Kittiwakes from the northern cliffs of Flamborough Head. The seabird city of the great white cape was first noted in 1725 and was lyrically described by Thomas Pennant in 1769. The start of the hazardous egg harvest is not accurately dated but may well have been centuries before the earliest deduced year of 1731, when a sugar refinery in Hull known to use eggs in its process was opened. In 1830, the egg thefts were seriously compounded by the aimless shooting of the breeding birds from boats, and by 1867 a third toll was fully apparent. This was the taking of the wings and feathered breasts of birds for the millinery trade, the main target being the Kittiwakes which were taken not only on the cliffs but also in the fields where they gathered nesting material. The kills were of a high order, for example 150-200 seabirds in one sportsman/day and 4,000 by one shooter supplying traders in one summer.

In 1865, Commander H. H. Knocker of the Coastguards calculated that over 230,000 birds and eggs were destroyed between mid-April and early August every year. This estimate has beein considered excessive but, importantly, it added substance to the clamour of voices raised in defence of the seabirds which included those of Charles Waterton (an early true birdwatcher who considered collectors "closet naturalists"), Professor Alfred Newton, the Vicar of Bridlington, the Reverend Francis Morris and John Cordeaux. A Sea Birds Preservation Bill was introduced into the House of Commons in February 1869. Although a House of Lords committee struck out the clause that would have stopped egg-collecting (and the climbers' local profit) and the St. Kildans were also allowed to continue their truly life-supporting harvest, a long list of breeding seabirds (to the tune of 33 local names but not that many species!), Oyster Catcher and Chough became protected by law on 24 June 1869. The penalty levied was up to £1 (£57 today) a bird "killed, wounded or taken". Sadly the historic Act did not stop the slaughter, merely switching its timing into the pre- and post-breeding periods. As late as 1902, a London milliner placed an order with a Bridlington entrepeneur for 10,000 Kittiwakes and Little Terns. Finally, however, local politicians took action and in 1905 an extended close season and the banning of boat-shooting and rocketing brought peace to England's greatest seabird colony. Although reduced, climbing continued until the summer of 1954 but the passing in that year of the first comprehensive Wild Birds Protection Act meant that thereafter only Herring Gull eggs could be taken. The last of the courageous climbers turned away from the cliffs and the famous local 'seabird pie', a round of eggs baked in pastry, was no more. Oddly for all the frenetic activity, no marine rarity fell to sportsman or climber. Four Brunnich's Guillemots were claimed from the general area of Flamborough Head between 1895 and 1909 but all have since been rejected. In terms of species, the only lasting loss from the centuries of depradation has been the Black Guillemot, which was always scarce and no longer breeds anywhere along the North Sea coast south of Kincardine. Much more significant than such minor ornithology was the achievement of lasting seabird protection, hardly the original intent of the brave climbers and the uncaring sportsmen and collectors.

Happily today the Kittiwakes of Flamborough Head, Yorkshire, are left in peace. Nearly 84,000 pairs were counted there in 1986.

Expedition to St. Kilda, 1956. Aboard the *Lochmor* between Mallaig and Loch Boisdale, 2nd July. From left to right, Dougal Andrew, John Arnott, author, David Wilson. Note size of what was then the BBC's smallest tape recorder which weighed "half a ton"!. JC.

As the *Maid of Harris* begins to pitch, apprehension grips the party. From left, Dan Bateman, Dougal Andrew, David Wilson and author. All but David were totally seasick during the nine hours voyage to Hirta.

Photographing Leach's Petrel by the Manse, Hirta, 5th July. From left Peter Naylor, David Wilson, Dougal Andrew, Dan Bateman. Note 8mm cinecamera. JC.

Author releasing Leach's Petrel after photography. Village Bay, Hirta, 5th July. JC.

Village, Hirta, St. Kilda, July 1956. Only the Manse (by shore) remained undamaged after two and a half decades of desertion.

Fraternising with Spanish fisherman on the quay, Hirta, 10th July. On left, John Arnott receives gift of huge bottle of red wine.

Birding in the Scottish Highlands, summer of 1958. Robin Emmett cooking English breakfast just before a Scottish gust blew the bacon out of the pan! REE.

Bob Emmett experiencing cold bum for hot eagle. PRC.

Peter Colston triumphant on a Cairngorm summit cairn. REE.

Red Squirrel breakfasting on Roy Dennis's wrist. REE.

Imperial Pursuits

*The British imperial writ of ornithology – in a colony - in an empire –
international and regional bird recognition - an astronomical measurement of bird
slaughter – three unforgettable authors forgotten – the apocalypse of* Stray Feathers *–
a survivor of rhino and leopard attacks – the first pocket guide – Henry Seebohm, not
of Japan but surely of Siberia –* his assistants in terra incognita *- the collecting zenith
– the robbing or killing of eight Emperor Penguins – the fates of their assailants*

One of the comforts formerly provided by Bartholomew's Atlas was the many
lands on all the continents that were coloured pink for Britishness. In the greatest
spread of the colour and particularly during the two World Wars, it entailed upon
us maximal geographic reassurance. The free range afforded and the duty deferred
to the parent nation's passport holders was amazing. Particularly in the 19th century
with no hint then of the unwinding of economic and political connections that
would follow, the writs of the second and later sons of the British gentry in their
service as military officers, administrators and world merchants were almost
boundless. From these classes and their many leisure hours sported an amazing
number of sportsmen and naturalists, not unique within European nationalities
but inevitably more numerous and much further scattered around the globe. Not
content with unravelling British nature, we took on the description and classification of most of the world's other animals as well. Here a stuck pig, there yet
another new species. As the most overt class of all creatures, birds were as usual
the most obvious subjects of our study.

Lammergeiers inhabit the finest montane wilderness of the Old World. Here one sails over Mackinder Valley on Mt. Kenya. In the foreground, an Alpine Chat and a Scarlet-tufted Malachite Sunbird make the long climb even more worthwhile.

To imply that the gifts of national, even sub-continental ornithologies to the pink lands were acts of British altruism would be to go far too far but our territorial acquisition and economic rapacity was tempered by some strands of conservative habitat management and a huge volume of scientific exploration and description. To illustrate the past peaks of our international ornithology, I shall address particularly one colony, one empire and one climatically harsh merchant destination, respectively Kenya, India and the sable lair of Siberia.

I was dead lucky with my National Service. On the date of my commission in 1952, the subaltern cadre of the Royal Scots was already full. "So sorry, Ian" said the Colonel, "you can't come with us to Korea". Stifling the total release from terror that his apology provoked, I muttered a vast lie of disappointment and proceeded to Kenya to experience the far lesser dangers of the Mau Mau Rebellion. Like many other birdwatchers in my age class suddenly sent to the ends (and endings) of Empire, I faced a magical but indecipherable field of birds for which there was no guide and apparently no handbook other than Austin Robert's *Birds of South Africa*, half a continent out of sync.

Yet everywhere flew the most marvellous feathered beings. As I made first military and then social converse with long-serving District Officers and white Kenyan families with love of their land and its nature, I was delighted to find piece by piece an already considerable faunal library. The most immediately practical reference to birds was a little green booklet by V. G. L. van Someren on Eurasian winter visitors to Kenya but the real stunner was the discovery, during a quiet time guarding an isolated farm on Mount Kenya, of the luxurious three volumes of Sir Frederick Jackson's *The Birds of Kenya Colony and the Ugandan Protectorate*, compiled by W. L. Sclater and published in 1938. To turn the latter's pages was to be transported back almost to primeval African bird paradise. From these and other references, I found out that within eight years of the British adoption of Uganda and Kenya between 1894 and 1896, ornithological writings had begun to surface to be first summarised into a national statement in 1925.

Clearly ornithological recreation and purpose had not been slow to follow the general colonial imperative. Alas my duties allowed me no time to enter the small circle of birdwatchers and ornithologists then clustered at the Coryndon Museum in Nairobi. My one accidental meeting with such a person came in the aftermath of a hopeless forest sweep for black guerillas above Meru on the east flank of Kirinyaga, to give Mount Kenya its proper name. He was Con Benson, one of two famous brothers, and his knowledge was truly African. Yet my most lasting memory of Kenyan natural history skills came not from any expatriate but from my exuberant askaris. The Samburu and Elgeyu especially could track any beast with feet and had their own tribal names for any sizeable bird. I do not know of any research into native bird lore, other than in the oft repeated wonder of the Honeyguide symbiosis, but no wonder so many true Kenyans have made good naturalists and conservators. Oddly I came upon little trace of the enigmatic Richard Meinertzhagen who 50 years earlier had been sent on a punitive expedition against the Nandi of Northern Kenya but we shall return to him later.

British rule and the accompanying early ornithology of Kenya ended in 1963,

The pheasants of south-central Asia much attracted the sporting gentry of the Raj. Of those that they introduced to the British countryside from the early 19th century, the community of Lady Amherst's Pheasants at Woburn, Bedfordshire, is the most appreciated by birdwatchers.

16 years after the Indian sub-continent underwent partition and returned to the two main religious camps of its indigenes. There, the British involvement had been three centuries long with rule through the so-called Supremacy being established as early as 1763. Astonishingly, the first signs of our ornithological imprint in India had begun in the late 17th century – the era of Willughby and Ray – when for example the resplendent pheasants of the Himalayas got attention from Major General Thomas Hardwicke. After the watershed of power for which he, other military men and the East India Company fought so long, it took only six years for British-based authorities like Pennant and Latham – at a range of over 3,000 miles – to begin the scientific naming of Indian birds. This was accomplished with only vicarious experience, drawings of shot birds and the odd tattered specimen. In an investigation of the more difficult passerines, I found that for Indian chats, Linnaeus had been even quicker off the systematic mark, having named, for example, the Magpie Robin and White-fronted Redstart in 1758. From 1817, even smaller, less distinguished passerines were described by the early observers cum collectors. In an investigation of the nomenclature of 123 warblers and chats covered in the eighth volume of the *Handbook of the Birds of India and Pakistan, together with those of Bangledesh, Nepal, Sikkim, Bhutan and Sri Lanka*, by Salim Ali and S. Dillon Ripley (1973), I found that over 20 Britons, four Europeans and one American had been involved in the initial flood of their

The Magpie Robin was one of the first Indian species to be classified from early specimens in the mid-18th century.

recognitions from 1821 to 1874. Due to Ali's happy inclusions of local dialect names, it is also possible to glimpse the indigenous knowledge on which the British collectors drew. The fund was enormous; even of 36 skulking Palearctic migrants, 19 had attracted at least one local recognition and to mention one species that is still a tough target in the field, the Blyth's Reed Warbler had six. How had such familiarity arisen? In the search for food, came the answer. In a remarkable assessement of *The Struggle for Existence of Birds in India* (1920), Donald Dewar made "very conservative" estimates of the death toll along their interface with hungry human beings. He reckoned that hundreds of thousands of hunters used lime, nooses, nets and traps to harvest an annual crop of 1,000,000,000 birds. With such slaughter long dated, no wonder so many attracted clearly named identities.

The first British specimen of Blyth's Pipit lay misidentified in a museum drawer from 1882 to 1963. It took 50 more years for its modern field records to be taken seriously. In January 2002, British observers found two birds (second and third from bottom) with a Tawny Pipit and Citrine and Yellow Wagtails at Sohar, Oman.

In the description of the Indian avifauna, the efforts of British Army officers and doctors dominated the few contributions from administrators. The commonest rank of the first group was Colonel and its members often employed small legions of local hunters and skinners as they filled the maws of both local and distant museums and display collections. In India, the ornithological crucible was most concentrated in the Asiatic Society of Bengal. There in the mid-19th century, Edward Blyth presided over a collection that grew to be the largest in the world outside Europe and America. Gould accredited him with the title "father of India Zoology"; in Britain today, few will know that he did rather more than name one difficult-to-identify warbler and a similarly awkward pipit. The first full description of Indian avifauna to be leavened with considered observations, habitat notes and even field sketches came from an army doctor, Thomas Claverhill Jerdon. Coming from Roxburghshire, he brought the sensitive intelligence of a Lowland Scot to the writing of his *Birds of India*, published in 1862 and 1864. Drawing on wide travel, overcoming the eating by moths of his early specimens, working for 35 years at his task, Jerdon monitored the addition of nearly 800 species to the Indian bird list. The mind boggles at such an Herculean achievement; yet he was an expert mammalogist too.

In contrast to the general scientist in Jerdon, Allan Octavian Hume was the sub-continent's impresario of driven collecting. Like Gould, he used the purchase of others' specimens to swell further the drawers that his own small army of over 50 native hunters were filling. The outcome was a collection of 100,000 skins, nests and eggs and an Everest of associated knowledge. Yet in his wish to write an even greater work than that of Jerdon, Hume suffered an apocalyptic loss. In 1884, two years after giving up his duties as a senior agricultural and customs administrator in the North West Provinces, he took a winter break from his Simla home. In his absence, his unthinking servants sold his life papers for their waste value of a meagre £25. Hume turned away from ornithology to politics and, apart from two books on gamebirds and the nests and eggs of Indian birds, all that survives of his meetings with Indian and Asiatic birds is his personal journal *Stray Feathers*, published from 1872 to 1888. How uncanny that his choice of title should be so prophetic of the hapless disposal of 25 years' work. Once again an amazing personality is saluted, in Britain today, only in the form of an Asian vagrant, Hume's (Yellow-browed) Warbler. The descriptive ascent of the Indian fauna during the

The Bewick's Swan (nearest) was not separated from the Whooper Swan (behind and in distance) until 1830. The former was one of the mysteries that drew Henry Seebohm to Siberia.

Raj culminated in E. C. Stuart Baker's majestic *Fauna of British India* (1922-31). Birds alone occupied eight volumes from the eventual single hand of another Imperial naturalist who was trampled by a rhino and lost an arm to a leopard, yet still shot snipe. (Look upon such men and their works, ye twitchers, and quail). The tales from the ornithological Raj are endless; one wave of them swamps most adventures from the participants' home nations without trace but this is not the place for more. I turn back towards Britain again.

A man with a foot in both camps is another hero of mine, Colonel Robert George Wardlaw Ramsey. Yet another Lowland Scot but educated at Harrow, his boyhood birdwatching was fashioned into skilful adult ornithology by a legacy from his uncle, the Marquess of Tweeddale. This combined a huge collection of skins with a brimming library of bird books, no standing start. Wardlaw Ramsay soldiered for 11 years from Afghanistan to Burma, adding several new species to the Oriental list, but his legacy to British, European and North African birdwatching was even more innovative. This reserved but "singularly attractive" man, revered by Eagle Clarke and President of British Ornithologists' Union during the First World War, decided that the sequential, ever longer handbooks should have a concise competitor. So he began his "great desideratum", the wonderful distillation of plumage description, distribution, status and biometrics that is his *Guide to the Birds of Europe and North Africa*. Although it lacked illustrations, the store of knowledge in the 355 pages deserves appellation as the first ever pocket or field guide. Appearing in 1923, it was five years ahead of Wilfred Backhouse Alexander's *Birds of the Ocean* which is often accorded first place in the genre of bird identification guides but in my view is the runner-up. Sadly, Wardlaw Ramsay did not live to see his little encyclopaedia published but faced with the premonition of his imminent death, he had the sense to place the incomplete manuscript with Eagle Clarke. He and John Stenhouse did the rest.

My copy of Wardlaw Ramsay's fact-trove was bought on the advice of Dougal Andrew. In his Palestine service in 1947, it had already served him well and when he learnt of my imminent departure to Kenya, he ordered that it should be in my kitbag. So early in 1952 for 12/6 (£8.90 today), I bought the copy that had been previously owned by a Marion Broughton. She had added but one note to the text, the length of a dead Redwing. For 20 years, the book got no such light treatment from me; I updated its nomenclature and taxonomy and added first locations for new birds seen abroad from Austria to Malta, Sudan and Kenya. It has travelled over 20,000 miles with me and its delivery of ornithological value has rivalled that of the Witherby *Handbook*. Since no guru guide gives you complete systematics down to subspecies and potted measurements, I still refer to it. At 80, the book and its author serve on.

I mentioned earlier that some descriptions of species were totally vicarious, being based on the import of specimens and their labels. One man used this basis alone to describe the birds of another empire, the lands virtually closed to foreigners in the 19th century which we now call Japan. His name was Henry Seebohm and his was the ultimate act of what now seems incredible trans-Imperial, scientific gall. Where did he get the confidence to publish in 1890 from the medium of entirely bought collections *The Birds of the Japanese Empire*? From a grasping personality (unusual for a Quaker and disliked by Gatke who worried that Seebohm would steal his work) and another life driven by the pursuit of birds in far places. His most famous adventures took place in yet another empire difficult to penetrate, Czarist Russia. Fur-traders had penetrated its high Siberian latitudes as early as 1611, eager for the sable coats of wolverine, but visits by natural scientists had not begun until 1837. Although "no Englishman had travelled from Archangel to the Petchora for 250 years", Seebohm was driven by the fact that five Arctic waders and the Bewick's Swan still had breeding grounds "wrapped in mystery". Sensing that only by exploring east of Arctic Russia into Siberia would he find them, Seebohm used the profits of his Sheffield iron foundary to finance two riverine journeys into the wilderness of taiga and tundra. On the first, John Harvie-Brown was his companion and, about both, Seebohm wrote some of the most ripping yarns of British ornithology. In 1875, he explored the basin of the Petchora River and in 1877 he investigated almost the entire course of the Yenisei. The original titles of his books were aptly *Siberia in Europe* (1882) and *Siberia in Asia* (1884). Today they are most accessible in the combined posthumous volume reprinted in 1901, called *The Birds of Siberia*.

The wealth of Seebohm's narrative is astonishing. Birds and the exigencies of their discovery and procurement are the main subjects but his powers of observation are reflected in descriptions of items as varied as ikons to native dress and of people who ranged from sterling boat captains to corrupt local politicians. Fighting particularly against the fickle climate, Seebohm led and organised an amazing harvest of discoveries including the top prize in his eyes, the eggs of the Grey Plover. Even now, as I discovered when I wrote the British history of the Scarlet Grosbeak (Common Rosefinch) in 1998, there is prime ornithological data in his tales. Like all the best collectors, Seebohm was also a gifted observer by eye and ear. He was the first to detect the differences given in wingbeat sound by rising snipes. Seebohm's accounts do not drip the names of people as fast as those of discursive authors like Morris but I also searched for what history and human resource he tapped in the preparation and the execution of his two journeys. Seemingly, only two Britons, Josias Logan and William Govedon, had got through to the Pechora in the early 17th century, respectively braving the full Siberian winters of 1611-12 and 1614-15, but they had left no hint of the area's birds. Nor were the Arctic Lapland explorations of five Britons, including the pioneer John Woolley and Seebohm himself, that preceded his eastward explorations particularly instructive. So Seebohm looked for foreign experience of Siberia and found that only six Russians and Germans had made any prior exploration of the Pechora and Yenisei, beginning in 1853. He faced almost *terra incognita*. On his first journey, he had the sympathetic and skilled assistance of Harvie-Brown and recruited three helpers, a Pole soon nicknamed 'Cocksure' due to his headstrong behaviour and two Samoyedes to increase his collecting and skinning capability. Otherwise he found only one priest and three other educated persons with bird knowledge.

On the second journey, two months were needed just to reach the Yenisei, by train and sledges drawn in turn by horse, dog and reindeer. Captain Wiggins

"Seebohm was a Quaker iron-founder and therefore no gentleman, especially since I feel he was usually right (e.g. in promoting trinomials), which was of course particularly unforgivable. Read between the lines of the introduction to the Dictionary of Birds *for some prize Victorian back-biting."* Bill Bourne, *in litt.* (2001)

The *Dictionary* here is the original *co-magnum opus* of Alfred Newton (1893-6).

The Common Rosefinch is expanding its range west across Eurasia from its winter home in northern India. In 1992, it made an attempt to colonise the east coast. Shown here is a classic trio of a mated pair with nearby a hopeful young male.

made a good but ornithologically unskilled companion but otherwise he was helped along 15,154 miles by only one priest and five other persons, one another Pole and one a Jew. Local observers met by accident numbered only three, a wandered Scot, the mate of a schooner and, in of all places the ill-fated Ekaterinburg, an "amateur ornithologist" working in a telegraph office. On both journeys, his regular sprinkling of roubles quickly produced live and dead returns from native peasants but Seebohm noted that only one native tribe, the Samoyedes again, had their own names for a few species. When writing up his journeys even in the more popular account that I have reviewed, Seebohm referred eventually to 18 direct authorities on identification and status and 11 indirect sources on nomenclature. Within all this, it is the apparent poverty of the local interface between man and bird that is so striking. Yet just over a hundred years on, the first modern bird tours were to find "State ornithologists" in every regional city. Once educated and organised, the Russian mind had not been slow to catch up with the west European appreciation of ornithology. Before (or after) you go on any bird tour to Asia, read Seebohm's books.

This chapter is becoming overlong and the above vignettes of British Imperial bird enthusiasts in one colony, one possessed and two foreign empires must suffice to show the relief of their efforts. In their exploration of the world's major bird faunas, they were rivalled in any large volume only by the Germans and the Americans, the latter still justifying active specimen collection. Only South America was largely shunned by our forebears. The start dates and periods of the greatest collections are given in great detail in *The Bird Collectors* (Mearns and Mearns 1998). This fascinating book demonstrates that their initiations peaked in the 1880s and that their cessations were most influenced by the preoccupations of the Second World War. No British museum is currently proactive in a bird-killing science that proved avian systems, fuelled some of our greatest adventures, gave employment to hundreds and, through the displays of the great houses and museums, satisfied the curiosity of hundreds of thousands in a civilisation without radio or television.

One more tale merits re-telling. Dr. Edward Wilson, its chief character, summed up its events as "the weirdest bird-nesting expedition … ever made". In retrospect, it was also the most lunatically brave and a terrible price was paid for just three skins and three eggs. Dr. Wilson served as zoologist and head of science to Captain Robert Falcon Scott's second Antarctic expedition from 1910 to 1912. For six years, he had fretted over his pet theory that the Emperor Penguin was the most primitive bird in the world and that its embryos would show it to be the missing link to reptiles. Against Scott's instincts, Wilson, Henry Bowers and Apsley Cherry-Garrard were allowed to leave Cape Evans on 27th June 1911 on a straight line 67-mile trudge to the penguin rookery at Cape Crozier. It was the Antarctic mid-winter with hardly any daylight but the three men, hauling two sledges joined together, moved out in high spirits. Within three days, soft snow prevented the simultaneous towing of their sledges and only Bowers was not frost-bitten. With the necessary repeats of single sledge pulls, the remaining journey length trebled; sleep became almost impossible; only their hot breakfast gave any comfort and

Emperor Penguins and their assailants below Mount Terror on 21st July 1911.

even after it, they took five hours to break camp and start the next leg. Only the magnet of the penguin eggs and stoic companionship kept them from turning back.

After 18 days of frozen hell, they reached a 800-foot col between The Knoll and the aptly named Mount Terror. Unable to burn their secondary fuel of seal blubber in the tent, their next task was to build a bothy from boulders and packed snow. Two more days saw the walls up but a fierce blow on 18 July prevented roofing. The next day the weather cleared and calmed; it was relatively warm at only minus 37°F. With the adrenalin of renewed hope, they dashed for the rookery on the cape and got close enough to hear the crowing of the adult Emperors. Cruelly with the dim midday light going fast, all that they could see was a series of pressure ridges too high to cross. So they returned to their tent and roofing task. The next day's effort saw the bothy sealed. They ate well and at first light, they went back to the ridges and found a way through and down. In a rising wind, Wilson and Bowers assaulted the colony of about 100 birds. With five eggs in their fur mitts and three ripped-off skins under their belts, the pair hurried back to Cherry-Garrard and the rope that would haul them back up. Clear of the ridges, darkness fell; Cherry-Garrard stumbled, breaking two of the eggs; worse still, for a terrible time, they could not find the hut. At last safely inside, they relaxed, their goal achieved; the wind dropped.

But the lull was the eye of a hurricane. Suddenly the wind returned to shred their canvas roof and blow their tent they knew not where. They faced death, hunched on sodden sleeping bags with only their boulder walls for windbreak. It was Wilson's 39th birthday. Another agony of freezing time was passed to the singing of any remembered song or hymn and the fantasy of sweet food. At last, the wind had less force and mercifully laid in a hollow only half a mile away was their life-preserving tent. Within it, they revived and buoyed up by the retrieval of the remaining six relics of eight parent penguins from the ruined bothy, they found the reserves to resume sledging and so reach Cape Evans.

English optimists will have it that "nothing is written" but the three were still to fall foul of the Antarctic fates. Recovered, undaunted by their near death, Wilson and Bowers went to the South Pole with Scott but this time their return journey ended in a tent that did become a shroud. Cherry-Garrard did not take part in the fatal attempt and eventually returned to London where he announced that as the only surviving collector, he would deliver the eggs with their precious embryos to the British Museum of Natural History. He duly did so only to have to depart without ceremony over, or thanks for, the hardest worn specimens in all of natural history. So with this final slight of human manners, the efforts of the world's most heroic birders ended. They will never be matched.

Today, world birders fly in to penguin rookeries by ski-clad plane and much more importantly the Emperors are disturbed by nothing more aggressive than camera lenses.

No penguins in the North Atlantic but plenty of Fulmars. Peter Naylor contemplates the fulfilled promise of St. Kilda in summer. Summit of Mullach Bi, above hundreds of Shearwater burrows. July 1956. JC.

The First Marshalling of Knowledge

*The magic of migration – early references – Gatke and Heligoland – John Cordeaux –
the first migration surveys – island collecting – a St. Kilda waif – lighthouse records
in Ireland and Britain – the collectors of Fair Isle – a Norfolk prize – inland rarities
– beginnings of local ornithology – succeeding national works – profiles of early
authors and bird finders – Horace Alexander and a Yellow-headed Wagtail*

In Britain it is difficult not to see the migrations of birds. Today many are on a lesser scale than those of 40 or 50 years ago when the near European as well as the British communities of migratory species were much more substantial, but the seeming miracles of invisible overnight fall or visible daytime passage still strike the eye and heart. The general thrills of such pageants, and the particular hope that within them will appear a rarity, tug many a birdwatcher from inland bed to coastal and even island dawn or send a few ever more eagerly to their local patch.

Brought up in a family that followed the pelagic herring around Scotland and up and down the east coast, I was no stranger to animal migration. I first fully recognised bird movement in the early 1940s when the position of my prep school south of Edinburgh presented incoming skeins of wintering grey geese. What turned this brush with a wonderful phenomenon into a lifelong addiction to it was the fortuitous association of my second boarding school with the Edinburgh members of the Scottish Ornithologists' Club (SOC) who pointed the way from Musselburgh to the Isle of May, Fair Isle and the latter observatory's inspiring first Director, Kenneth Williamson. At a time when I rashly assumed that Scotland's national game was rugger and not golf, I also made the mistake of thinking that my country had taught all of Britain and Europe about bird migration. The truth, of course, was other.

"I have been travelling to and from London most days, in order to pore over skins at the British Museum. One day, I tried to find your mother's so that I could scrounge a cup of tea – but I got completely and absolutely lost in a maze of small streets around Marylebone and gave up in disgust. The more I have to do with London, the more I realise what a bird feels like when it gets mixed up with an east wind on migration." Ken Williamson, in litt. (1959)

An example of unsuccessful human drift.

Skeins of Pink-footed Geese, heading for the Moorfoot Hills, are often seen in late autumn from Dalhousie Castle, near Bonnyrigg, Mid-Lothian.

Although references to migratory behaviour occur in ancient literature as far back as Homer, observations linking it to seasonal weather did not reach the printed page in Britain until the mid-17th century. Norfolk's first ornithological chronicler, Sir Thomas Browne, was their author but it took more than a century to pass before the first considered discussions of the phenomenon, by Gilbert White and

Dr Edward Jenner, appeared. These were still inaccurate and indeed as late as 1817 in Europe, Aristotle's myth that summer visitors hibernated was still being repeated. What finally took the scales off our eyes was the initially fortuitous combination of an increasing, gainful collection of unusual migrants, as specimens or display pieces, with a mounting recognition of the weather systems that caused their arrivals. The seemingly magic formulae of spring rush and autumn fall, caused essentially by locally adverse weather interrupting overhead movements, were first recognised on a then British islet in the south-eastern North Sea, Heligoland. (It was swapped for Zanzibar in 1890.) Earlier in 1837, a 23 year old German, Heinrich Gatke, became secretary to the British Governor and began a long association with the islet's trappers and shooters. From their harvest of migrants, he was provided with the rarest birds and greatest gains. Although regarded as the European prophet of bird migration, Gatke did not elucidate the origins of the arrivals that gave the isle its ornithological reputation but nevertheless this spread to Britain in the 1850s, igniting other more imaginative minds. Meanwhile further east in Imperial Russia, another observer had gone a large stride further and organised the first co-operative watch of bird movements in European Russia. By intelligent correspondence, A. von Middendorf was able to correlate the advances of spring migrants, producing maps with isochronal lines. When these were presented to the Imperial Academy of Sciences in St. Petersburg in 1855, bird migration in Europe had a first description of broadfront movement to go with the more individual astonishments of the Heligoland records. Who would take the study of migration on?

In 1972 I re-entered the fishing industry, finding myself in Hull in time to witness the virtual end of the British distant-water fishing fleet and yet another disastrous crash in herring stocks. My solace was the Yorkshire coast and its teeming birds, especially the migrant ones. So it was with great expectations that I approached the gate to Spurn Point on 15 August 1972. Alas as small dogs, even my peerless Sheba, were *verboten*, I was turned away and so I drove straight to the free range of Flamborough Head to begin my long association with England's great white cape. It had always caught my eye on the map but had been rather shrouded ornithologically. As Spurn hosted BTO ringing courses and Filey Brigg attracted early sea-watchers, only a few observers and the odd ringer were working the head and not from dawn to dusk. An irresistible challenge beckoned and so began the Flamborough Ornithological Group and the re-discovery and development of the head's potential for bird observations. I write 're-discovery' because we soon found out that in 1872 and 1894 a certain John Cordeaux had co-operated with the leading Bridlington taxidermist, Matthew Bailey, and compiled from specimens and field observations a fascinating inventory of Flamborough's birds in the 19th century. The relative neglect of the place for the ensuing seven decades puzzled even more.

John Cordeaux was a Lincolnshire farmer who lived at Great Cotes, opposite Spurn Point at the mouth of the Humber estuary. From 1860, he adopted what he called the Humber district (effectively the ill-fated county of Humberside) as his patch, exploring it for migrants as no other British observer had done before

The great white cape of Flamborough Head, Yorkshire, was one of the places that inspired John Cordeaux to study bird migration. Today it has become the foremost seawatch point facing the western North Sea. Here a Little Shearwater (nearest), a Sooty Shearwater (extreme left) and six Manx Shearwaters fly south.

anywhere. Initially attracted to Flamborough Head where he was a member of the group that argued successfully for seabird protection, Cordeaux found only five other local like-minds and turned to correspondence in his quest for an understanding of migration. With no library of bird books, his contacts with Alfred Newton were crucial and it was the latter who suggested that Cordeaux should compare his own findings with those emanating from Heinrich Gatke on Heligoland. Bonding immediately with Gatke, Cordeaux became the first Briton to visit him and experience Heligoland's embryonic bird observatory techniques. He spent five days there in mid-September 1874, a year before Henry Seebohm's more publicised stay. So was formed the triumvirate of great migration students (and friends) that would add the basis of a new understanding of bird movement to western European ornithology, by means of an amazing saga of exploration and co-operative study.

A Heligoland–Humber axis of comparable observations was soon forged. Cordeaux publicised Gatke's findings in *The Ibis* in 1875 and he and Newton became almost weekly correspondents with a man who wrote, "Don't hold back with any queries you fancy I could answer. I shall be too happy to give all and any information at my command on this our mutual hobby". In 1878, Cordeaux also joined forces with John Harvie-Brown, Scotland's powerhouse of faunal stocktaking, to launch his concept of the first-ever nationwide survey of coastal migration. From that year, lighthouse keepers and lightship crew members were to have less boring days and nights as eventually 157 stations around Britain and Ireland received schedules of instructions on how to record bird migrants. The returns provided the first really comparable migration data, and reports on the observations followed smartly from 1881, covering by 1888 all the eight years' work administered by the scheme's two inventors and their fellow members in a committee set up by the British Association of the Advancement of Science. They had shunned the universities, given Newton's advice that an approach to them would not find a similar questing spirit. The rest of the wonderful story needs illustration which is given in the following map and schedule and their captions.

The distribution of the so-called "coastal lights" where migrant observations were solicited by the Committee of the British Association of the Advancement of Science (from Pashby 1985). Its initial members were J. A Harvie-Brown, J. Cordeaux and Professor Alfred Newton; later, they co-opted P. Kermode for the Isle of Man and A.G. More and R. M. Barrington for Ireland. When Kermode proved a dilatory correspondent, he was replaced by W. Eagle Clarke. Cordeaux was the first Secretary, responsible for the correspondence with stations (all crammed into three evening hours of a farmer's day), handing this task over in 1896 to Rev. E.P. Knebley who had replaced More. By 1885, the Annual Reports were

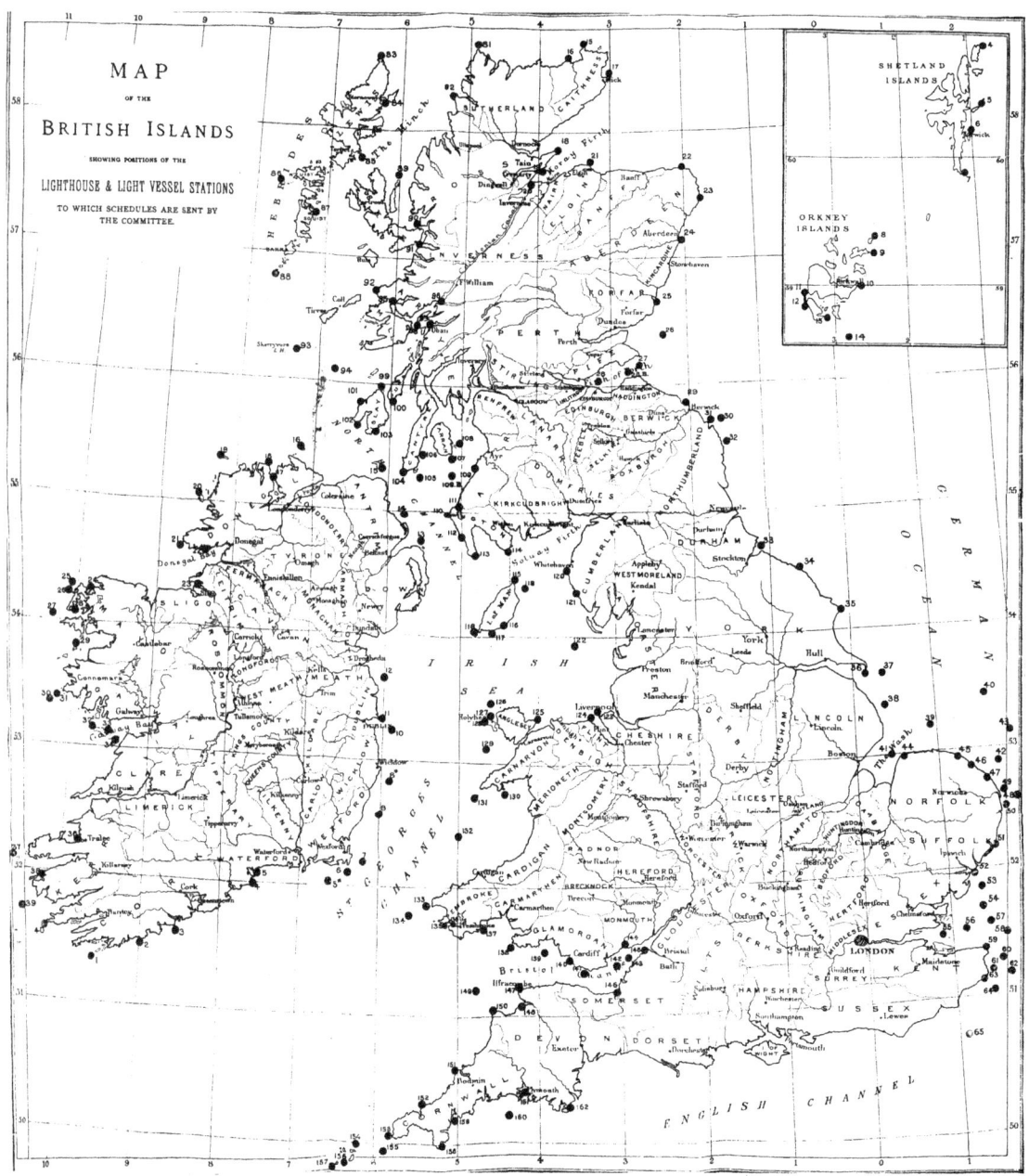

popular reading. After some hesitation from Ireland (over the ownership of the Irish Station schedules), a central archive was formed; the preparation of a full digest was begun by Eagle Clarke in 1888. This mammoth task took eight years, Newton insisting that it should be "written as much for the men (keepers) as the scientific (community)". In addition to the 157 stations on the map, nine others as far apart as Iceland, Sweden and the Channel Isles were canvassed. Up to 1895, Cordeaux's correspondence with Heinrich Gatke added a constant North Sea (or German Ocean) axis of thought. Sadly, Eagle Clarke was prevented by illness from presenting his digest but appropriately that final task fell to Cordeaux, the scheme's amateur architect for 20 years, at the BAAS meeting in Liverpool on 19th September 1896. Gatke's doubts about "how solutions possibly could be arrived at" and his feeling that "man would never understand the riddle" were superseded.

The return from the Pentland Skerries lighthouse keeper for 17th April to 10th May 1889 (from Pashby 1985). A fascinating example of the recording effort from one of over 100 responding stations. Twenty-six species are listed for the 14 separately dated arrivals within the 24 days watch; after the wind went east on 24th April and local occlusion persisted, the numbers and diversity of 'drifted' migrants increased. Of particular interest are the forerunning Black Redstart which recuperated for five days, the Wren which has since been proved by ringing to cross the North Sea, the Grey-headed Wagtail, the 'Grey Owl' (a name coined by Ray in 1678 for what proved later to be the grey phase Tawny Owl), and the classic

The Good Ladies of Scottish ornithology Miss Leonora Rintoul and Dr. Evelyn Baxter gave separate witness to the enthusiasm of the Pentland Skerries watcher. He sent them bird corpses, and was rated "a very good man, tremendously keen".

Observations on the Migration of Birds.

Name of Station, Pentland Skerries.

I. Date.	II. Name or Species of Bird, and Number.	III. Time when Seen.	IV. Force and Direction of Wind.	V. Weather, Clear or Fog, Rain or Snow, etc.	VI. Number of Birds striking glass of lantern, and whether killed or otherwise.
April 7	1 Heron	about 7 p.m.	Fresh W.	Haze	
" 20	Puffins	all day	Mod. S.W.	Showers	
" 22	2 Fieldfares	"	Breeze N.W.		
" 24	Flock of Fieldfares	"	Strong N.E.	Haze	27th. An occasional Black-bird pass now & again. Also 2 Redstarts have been seen for 2 considerable time.
" "	1 Black Redstart	Afternoon	"	"	
" 25	Song Thrush	Forenoon	Fresh S.S.E.		
" 27	Chaffinch & Ring Ouzel	all day	" S.E.		
" 28	2 Pied Wagtails	Forenoon	Strong	"	30th. Black Redstart still remains on island.
" 29	Brambling	All day	"	"	Rain & "
" 30	1 Com. Wren	Forenoon	Fresh	"	"
" "	Arrival of Dunlin	—	"	"	
May 1	Fieldfares, Ring Ouzel, Blackbird, Redbreasts, Reed Sparrow (female) & the Whitethroat (In addition to the above)	All day	Strong E.	Rain	note on page 2
" 3	Green Linnet, Whin Chat & the Grey-headed Wagtail	"	"	S.S.E. Haze	
" "	2 Curlews	Forenoon	"	"	
" "	1 Grey Owl	Afternoon	"	"	
" 4	4 Grey Crows	Morning	"	"	
" 5	Small flock of Turnstone	all day	Fresh S.E.		
" 8	Chaffinch, Blkbird Song	"	"		
" "	Thrush, Blk's Header	"	"	Fog	
" "	Bunting (male) + a few	"	"		
" "	Fieldfare & Ortolan Bunting	"	"		
" 10	Arrival of Corn Crake	Forenoon	Strong E.S.E.	Haze	
" "	Com. Redstart	p.m.	"	"	

simultaneous overshoots of the Black-headed Bunting (male) and the Ortolan on 8th May, of which the former is strangely denied and demoted amazingly to "C. Bunting" by another (editing) hand. As a Reed Sparrow (female) is listed earlier, the keeper can hardly have mistaken a male Reed Bunting for the Black-headed Bunting. The four or five extraordinary vagrants may have seemed too prominent (in a migrant flow of only 38 precisely counted passerines and perhaps 40 other birds) to the contemporaneous judges but modern observations frequently present similarly high incidences of uncommon birds in a spring rush.

The Grey-headed Wagtail breeds in Fenno-Scandia, regularly overshooting to Britain in late spring.

With the publication of R. M. Barrington's digest of Irish results in 1900, the study of migration around what could then be called the British Isles gained entirely new levels of published data and debate.

In 1905, six years after Cordeaux's death, the British Ornithologists' Club (BOC) decided to relaunch the enquiry into migration and formed another Migration Committee, of which Dr. Norman Ticehurst was Secretary. Unlike the earlier schemes which had produced records from pinpointed coastal stations, the new investigation solicited a wider effort, for example from the first few RSPB wardens, then called watchers, and even amateur observers inland. A study of still unrivalled perception, the BOC survey dealt first with 34 scheduled species during their incoming spring migration but later asked also for data on their autumn movements. These were soon recognised to be truly complex, presenting waves of passage as well as mere arrivals and departures. This third phase of co-operative migration study ended in 1914 with the outbreak of World War I.

Through most of the decades just discussed, the collecting of wild birds for gainful sale to game markets, museums and the country houses of specimen display gave employment to the fowlers of the countryside and particularly the east coast of England. By bush shooting, netting and liming, multitudes of birds were scrutinised in expert hands, collection drawers and display cabinets and many discoveries were made. Increasingly, the miracles of vagrancy from Asia, let alone passage through Europe, were demonstrated.

Until 'obtained' on 14th June 1894 as the first British specimen, a Subalpine Warbler would have shared the garden of the St. Kilda Manse with the sturdy local race of the Wren and a Rock Pipit.

On 13th June 1894 following a savage south-westerly gale, a male Subalpine Warbler was found in the garden of the St. Kilda Manse by the 23-year-old Jannion Steele Elliott from Dudley in Worcestershire. An egg collector and keen walker, he knew little of the full zeal of a West Hebridean Sabbath and his enforced inactivity on that day prevented the bird's immediate death. It recuperated in a row of young peas and a nearby seeding parsnip complete with insects. Came Monday, however, and fowling of Fulmars and tiny vagrants alike recommenced. In the presence of Messrs. McKenzie and Fiddies, the bird was shot by Steele Elliott and immersed in spirit for initial preservation. It was sent to a Durham taxidermist, John Cullingford, for preparation as a specimen skin and eventually, on 19th December 1894, this was exhibited by Dr. Richard Bowdler Sharpe to the assembled gaze of the British Ornithologists' Club as an example of the nominate Iberian race. This tale is exemplary of many of the records of vagrants and scarce migrants made as the exploration of other 'potential Heligolands' proceeded. What was hit proved time and again to be unusual history. The ornithological exploration of isolated islands increased as the lighthouses continued to deliver not just a repeated substance of migration but also other first records, for example the Arctic Warbler that struck Sule Skerry light on 5th September 1902 and languished for seven years as Britain's second Greenish Warbler, before Eagle Clarke shot a similar bird on Fair Isle on 28th September 1908 and realised his error.

In retrospect the total product of the lighthouse and lightship observations, which continued beyond the initial schemes in all countries, is unique even in world terms. Clearly the confidence of the initially untutored observers in recognising unusual visitors and casualties grew as the exchanges of confirmed identifications multiplied. The number of truly significant records became sufficient to confirm that European and Asian vagrants were reaching almost every part of the British Isles. Oddly, however, the incidence of such records was much higher around Ireland than around Britain. In the former country, the energetic solicitations and encouragements of R. M. Barrington led to a published 70-year series of over 150 records from 36 stations. These featured 120 rare or very unusual birds, of which 50 were concentrated at south-eastern locations; one was the first ever Pallas's Grasshopper Warbler on Rockabill on 28th September 1908. In Britain, the three successive commands led to a series of nearly 60 records of unusual vagrants from 16 Scottish, 10 English and one Welsh stations. Surprisingly 25 were made along the northern and western periphery of Scotland. Such bias behind the published selection of these records cannot now be examined but interestingly the length of the main British series was only 35 years, with only three dated records after World War I. With lighthouse and lightship men much reduced in number following the automation of the equipment that was once so much their time-consuming charges, one thing is certain. Even with the modern bird observatory system, the amazing exercise will never be repeated and the enthusiasm of its scientific achievement outclasses totally the energetic but essentially repetitive pursuit of birds that is the competitive hunt for modern-day rarities.

Among the isolated isles that were searched, Fair Isle was pioneered by Eagle Clarke from 1905 and soon became recognised as the *locus classicus* of rarities in Britain, rivalling Heligoland and so quickly attracting a small caravan of distinguished bird hunters. They made annual pilgrimages there to share the migrant searches with Eagle Clarke and his islander assistants. Hence in sailed, on the private yacht 'Sapphire', the Duchess of Bedford who rented the Pund croft and keenly observed human breeding rates and sheep dipping by teapot between rhapsodising over the beauties of the western cliffs and small migrant birds. Another notable visitor of Fair Isle's first ornithological era was Dr. Norman Kinnear who went on to direct the British Museum of Natural History. Chief among the islanders who lent extra eyes and ears to these early collectors were George Stout (of Busta), his brother Stewart and cousin Jerome Wilson. Eagle Clarke ended his main series of visits to Fair Isle in 1911 and published his *Studies in Bird Migration* a year later. In an echo of Gatke's incomplete understanding of the phenomenon, the book was flawed by too much account of the theory of concentrated flyways. It was not until 1917 that Eagle Clarke and Leonora Rintoul and Evelyn Baxter, with their experience of the Isle of May, recognised fully the important correlation of east or south-east wind to the arrival of migrants and vagrants from the North Sea. Eagle Clarke came back to Fair Isle after the end of World War I but soon handed the baton of observing and collecting to Surgeon Rear-Admiral John H. Stenhouse who continued the studies until 1928.

It would be wrong to leave the history of the early study of migration without an east coast tale. Today the north Norfolk coast is subject to many conservation regimes and access to much of the littoral wilderness is rightly rationed. A century ago however, it was the hunting ground of three particular tribes of shooters – the Royal and landed gentry, the gentleman naturalist and the usually more native

One of the delights of recent observations in Donegal has been the rediscovery of Greater Redpolls which have flown the North Atlantic to recuperate with local Twites (behind) and a Lesser Redpoll (nearest bird). Their trail in Ireland had been lost for 35 years.

Duchess, a Sussex Spaniel, and Edward Ramm pursue Britain's first ever Pallas's Warbler at Cley next the Sea, Norfolk, on 31st October 1896. Nowadays the 'seven-striped sprite' comes in scores every year, crossing over 3,000 miles of Eurasia in reversed migration.

"When I was working on Fair Isle in 1997 I was lucky enough to find a Pallas's Grasshopper Warbler, definitely a megatick for most of the birders staying on the island. A friend of mine overheard a conversation between two of these birders, which he reported back to me with some glee.

"Guess what?"

"What?"

"A WOMAN found it! Guess what else?"

"What?"

"She IDENTIFIED it!!!"

Hopefully there will be more women birders in the next century, but I don't quite see where they are going to come from …". Jane Reid, *in litt.* (1999)

Yet 90 years earlier, an English Duchess and two Scottish ladies were sharing the lead in vagrant hunting and shooting!

bush-shooter. To the last, a good find and an accurate shot could deliver a gain at today's values of hundreds, even thousands of pounds. As the securing of valuable rarities also removed all doubt of identity, ornithology advanced but at total cost to its avian subjects. The most famed story concerns the first British Pallas's Warbler, obtained at Cley next the Sea on 31st October 1896. Following a month of easterly wind, two 'bush-shooters', Bob Pinchen and Edward Ramm, had gone out again in the search of rare prizes. With them were their canine partners respectively called Prince and Duchess, both 'bush-dogs' of great skill in the finding and retrieving of small birds. Tough but not totally insensitive to icy squalls, the quartet took shelter from one furious spell of rain by sheltering in the lee of an ancient railway carriage. After a while, Duchess pricked up her ears; alerted, Ramm saw a small bird flutter past; it sought shelter in a corner of the sea-wall. Pinchen wrote it off as just another 'titty-wren' or Goldcrest but seemingly the Sussex Spaniel knew better. She followed the target and came to a big grass tussock. Ramm brooded on his initial reaction that the bird had indeed been more than a 'crest but it took several imploring, even reproachful (it was said) 'speaks' from Duchess for him to take up the hunt himself. In the next lull, Ramm braved the gale and approached the tussock. Out darted the sprite to meet a discharge of small shot. Ramm bent down and picked up a warbler new to him; it had a lemon-coloured rump and was eventually confirmed as the first British and second European specimen of what was then known as 'Pallas's Barred Willow Warbler'. Ramm gave due credit to Duchess for her perception which made him a tidy profit. The price paid for the prize of the newest British bird by a E. M. Cannop was then £50, equivalent 104 years later to about £3,800. Intriguingly, Pinchen who had ignored the then ultimate prize in all bush-shooting history went on to become a warden!

So tame was the only modern Houbara at Westleton, Suffolk, from 21st November to 29th December 1962 that it could be adored from car windows.

The best place to muse on the old dealings between Norfolk man and bird is the raised lawn of the Dun Cow hostelry at Salthouse. Where today's birdwatchers trudge past with their triffid-like telescopes and tripods, the ghosts of the old bird hunters still flit amongst the backdrops of wide marshes and shingle dunes. North Norfolk's bush-shooters did not, however, have a monopoly on ancient rarities. Other localised bush-shooting occurred in coastal Yorkshire, Lincolnshire and Suffolk. Elsewhere in Norfolk the Yarmouth Denes were favoured for trapping, as were many localities along and above the south coast as far west as Eastbourne. In the main 150 year period of collection and secure identification which lasted up to about 1925, the east coast from Fife south to Kent provided over 120 records of glamorous near-passerines like the Roller and nearly 200 occurrences of rare passerines.

The successful chase for rarity can be a feast in more ways than one. Britain's third Houbara appeared on Clubley's Field near the Spurn Peninsula on 17th October 1896 and survived a poorly aimed first shot on that day. Alerted presumably by early telephone, W. Eagle Clarke and H. F. Witherby travelled up to 240 miles by unrecorded means to watch it through binoculars on the following day but made no attempt to prevent an eventual lethal fusillade from G. A. Clubley. Witherby went home but Eagle Clarke and John Cordeaux further celebrated the event by dining on the body. They "found the flesh dark and tender, tasting of wild goose with a savour of grouse" (Mather 1986). No wonder the Houbara is so prized by Arab palates. Happily Britain's fifth and last individual, in Suffolk in the early winter of 1962, was only shot in black and white. Now separated from the Houbara as MacQueen's Bustard, it remains a classic 'blocker', that is a bird for which modern twitchers eat out their hearts.

Any wide survey of the historical geography of uncommon target species shows that, just as their mentions in the old books indicated, countrymen of all classes had eyes for unusual birds and often had guns handy to prove their perceptions.

Sir Hugh Elliott was one of the last colonial administrators who leavened civil service with enthusiastic field ornithology.

My favourite character is the wherryman who must have performed a small miracle in barge-handling to be able simultaneously to shoot two Bee-eaters, as they hawked insects over the River Yare east of Norwich on 3rd June 1854, but what is of more lasting significance is the apparent ubiquity of such chance meetings in the late 19th century. My own last two study areas have been 28 and 80 miles from the nearest sector of coast. Yet in the second half of the 19th century the former inhabitants of these patches and the 10-mile circles around them recognised and then secured therein respectively 13 and 12 rarities. Remembering that not even a handy telescope was available before about 1870 and that binoculars lacked central focusing until fully developed during the two World Wars, the visual acuity of the early observers must have been of a high degree. Conversely, the incidence of valuable prizes gained to the number of cartridges expended has been variably assessed as between 1 in 500 and 1 in 1,000. So clearly a large number of bows were drawn at ventures, as exemplified by E. C. Arnold who shot Willow Warbler after Willow Warbler in the hope that the next would be rare!

The modern equivalent of an obtained specimen is a caught or closely watched bird but its identity may remain elusive. This leaf warbler spent nine days in the Parsonage, St. Agnes, Scilly in October 1969. With a cracked voice and changing plumage tones, it brought two world twitchers and even luminaries like Peter Grant to their knees in frustration. It died un-named.

While early bird hunting provided sport for dogs, and sport, gain and a scatter of fascinating specimens for men, ornithological gathering was also being exhibited. David Ballance accredits the first attempt at systematic local ornithology, at least in the form of a partial county list, to Sir Thomas Browne. In about 1662, he made a manuscript contribution to Christopher Merrett's *Pinax Rerum Naturalicum Brittanicarum* which was published in London in 1666. It is known that Browne lived in Norwich for 46 years and his *Notes on Certain Birds found in Norfolk* and other letters contain many glimpses of the rich avifauna of old Norfolk. In his time, duck decoys of the Dutch model were scattered all the way from Norwich to the sea; wildfowl abounded "very much". Hoopoes often came to be described with a hint of aestheticism as "gallant marked"; they were "not hard to shoot". Avocets summered in marshes and Spoonbills had formerly nested with herons at Claxton and Reedham. The next real attempt at lists with lasting value and even dated records came from Northamptonshire in 1712 and coastal Essex in 1730.

Following the appearance of Linnaean nomenclature from 1758 and a first full British list in the first supplement to Latham's *Synopsis* in 1787, a much improved discipline was available to local authors. Lists from Northumberland in 1769, the Sandwich area of Kent in 1792 and Derbyshire and the Selborne area of Hampshire in 1789 all showed an increased focus from reporters and writers alike. Within 15 more years, the readers of Gilbert White's diarised letters, Thomas Bewick's evocative woodcut images and George Montagu's *Ornithological Dictionary* lacked no example necessary to an attempt on written and illustrated ornithology. The national checklist of 1802, combined into the dictionary, invited more local editions and although initially often buried in wide-ranging local historical descriptions, lists of birds for individual localities and counties multiplied.

In 1843, the national discipline in ornithology was further moulded by the completion of Yarrell's *History of British Birds*, a work that was to become through succeeding editions the standard reference for over 60 years. In addition, 1843 also saw the birth of a first conduit for topical ornithology. This was the *Zoologist*,

In the 17th century when the marshes east of Norwich, Norfolk, were almost true wilderness, Spoonbills nested at Claxton and Reedham. From 1998, the fine birds have made breeding attempts at up to six British localities.

a journal launched by another Quaker naturalist, Edward Newman. He was also natural history editor of *The Field* and thus had a unique central viewpoint over both august and popular bird lore. One of the most valuable early roles of the *Zoologist* was to provide space for more county or area records. Attempts at major county avifaunas were made as early as 1809 but only Richard Lubbock's *Fauna of Norfolk* in 1845, of which two-thirds dealt with birds, and A. E. Knox's *Ornithological Rambles in Sussex* in 1849 were effectively such books. According to David Ballance, the day of the county bird book, complete with precise geographic title, arrived in 1863. In that year, an illustrated pamphlet on the *Birds of East Lothian* by W. C. Turnbull was published in Philadelphia, USA, and re-issued from Glasgow in 1867. Much more worthy progenitors of the genre were, however, James Harting's *The Birds of Middlesex*, which appeared in 1866, and Henry Stevenson's first volume of the most impressive *Birds of Norfolk* which came out in 1866. This work was not completed until the third volume was published in 1890.

Intriguingly the *Zoologist* review of the first of these books condemned the concept of county lists as "entirely useless". How strange such comment looks today and in the light of the flood of similar works that followed and the now almost universal genre of bird atlasses. By the First World War, more than 50 avifaunas had been published for English counties and ten parts to the Scottish faunas had appeared. The engine for all this description and stocktaking was the growing affection of a large leisured class for scientific discovery. This was partly fired by the challenges put out by Charles Darwin in his books on the *Origin of Species* (1859) and *The Descent of Man* (1871) but was more sustained by the solid advances in ornithological knowledge that illuminated the inherited affections for and the physical involvements with birds, fuelling further the beginnings of conservation conscience.

Having cut many of my birdwatching teeth in Mid- and East Lothian and London, I sensed the long history of ornithology early in my experience of birds but the images of the early authors remained hazy for 50 years. My tardy investigation of them was greatly aided by the biographical portraits included by Simon Holloway in *The Historical Atlas of Breeding Birds in Britain and Ireland 1875-1900* (1996). Forty-six of these featured partly comparable information. There was no marked correlation between age and the output of a major work. The youngest compiler was an Eton schoolboy of 16 called Alexander Kennedy but he was greatly aided by other senior authorities. The rest of 35 aged authors fell into three year groups, from 25 to 35, from 42 to 51, and 56 to 71, with one straggler at 77. Along 32 educational paths, universities and the foremost private boarding schools were clearly important stations. Experience of Cambridge-led thought was the fortune of ten; of the 13 other institutions, Oxford had nurtured four minds and Eton, Harrow and Uppingham together seven more. Of 46 occupations, only 12 were in commercial trade or banking; 31 were in the professions or service to the community, with eight museum or collection curators the foremost and six curates and five scholars the other leading groups. Only two were taking any living from their subject. In terms of writing effort and prior research, the most industrious author was John Harvie-Brown, solely or partly responsible for seven books, and T. E. Buckley, Harvie-Brown's collaborator, for six. Eight other authors wrote two books each. With so much personal investigation required, the years of preparation noted for nine works ranged from two to 30 and averaged 13. Yet between 1866 and 1912, the determination of local nature and particularly birds attracted an unstoppable fraternity of observers and authors. The full temporal pattern of 59 main publications was mapped in Fisher (1966), although they can be most easily traced through Holloway (1996).

At 21:00 hours on 22nd September 1954 after a cold crossing on the 'Good Shepherd I', I stepped onto Fair Isle for the second time in my birdwatching apprenticeship. Among those who greeted me and the rest of the incomers was a notably thin man with a broken arm strung across his chest and an oddly crumpled face but yet possessing the unmistakable air of a gentleman. His name was Horace Gundry Alexander. As we thawed out in the observatory, Ken Williamson told us of a mysterious wagtail that was tantalising him and the incumbent visitors. Horace

The handwritten card reads:

SPECIES — WAGTAIL, YELLOW-HEADED

Trapped — Gully — Date 20.IX.1954 — Time 1745 GMT.

WEIGHT	WING FORMULA	COLORIMETER	
18.09	1st 2nd ⎫ = longest 3rd ⎭ E ⎰ 4th 1 shorter ⎱ 5th 3 " 6th 12 " 7th 17 "	Red Yellow Blue Dial	Hind claw 11 mm. Hind claw within toe 20 mm.

RING		Wing 85	mm.	Tarsus 27	mm.
B88594		Bill 16½	mm.	Tail 82	mm.

PARASITES — One Ornithomyia fringillina

Examined by ———— 1677

The visual record of Britain's first Yellow-headed or Citrine Wagtail began with these field sketches, which were later worked up to a finished line drawing. The card documented the biometric details taken as it was examined and ringed.

or H.G., to give him the couplet of initials that were his ornithological binomial, made no comment. Sleep was difficult; my first search for the bird on Buness was fruitless and a glimpse of it by the observatory lounge was infuriatingly just that. No other member of the heavyweight incoming team which included Guy Mountfort and James Ferguson-Lees had any luck at all. Rather sneakily, the ornithological fates had decreed that the bird would be re-trapped by the observatory maid on her way to breakfast duties. That meal over, Ken put us out of our miseries and announced the bird's presence in the laboratory. Immediately it was surrounded and "heavily examined", as my diary recorded. With nothing like it in the Witherby *Handbook*, even the experts were scratching their heads. "If I was still in Kenya", I ventured, "I would be thinking of Mountain Wagtail but that's impossible". "If I was in India", countered H.G., "I would be thinking Yellow-headed Wagtail. Is there not something about that species in Gatke?" Ken turned to the bookshelves only to find that the observatory's copy of the book on Heligoland's 19th century rarities was in German. Entered Dad who in the course of exporting salt herring to Danzig before the war just happened to have become fluent in that tongue. "Go on, Dad", I said, "you do the sentences and I'll give you a hand with the *unds*". My sally brought a relaxing laugh to the assembly and the decoding of Gatke's account went well. A strong possibility emerged that the

SEVENTY YEARS
of BIRDWATCHING
H.G. Alexander

Robert Gillmor's cover design for Horace Alexander's entrancing reminiscenses.

wagtail was indeed a Yellow-headed, now called Citrine. H.G.'s smile went off the sides of his face. The wagtail spent the rest of the day near the observatory and my diary note stated (in the first of my many repeated worries over excessive bird-handling) that it "seemed to have recovered from the bashing it received early on". It finally left overnight on the 24/25th and its identity was fully confirmed in a skin examination by the senior observers.

I never became as close to H.G. as several of my peers but he never forgot the Wallaces' part in the wagtail debate. Bouts of later correspondence from within and without the Rarities Committee touched particularly on the then ultimate challenges in leaf warbler and stint identification. My respect for him soared in 1974 when his ornithological autobiography appeared. *Seventy Years of Birdwatching* is a remarkable testament not just to his good self and Quaker family qualities but also to the springs of the purposeful but joyous hobby that is birdwatching. From 1897, he was a direct witness to much of what I have been trying to describe in this chapter, the first full marshalling of observer cum shooter effort and local knowledge.

William Eagle Clarke

William Eagle Clarke was born in Leeds in 1850. From his schooling at its Grammar School and Yorkshire College came general nature study and then a growing predilection for birds.

His first ornithological paper appeared in 1879. After a curatorship at Leeds Museum, Eagle Clarke migrated north to Edinburgh where from 1888 he worked in the Natural History Department of the Royal Scottish Museum. He became its keeper in 1906, having developed into a "fine editor" and "thoughtful faunist". He is best remembered for the originality of his own intrepid "island going" after rare migrants, especially on Fair Isle which he pioneered from 1905, and his share in the distillation of the remarkable lighthouse observations. He wrote the two famous volumes entitled *Studies in Bird Migration*, published in 1912 (and effectively the brief for the bird observatory movement that began in 1933). Two of his other favourite places were the Pentland Hills and Aberlady Bay, still enjoyed by succeeding generations of fellow hill walkers and birdwatchers. He died in 1938.

In this photograph taken on Fair Isle in October 1912, Eagle Clarke is the epitome of Victorian gentleman cum specimen collector. It is the shotgun in his hands that takes pride of place as an instrument of record. His binoculars are tucked over his oxter, hardly in a ready position. Even so his eyes are sharp enough to spot Britain's first Thrush Nightingale and Dusky Warbler and his aim good enough to consign them to museum drawers where they can still be inspected. Truly what was hit is history.

Left: passengers 'zipping up' for the crossing to the Isle of May, 21st April 1950. WJW.

Below: passengers 'marooned' KY112 leaving Kirkhaven for Fife. WJW.

The Low Lighthouse, Isle of May, April 1950. Home of the second oldest bird observatory in Britain. WJW.

In the 1950s, anyone thrilled by birds and keen to get close to them went to bird observatories. Equipped with traps and helpful expert wardens these magic places provided the platform for many a field ornithologist's takeoff.

Four hopeful observers on the Isle of May, April 1950. From left, Jim Gibson (opera glasses in case, small stalker's telescope), E.L. ('Eek') Turner (French master and leader of Loretto School Ornithological Society), Nigel Turner and author (Zeiss 6x30 binoculars, ex-German Navy, without central focusing but many observers' start in post-war optics). WJW

The Ornithological Cusp of the 19th and 20th Centuries

The importance of higher education – the dominance of Cambridge-led thought in the late 19th century – Alfred Newton, Victorian professor and mainspring of ornithological change – his part in early migration study – the lost history of the Cambridge Bird Club – a trio of its members before the First World War – three Quaker brothers and their brilliance – a salute to Newton's ghost – Oxford's limited contribution – Darwin's lack of ornithological rank

God is reported to have delivered "the fowl that may fly above the earth in the open firmament of heaven" on the fifth day and, after an actual time lapse of 140 million years, man on the sixth. It was His early saints and ministers that first took sensory and sensible stock of the earlier class of being. As I have already shown, however, their partly apocryphal hold on early bird recognition was not a quotient of Holy Orders. As in all other early sciences, the associated privilege of education was the *vade mecum* and, as we have just seen, that portal with sufficient leisure for the close pursuit of animals was the social formula for the growth of their study throughout the wen of the British Empire and beyond.

So I investigate next the two oldest and greatest seats of learning in England, Oxford founded in 1249 and Cambridge with its earlier start uncertainly dated but within the 12th century. In the girding of modern ornithological loins, what did the ancient rivalry of the two universities produce? Undoubtedly it was two 'light blue' clerics, Turner and Ray, that first took up the baton of British ornithology. Their start is agreed by Fisher, and, more pertinently, by a Royal Chaplain, Charles Raven, who was Vice-Chancellor of Cambridge after the Second World War (and the Master of Christ's College when Dr. W. R. P. (Bill) Bourne pursued there his medical and zoological studies). Never fraternising with younger birdwatchers, Raven nevertheless wrote splendid popular ornithology in the 1920s and important histories in the 1940s. Both series of books confirm that the fundament of British ornithology was served directly by Cambridge learning. Moving on to the mid-19th century, Cambridge-educated minds appear again to have formed the foremost intellectual group among the authors of early county avifaunas but where did the beginnings of the modern system of bird study and bird caring appear? Outbursts of geniuses like Gilbert White and an astonishing teenage lass called Margaret Emily Shore, who made even more endearing and thoughtful observations before her death at only 18 in 1839, were beacons of remarkable spontaneity. For the gift of a framework of initiatives that would give academic ornithology and joyful birdwatching a closely parallel set of rails, we have to thank another Cambridge and Magdalene man, Alfred Newton.

Newton ended his life as a stooping, ancient-habited, top-hat-toting, utterly misogynistic, white-haired and -whiskered senior don, shuffling along the Cambridge streets aided by two sticks, his mind fretting still on the extinction of his favourite other being, the Great Auk. He was hardly the epitome of a thrusting Professor of Zoology and Comparative Anatomy. His father was a West Indian planter who reinvested his profits in a squiredom near Thetford and the education of Alfred and his brother Edward. The latter opted for the Colonial Service in the Malagasy Region, becoming an important collector there, but the former stayed on at Cambridge to become the university's foremost natural scientist of the 19th century, enjoying ten years of travelling fellowship and dispensing four decades of full professorship. His early field experience came from the high latitudes of the Holarctic 'bridge' and his writings included Britain's second great *Dictionary of Birds* (1893-1896) and other works on subjects as diverse as evolution and extinction. A high Tory opposed to any abrupt disruption of the scientific continuum, his behaviour towards young people has been variously reported. Fisher gave him credit for finally bringing the hobby-pursuit of birdwatching into the fold of biological science but a closer witness – another collector, Cambridge don and Richard Meinertzhagen's brother-in-law, Dr. A. F. R. Wollaston – took absolutely the opposite view. Writing Newton's life story in 1921, Wollaston decided that his extreme conservatism had put the more academic lobby of Cambridge science off ordinary birdwatching for good.

Whatever the schisms of his commentators, the achievements of Alfred Newton in terms of legacies still benefiting British birdwatchers and ornithologists and ordinary society are outstanding. They range from the assembly for the University of a leading library of bird books and a great collection of skins (including those of his brother), the foundation of the British Ornithologists' Union – he, Edward and ten other gentlemen "attached to the study of ornithology" gathered in his college rooms in 1858 – the drafting of the earliest general statute on bird protection in 1880, 13 years after his first public denunciation of Victorian bird slaughter, the secure management of the second stage of the early study of bird migration in 1880, and the payment of the first guinea subscription (about £77 today) to the precursor of The Royal Society for the Protection of Birds in 1889.

Whether any pair of Great Auks ever bred on a Scottish rock shelf is now doubted. Only the ghosts of the St. Kildans know. The last southern specimen was taken on the Waterford coast of Ireland in 1834, ten years before the last two of their kind were bludgeoned to death on Eldey, Iceland.

Let us return to the reporting schism, however, and see if the Professor ever enjoyed and shared birdwatching adventure in his later years. Undeniable was his attraction to the mystery of bird migration, this having been particularly stimulated not by news from Heligoland but by the deduction of a Swedish poet Johan Ludvig Runeberg that birds must migrate south in winter in pursuit of a still rising sun. This appeared in *The Times* on 8th September 1974 (the year of John Cordeaux's first experience of Heligoland). Newton and that other champion of evolution, Alfred Russell Wallace, were stimulated to discuss the phenomenon of migration in *Nature* and in 1875, Newton contributed an article on it to the *Encyclopaedia Britannica*. Furthermore, he went out into the field to see movements for himself. At Hunstanton on 14th October 1877, he was the first observer ever reported to witness and surely be thrilled by the westward coastal passage of autumn diurnal migrants along the north Norfolk coast, as noted by Cordeaux in his review of migration on the north-east coast of England in 1878. The passage, although depleted, still inspires observers today. In 1880 came Newton's appointment as the chairman of the general migration enquiry of the British Association for the Advancement of Science, the sequels to which I have already described.

Along the wide horizon of the north Norfolk coast, diurnal passage can still thrill the soul. At Holkham, a migrant Sparrowhawk flushes two Snow Buntings. Above them, Lapwings, thrushes and Starlings move into an autumnal zephyr.

The Cambridge crucible of ornithology needed an injection of hot youthful metal, however, and the odds are that once again Newton may have seen to this, if his celebrated Sunday ornithological soirees attended by young people qualify as the first-ever gathering of birdwatchers in Britain. The loss of old records prevent full proof of this and the dining club of Edwardian students that replaced the soirees, its members soon lost to the First World War, but it is surely significant that the first-ever list of Cambridgeshire birds, however scant, appeared in 1904, three years before Newton's death and one year before that of the Edwardian reconvening. If 1905 was indeed the start of shared birdwatching at Cambridge – Bill Bourne will have it that this was verbally agreed at the current club's '50th' anniversary by no less authorities than Dr. David Lack and Tom Harrison – then the club's metamorphosis began before any other in Britain, although even this is only arguable if you ignore the earlier births of the general natural history societies which began in London in 1858.

As the retrospectively hapless organiser of the '50th' anniversary dinner in 1955, I committed several sins, for example forgetting until very late in the process to invite Dr. W. H. Thorpe who was either the re-constitutor or, in his own view, the founder of the club in 1925. As, however, I was never in charge of any old papers, I plead not guilty to Bill Bourne's recent charge that I lost those of the Edwardian era. It also seems odd to me that David Lack did not refer to the club's full ancestry in his *The Birds of Cambridgeshire*, published by the Club in 1934. It is certain, however, that true birdwatchers were combing the county from 1882. One at Royston used the pseudonym of 'Rambler', a nice change from the normal collector image of the period, and a noticeable grouping of passerine discoveries from 1893 to 1915 proved real application by the three observers most responsible for them. They were N. F. Ticehurst, Dr. A. F. R. Wollaston and the most energetic of three Alexander brothers, Horace Gundry. What a trio they made. The first became the ornithological sage of St. Leonards and a future Editor of the Witherby *Handbook*. The second was a collector who had given up the gun, was to write Newton's biography and appallingly to be murdered by an undergraduate gone mad. The third was the most energetic bird recorder of his and closer generations and quite possibly of all time. Whatever the age of its bird club, such men saw to it that Cambridge became the *alma mater* of fact-gathering birdwatching as the 19th century slipped into the 20th but they also had to watch its early promise of a responsive field ornithology dashed by the outbreak of the First World War.

The Alexander family merits fuller examination. Like the other great Quaker ornithological family, the Gurneys of Norfolk, its members were denied many normal pleasures. Brought up on the Kent/Sussex border at Tunbridge Wells, there was no drama nor music for the four brothers in order of age, Gilbert, Wilfred, Christopher and Horace. So the three youngest brothers took up not the gun and specimen collection, as had John Gurney, but the map, the diary and the chart and charged off into their local countryside and nearby wilder regions like the Romney Marsh and Dungeness. Conscious perhaps of Gilbert White's injunction to observers to use the discipline of a diary, Gilbert reduced the pestering of his youngest sibling Horace by giving him at the age of only eight *The Naturalist's Diary: A Day Book of Meteorology, Phenology and Rural Biology*. (Phenology is the study of recurring phenomena.) With Wilfred initially more interested in plants, it was the bonding of the field aces Christopher and Horace that was to produce a truly prophetic piece of field ornithology. Only 20 and 22, they published in 1909 – before Horace's Cambridge years – "a plan of mapping migratory birds in their nesting areas", so presaging the fashion for study into the avian territorial imperative of the 1920s and even the numerical and spatial discipline of the Common Birds Census of the 1960s. What I like so much about their early brilliance was its causality of daily observations prompting a flood of questions and a trickle of answers. Their studies would today be called 'patchwork'. Another of their remarkable achievements, marshalled by Horace, were the birdsong charts published in *British Birds* and the Witherby *Handbook*; these were the distillation of over 40 springs and summers of dated songs. Add to such harvest, Horace's skill in bird identification, which was aided particularly by his experience of Asian passerines in India, and the auras attached to him and his brothers as early 'total birdwatchers' are fully merited. Tragically, Christopher went down at Paaschendale in the 1917 slaughter. I return to Wilfred in my next chapter but, for the moment, a proper farewell to Alfred Newton is overdue.

In my three years at Cambridge from 1954 to 1957, my only contact with Newton's ghost was through visits to his collection and the odd glance at his portrait which hung on the wall of the vestibule. Horribly precocious with the

I'm not easily impressed – but the work of the early ornithologists was incredible. Many people today are just 'playing at it' in comparison. Frank Moffatt, east Yorkshireman, *in litt.* (2000)

Closely related to two other ground finches of the Galapagos islands, *Geospiza fortis* is distinguished only by the shape of its bill.

"Darwin is sometimes supposed to have been an ornithologist; actually (his practice as such) was limited to shooting birds and passing them to his assistant to skin. The variation of animals on islands first appears to have been noticed by Captain Fitzroy in the Falkland Isles and Darwin remained unimpressed until he also noticed it in the mocking birds and tortoises (not the finches – so often associated with him, even in a book title – which he was unable to tell John Gould about) of the Galapagos afterwards". Bill Bourne

performance of David Lack in the 1930s our guiding example, my peer group had little time for current or past ornithological dons as we charged imperiously over Cambridgeshire and large chunks of Lincolnshire and Norfolk to boot. To reduce my retrospective guilt at ignoring a great man, whatever his foibles, I salute him now as a migration student and especially the original 'Architect of Works' for the system of and constituency for bird protection that now exists in Britain . Whether the productivity of the British Ornithologists' Union has done him equal credit is debatable but I shall return to that issue later.

So, where in all this are the minds and boots of Oxford educated ornithologists? Amazingly as Newton worked on the future framework of both professional and amateur bird study, the spires of Oxford seem almost to have been dreaming. Neither in Fisher's book nor in that of the Mearnses is there any reference, let alone compliment, to Oxford ornithology in the late 19th century. All that I have found are five Oxford graduates, the egg-collecting Reverend F. C. R. Jourdain, of whom I write more later, three other cleric authors and one lay editor of county avifaunas. That for Oxfordshire itself, published in 1889, is regarded as "one of the most detailed and accurate" of the early genre (Holloway 1996) and its author O. V. Aplin's other achievements consist of an investigation of the birds of Uraguay and many papers on the distribution of scarce British breeding species. Otherwise in the zenith of collection and in the first real field studies, the influence of Oxford education was not substantial.

A quarter of a century later, it would all be very different. Cambridge would become introspective, turning to more local objectives and serious, intricate research, with Dr. Thorpe dismissing the collectors as "contemptible" and the drive of his fellow dons losing the Olympian aspect of Newton's legacy. Cambridge dropped the baton and the 'dark blues', native or adopted, picked it up. Oxford was to create most of the rest of modern field ornithology but I come to the coincidences of that remarkable achievement later.

Before leaving the 19th century, one other member of that epoch's scientific society should be noted. The more forward of the two great independent minds that shattered the Biblical view of the one week creation, Charles Robert Darwin was 20 years older than Newton and, disliking his excessive interests in birds, refused to support Newton's professorship. Not one to forgive ornithological slight, Bill Bourne returned the disfavour in *The Study of Bird Migration at Cambridge* (2000), denying Darwin any real ornithological merit. A mite suspicious, I looked for corroboration of such low estate and there it was in J. Weiner's *The Beak of the Finch* (1994). The scribe of evolutionary theory had indeed not separated the finches, so often considered as his sharpest perceptions, in the field. Worse still, he had not even labelled separately those that he shot on the first two of the Galapagos islands that he visited, consigning both samples to the jumble of one bag. No wonder Bill Bourne gave the verdict set alongside on Darwin's ornithological status.

The content of the two characteristically long, convoluted but pregnant sentences is explored in a fuller statement entitled, *Fitzroy's foxes and Darwin's finches* (1992). It seems that our greatest conceptual naturalist was not perfect.

The Architects of
Field Ornithology

*The birth of a master journal – British Birds – some of its early contents –
Harry Forbes Witherby, prophet, policeman and publisher – teenager memories of him –
his egg-collecting Reverend colleague – an end to feuding – other springs of
birdwatching – Bernard William Tucker – the establishment of field identification –
the development of 'economic ornithology' – Wilfred Backhouse Alexander, from
prickly pear to first photographic guide - "the odd Oxford circus" –
Max Nicholson, scribe of the new creed and opportunist of world conservation*

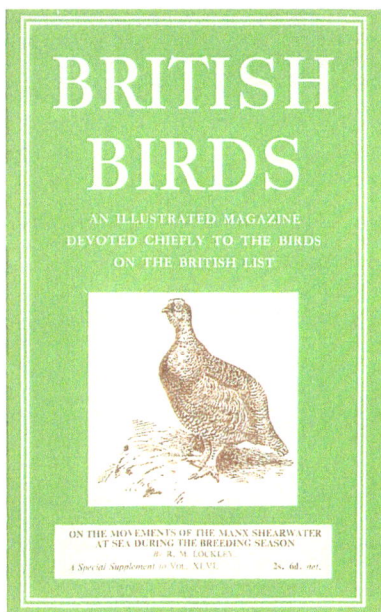

The original cover of *British Birds*,
unchanged for nearly half a century.

A week before Newton died in 1907, another exceptional but this time corporate being was born. It was and is the 'master journal' *British Birds*, known fondly throughout the ornithological world (and especially to those who owe it their best training) by its initials *BB*. It was the much-cherished brain-child of a London printer and publisher Harry Forbes Witherby. According to Tom Harrison, one of the stars of the Cambridge Bird Club in the 1930s, the man and his journal came just in time to end the fiddling with Victorian trivia, to attract young observers to real purpose in birdwatching and so to revitalise and re-orient Western ornithology. Witherby's own thoughts for the journal were astonishingly prophetic of the professional and amateur developments in bird study of the 20th century. When at the January 1907 meeting of the British Ornithologists' Club (then the *sanctum sanctorum* of the rearguard), even the ultimate museum maestro, the Honourable Walter Rothschild, gave the plan his support, a new die was cast.

BB might have had a stuffy title; both *The Magazine of British Ornithology* and *The British Ornithologists' Magazine* were considered but happily the birds themselves were chosen. The journal has reappeared almost every month for nearly a century, dressed in editorial news and edited thought that still sets the rest of birdwatching media a high standard. The first issue was dated 1st June 1907; it cost an old shilling (£3.40 today) and for that, a target group of 200 to 300 readers got the details of the 20 new species to arrive in Britain since 1899 and a lead paper on the 'home life' of the Osprey. These and the other notes had come from the pens of the leading observers that Witherby had attracted to the journal's id. In the second issue came a significant milestone in resident bird recognition – proof that both the Marsh and Willow Tit inhabited Britain – and the concept of the first systematic national scheme for ringing.

In January 1908 came the first of rare errors. For a breeding pair of Hen Harriers in Surrey, readers later read Montagu's. In February 1908, mentions were made of Miss Leonora J. Rintoul and Dr. Evelyn V. Baxter, accorded the accolade of being "very keen and competent … spending a month … on an island (the Isle of May)" and later known as the 'Good Ladies' of Scottish ornithology. In the same month, Fair Isle was noted as fast "becoming a second Heligoland under Mr. Eagle Clarke's able management". In March 1908 appeared the first public appeal of the Royal

The British subspecies of the Marsh (upper) and Willow (lower) Tits were not formally separated until 1900 in the last addition of a resident taxon. In the field, the easiest way to identify them is by their commonest calls!

Society for the Protection of Birds in favour of the Watchers' (early wardens) Fund. Interestingly, the money was to be spent on the defeat of "destructive ignoramuses", the utterly selfish last collectors of the by then scarce and over-prized British-taken specimens. Such people had been first labelled as criminals by the great wildfowler cum observer of the Borders, Abel Chapman, in 1889. In the same issue, there was even an optimistic forecast that the Spoonbill would soon return as a breeding species, an event that did not occur in fact for another 91 years. Hopefully, these examples and those inset will suffice to demonstrate how Witherby and his immediate circle guided *BB* to a successful landing, on its feet and running. Its themes were and are still the stuff of birdwatching and field ornithology. The international applause that flooded into the 50th Anniversary issue in June 1957 said it all. *BB* is respected and enjoyed around the world, belonging now as much to its successive generations of eager, mostly amateur contributors as to its original creator.

Witherby was subject to two ornithological cultures. His book imprint carried the Old Bushman's motto of the collectors' age: 'What's hit is history, what's missed is mystery'. Yet as his heart and mind divined a new non-lethal set of study disciplines, Witherby became the foremost planner and manager of birdwatching and field ornithology in the first four decades of the 20th century. Born in 1873, he was brought up in Blackheath and at Burley in the New Forest, the latter forming the test bed for his lifetime's study of birds, and educated at Sherborne, where his favourite game was Rugby football.

Witherby had the sense to marry Lilian Gillson, who on their honeymoon in Algeria was promptly introduced to "the unromantic art of taxidermy" and who supported him throughout his execution of the three elements of the British bird bible that in addition to the children he fathered. The first two were *A Hand List of British Birds* (1912), the first reference to link sound nomenclature with accounts of each species' distribution and migration, and *A Practical Handbook of British Birds* (1924), effectively the *Handlist* expanded by texts on field characters, habitats and breeding and still wonderfully accurate descriptions of plumage and moult. The latter was a part-publication interrupted by the vicissitudes of the First World War, in which Witherby served abroad in the Royal Naval Volunteer Reserve from 1917 to 1919. The war period also included the demise of *The Zoologist* journal after "an honourable career of 113 years", its ornithological component being then incorporated into *BB*. The third element and final station of Witherby's remarkable continuum of ornithology was *The Handbook of British Birds* (1938-41) but we shall come to that later.

Witherby's drive towards a symposium of British bird learning attracted collaboration and assistance from the best of his contemporaries. The efforts of Dr. Ernst Hartert, Rothschild's museum curator, and "that clever and charming ornithologist", Mrs. Annie C. Meinertzhagen, drew particular praise. Witherby saw too that after the completion of the *Practical Handbook*, any last academic dam restricting amateur interest and observation had been broken. Seeing the formation of the British Trust for Ornithology as the beginnings of a new era, he clearly sensed that the volume of prior work on British birds would soon be swamped by a whole new literature not just of feathers and measurements but of their full biology. Witherby's own printing and publishing company would do much to facilitate this flood but what sort of person was he?

I owe the late Tom Harrison and another of my own great mentors, Phil Hollom, for clear memories of the God-like controller. Together they had conceived a plan for a national census of the Great Crested Grebe, the once much-persecuted source of 'avian ermine' for ladies' hats, that was doing its best to reoccupy British waters. Inspired by Max Nicholson's ground-breaking efforts to count Herons, the census was to be their contribution but there would be nothing without Witherby's sanction and support. So one November day in 1930, as a pair of "lusty boys" still in their teens, Tom and Phil were summoned to Witherby's Hampstead home to be interviewed by him and Max Nicholson. The latter was only 26 and already convinced of the merits of their scheme but Witherby was at 57 a tall, thin "awfully grim" elder. He first addressed the adolescent enthusiasts during a "sticky spell" with "reasoned suspicion" and "considerable probing" and then conversed with them in a "severely authoritative and rather dry" manner. Suddenly, however, the seemingly austere spell of his "bird- and book-stacked" study was broken by "the reassuring roar of a sports car" outside the house and the energetic playing of "jazz records" inside it. It was a normal family home after all.

Tom and Phil's trial ended with Witherby's agreement to provide the paper logistics and a *BB* launch and Max's offer of his prior experience and 'flying squad' coverage in the event of gap or crisis. All went well and the results were published as the first-ever national count of Great Crested Grebes in three autumn issues of *BB* in 1932.

Advertised widely, even in the *Times* and on the BBC 'evening news', the activity of counting a beautiful bird had appealed not just to birdwatchers but also to keepers and their bosses, landowners up to the rank of duke. Observations on 1,000 lakes had flooded in from more than 1,300 people to an undergraduate and an office boy. Sixty-seven years on, Phil Hollom has mused on the "remarkable response". He sees it as presaging the future enthusiasm for BTO surveys and atlasses but I sense that another of his thoughts that actually "people had a fuller appreciation of birds than had previously been thought" is nearer the real mark. It fits my earlier partial findings that a recognition, if not a love, of birds runs very deep in at least our rural society.

Hollom had actually met Witherby and his wife earlier in 1930. They and Phil and his mother alias chauffeur had independently decided on some early summer birding in Pembrokeshire and they found themselves together at dinner in a small hotel at St. David's. Later, Phil served Witherby as an amateur estate agent, being responsible for finding his retirement home. Remembering the youth from Pembroke and of grebes, Witherby had asked Phil to show him and a guest "a Swiss lad called Schifferli some nice birds". They went to Chobham Common and enjoyed its Woodlarks and Grasshopper and Dartford Warblers. Witherby took to the place and Phil spotted nearby the delightfully named Gracious Pond Farm which the great man duly bought. With growing confidence in their relationship and in spite of an nearly 40-year gap in age, Phil also persuaded

After the Grey Heron, the Great Crested Grebe was the second species to be made subject to a national census organised by two teenagers in 1931. This pair beautify the lake of the Hollybush Estate, Staffordshire.

A Greenshank flies up to yodel over the flow country of Sutherland.

Witherby to visit Fair Isle with him. It was no easy place to reach in the 1930s, with "communications too uncertain and accommodation too 'personal' and difficult to arrange, for the pilgrim trail to blaze brightly or regularly",. In return for Phil's additional service in two types of conveyance, Witherby let him watch the creation of the *Handbook*, while both he and his brothers were passed fit to take the Witherby girls "partying, etc". Such is the rest of life and from shared sinfulness, I have spotted that HFW must have really enjoyed tobacco. So proper and "very tweedy" – a tie removed in sunny Corsica looks very risque – the photographed presence of cigarette ends between index and second finger of left hand suggests a hint of frailty, at least in the case of nicotine. Occasionally, too, he would open up in a letter as shown in the inset piece received by Horace Alexander. Unlike most of those of the 19th century collectors and scientists, Witherby's works remain vibrant and accessible. The still so user-friendly texts of the triple symposium and the continuum of *BB* are living testaments to his and his close peers' gifts to British ornithologists and birdwatchers. He died in 1943, as the tide turned in the Second World War which, like the First, removed another batch of promising enthusiasts reared on his example and taught by his writings.

Eight years senior to Witherby was a cleric who has attracted a much shadier press. "He was an (expletion deleted) oologist" is the usual short epitaph of today's ignoramuses of his true worth. The Reverend Francis C. R. Jourdain did indeed assemble a huge egg collection but he also made it scientifically useful, since the field adventures that he had during its accumulation also gave him an immense knowledge of the breeding birds of the West Palearctic. Thus, although his *Pastor Pugnax* style grated on many and his egg-collecting became for nearly all first unfashionable and second unpardonable, Jourdain was a powerful collaborator with Witherby. His learning frequently crossed the Atlantic in contributions to A. C. Bent's *Life Histories* of Holarctic species. Horace Alexander rated his knowledge of birds as second only to that of the oracular Ernst Hartert. Unlike Witherby and the Alexanders, Jourdain was no dresser in the field, all baggy trousers and windjammer, but the Mearnses describe his fieldcraft as "superb" and his note-taking as "meticulous". In his spring expeditions of 39 years, he travelled light, slept rough, and worked tirelessly, blowing eggs and writing notes on the breeding cycles of his quarry often past midnight. Oddly his daughter disapproved of his egg collecting, not for its wronging of birds but for its excessive temporal "laying up of treasures for himself on earth", and was responsible for its neglect, if not its partial destruction by bomb blast. Finally I must note that from another determined egg-collector, Desmond Nethersole-Thompson, came proof that Jourdain did eventually concede the purposelessness of egg-collecting. It was he who advised Nethersole-Thompson to give it up and study the whole being and place of a bird species. Hence the latter's hard-won trysts with Greenshanks and his great monograph about them published in1961, which was followed by other works on Highland birds.

In retrospect, the manner in which field ornithology was advanced before and after the First World War was remarkable. In spite of the terrible loss and interruption of lives by trench warfare, the new spirit of purposeful birdwatching and co-operative investigation acquired real momentum. As time ran out for the often feuding authorities of the Victorian and Edwardian eras, new minds focused on the achievement of a fresh ornithological politic and a code for its practice. Skilled and disciplined observation of live birds began to supplant the shotgun and the endless stuffing or mere display of specimens. It was not enough any longer to name or own or simply admire; it was becoming necessary to qualify and quantify live birds in the fast-changing environment of Britain. So let us go back to Oxford and meet the young enthusiasts that picked up the baton dropped by Cambridge.

B. W. T. These were the seemingly magical initials that belonged to Witherby's chief ornithological heir, Bernard William Tucker. By all accounts a true gentleman, who never quite lost a schoolboyish manner and wrote in a beautiful hand, B.W.T. is remembered most as the apostle of secure field identification. That discipline was trailed in the *Practical Handbook* but it really took hold after his remarkable texts were published in the Witherby *Handbook*. The intervening period of some 25 years was Tucker's youth and yet by the end of it he had forged much more than identification criteria. Alas, I never met him, missing him by four days at Staines Reservoir in September 1950. The magnet to him and many others was the first Sabine's Gulls to penetrate the London area. On the 15th, the already ill B.W.T. was taken to see an adult; although exhausted by the effort, he "would not have missed it for anything". The pelagic gull with the three triangles on its wings was a great rarity in the years before seawatches and B.W.T.'s last new bird. On the 19th, a chance visit to Perry Oaks Sewage Farm and a meeting there with Cliff White got me a Baird's Sandpiper and later my shot at the adult gull and an immature that had also appeared. They were very distant, if not 'stringingly' so, and I waited another thirteen years for one at my feet on the Dungeness shingle. I also had no idea that I had trod in B.W.T.'s recent tracks. Earlier in 1950, however, a description of my and two other schoolboys' Yellowshank at Aberlady Bay on 13th May had come under his gaze. Although in one of his last acts as editor of

An adult (left) and an immature Sabine's Gull mingled with Black Terns at Staines Reservoir in mid-September 1950. The first gull was Bernard Tucker's last new bird.

British Birds and sole judge of British rarities he was not able to write personally, a letter from the assistant editor J. Duncan Wood informed us that B.W.T. had been "pleased to accept" the record. The joy of that first rarity acceptance pulses still. With Duncan's guidance and copying the style of other such notes, I crafted my first published sentences and, in spite of some ribbing from Dougal Andrew, the record of Scotland's second Lesser Yellowlegs (its modern name) has survived so far. The associated receipt of some sterling advice from Duncan on how to conduct more *purposeful* birdwatching was typical of courteous help that the editors of *British Birds* gave to teenage aspirants 52 years ago.

B.W.T. grew up in Somerset and was sent off on a classic journey of higher education, its main stations being no less lofty than Harrow and Magdalen College, Oxford. In an early photograph, he looks incredibly brainy and in his achievement of a first class BA Honours in Zoology, he displayed a penchant for plants and reptiles, which he kept as pets, as well as birds. This spread of interests produced employment by the University in a succession of zoological posts and a two-year spell in marine biology in Italy as an Oxford Scholar, during which time he found

British Birds
326 High Holburn
London W.C.1.
June 17, 1944

Dear Mr Barber

I am sorry not to have written to you at the time with reference to your letter about Rooks, Magpies, etc perching on sheep's backs, but I have been very much pressed with other work. I should not regard the perching of Jackdaws or even Rooks on sheep as particularly unusual, but personally I have never seen Magpies, which were primarily the subject of the correspondence, doing so, and although it is now clear that they do so more frequently than the available evidence suggested, the number of readers who wrote to say that they had observed it was by no means large. You are the only observer who has found it really common and it seems evident that the habit must vary locally. I am obliged for your data on the parasites taken. This is what one should expect, but yours is the first positive evidence and is what I wanted. I will publish a brief note on the matter if you can enlighten me as to what exactly is the 'big sheep louse' to which you refer. I cannot trace a big louse which occurs regularly on sheep. If you can give me the scientific name or give me any particulars which would enable me to identify it – or get me a specimen – I should be much obliged.

Yours truly
B. W. Tucker

A classic example of editorial care attached to what today would be seen as an obtuse observation, from the wartime period of *British Birds*.

his wife and work partner Gladys. Birds became his particular realm, however, and in 1946 B.W.T. became the first Reader in Ornithology at any British university. In that science, his finest evaluations were of the nature of species and subspecies but the reason why he played such a great part in the Oxford epoch of early field ornithology was his willingness to collaborate and his contagious enthusiasm for birdwatching as a hobby. Not without caution, he could deliver slapped wrists and so many excellent, balanced observers came from his student broods.

Although his own explorations were confined to wilder Britain, near Europe and Spitzbergen, B.W.T.'s perceptions of birds as beings were universally applicable throughout the Holarctic and Palearctic avifaunas. In the new discipline of field identification, he offered all who would follow his lessons a confidence in disciplined sight and thought that no shotgun exhalation had ever occasioned. Uncannily in retrospect, his Field Character texts for the Witherby *Handbook* were close-timed with the simplification of Nearctic bird identification that Roger Tory Peterson introduced in 1934 in his *A Field Guide to the Birds*, the first pocket book to carry the name of the later flooding genre. Both authors used italics to stress the crucial diagnoses but in the *Handbook*, the wonderfully accurate paintings of plumages stood unmarked. In the *Field Guide*, there were pointer lines on the plates that made learning the essential combination of fieldmarks even easier. Word for word, however, the quality and diversity of analogous image in B.W.T.'s texts was the better of the two. Although both Max Nicholson and Duncan Wood saw him as the least dramatic of men, his bequest of the first living vocabulary of field identification remains a fundamental gift to the modern birdwatcher. At long last, the guns could be racked.

Following so closely Witherby's example, B.W.T. had no taste for rash leaps and unconsolidated advances. What would he have made of these days of 'anyone can be a taxonomist or cladistician'? With DNA-driven separations unknown in his time, he was perforce cautious of splitting species. Yet perhaps unwittingly and in spite of what he himself wrote in *British Birds* in 1949 in an attempt to close the stable door, the bolting horses of subspecies were dramatically exposed by the old *Handbook*'s separate treatment of them. I for one have been subject to a lifetime's temptation to indulge in their identification since that magic day in May 1947, when I first opened the five great volumes, met the four great authors, read their science and coveted every last one of the 519 live taxa therein. According to the taxonomy of the 1940s, this total featured 423 species. About 40 new ones had appeared in the six decades since the *BOU*'s first official list.

Tucker's brief absence from Oxford ended in 1926, a year of great significance to the metamorphosis of field ornithology as it also saw E. M. Nicholson's arrival at the university. A self-confessed loner, owing most of his precocious skills to his study of White, Hudson, Kirkman and Coward, he professed rather anti-establishment attitudes at least towards the BOU and RSPB. Max Nicholson was also initially suspicious of the Oxford Ornithological Society (OOS), whose members he considered as keen on rarities as had been the collectors. Persuaded reluctantly to join the society, he found to his surprise that it contained a willing team ready to follow the blueprint for new bird studies that *British Birds* was exhibiting and his own *How Birds Live* (1927) had sketched. Even with the hindsight of today's many surveys, the outcome of the OOS's plunge into total bird study was astonishing. A small house-trap in Christchurch meadow preceded any ringing effort at an observatory, and bicycling members counted all the birds on the university farm at Sandford every week, winter and summer. Furthermore in 1928, with help of the readers of *British Birds*, Max Nicholson organised and delivered the first-ever census of a British bird, the Grey Heron. The era of monitored bird populations had begun.

Bernard Tucker recognised the importance of the new modes of ornithological enquiry to Oxford's zoological tradition but there were no university funds to support them. He and Nicholson had perforce to raise money elsewhere, persuading the then Ministry of Agriculture of the potential benefit of the Oxford Bird Census in settling the arguments over birds as friends or foes of agriculture. A three year grant was made and with other funding achieved by similar cajoling, the two chief architects of responsible purposeful birdwatching and counting had secured its embryonic organisation.

Max Nicholson's specification of a much needed manager for this remarkable diversification of field ornithology was someone "with the enthusiasm of an undergraduate and the skill of an experienced scientist". The hope was that this person would take on most of the preparatory work needed to form a permanent Institute of Ornithology, backed by a national organisation of field studies. Who else but an Alexander appeared to fill the role and hope?

Wilfred Backhouse Alexander, known as 'W.B.' in apposition to his brother 'H.G.', had gone to Australia in 1912, apparently a confirmed botanist and doing there some remarkable work controlling the invasive prickly pear. He returned from there, however, in the same sea change year of 1926, with his renewed interest in birds most obvious in his remarkable grasp of seabird identification. This he committed to print in *Birds of the Oceans* (1928). This was the second-ever pocket guide, or first-ever field guide if that genre is diagnosed particularly by the inclusion of helpful drawings and instructive photographs. One of the latter even showed the wing pattern of the Common Tern 40 years before it was finally elucidated by a 'modern' observer. Failing to get a curatorship at Cambridge, W.B. was free for selection as the man for the Oxford tasks and given the nick of time injection of hard-won funds, he was offered and accepted the new duties in 1930. By then 46, W.B.'s experience and world vision saw off any risk of a parochial chair in the new programme of Research into Economic Ornithology that Max Nicholson and Bernard Tucker, still only in their late 20s, had brought into being.

Although after his extra quarter-century of largely personal study W.B. was not a lover of administration nor co-operative studies, the complement of his and B.W.T.'s skills brought the whole project to fruition and continuing radiation. W.B. went on to complete 15 years in the "rather odd Oxford circus" (as Max Nicholson called it) that had replaced the 'Cambridge Olympus'. With Nicholson, Tucker, Jourdain and Guy Charteris, he contributed to the foundation of the British Trust for Ornithology in 1933, for which he particularly oversaw the early bird observatory initiatives, and the Edward Grey Institute for Field Ornithology in

W.B.'s little guide was subject to a" steady demand for the only book of its kind available" from 1928 to 1983.

1938, of which he was the first director and librarian, continuing in the latter post for a decade after his retirement in 1945.

It was a matter of chance which of the two remaining Alexander brothers you met in the booms of birdwatching that occurred before and after the Second World War. For me, it was H.G. on Fair Isle but for the majority, it was W.B. at Oxford. The latter's field characters in the 1950s were caught by Bill Bourne, who described him as "a genial, short, round, rubicund, arthritic and temperamental old soul", very different from "the thin, tall, quiet guru" that H.G. resembled. W.B. shared with H.G., however, the charming Quaker attribute of *liking* to chat with young people, taking particularly to Bill who found him not one but two new birds. In H.G.'s book, there is a lovely vignette of an early 'twitch' – from Oxford to the once famous Cambridge Sewage Farm – occasioned by brotherly communication. The target was the unlikely pair of Yellowshank and Water Pipit. W.B. bustled over in a car full of eager undergraduates; Bernard Tucker followed in a chartered taxi to enjoy particularly the more intriguing bird of the two, the then subspecies Water Pipit. So as long ago as 1934, given the prospect of new birds, there was no pulling of rank, only the urgent pursuit of new birds and increased experience. A lasting motivation!

It is time, however, to return to the once outsider the late Max Nicholson, finally lost from British ornithology in his 99th year. In my ornithologically fortunate life, I was privileged to observe him closely in five of the stations along his path through modern field ornithology and its increasing contribution to conservation. His was a difficult personality to identify, for although his written testament to birds and their interface with man is immense, there swirled about him a force-field of outward energy that did not allow easy access to the quite small and restless man that was its source. For example, I winced to see him 'dry up' disastrously in front of the arch television interviewer David Frost, in a rash attempt to be iconoclastic in front of a live audience expecting no more than undergraduate satire. I have also heard him choke on Desert Island Discs, as he remembered the still and simmering personal loss of close friends mown down in the First World War. These were signs of vulnerability and compassion. I found his writings informative and inspiring but a mite patrician. I served with him as an editor of *BB* and *BWP*, and I was constantly amazed by his application to work and duty. Somehow the essence of the long-lived, singular human being cum world engine of conservation policy and practice still eludes me, but as Norman Moore's obituary truly noted, he was one of the greatest men of our age.

Max, as I could *just* call him, does not command complete allegiance in the generations close to mine but I know that his trick was to extract exceptional performance from those who became subject to his exemplary leadership. Not just as one of Britain's best naturalists, to repeat Bert Axell's appellation, Max translated his innate intellectual ability into courageous opportunism and, beyond that, a blazing defence of the other beings that have the misfortune to be co-evolving with man. The essential defiance of Mammon in the International Biological Program was very much to his heart and mind. No slouch at delegation, he left Bert Axell to the task of forging the new reserve of the Coto Donana with a characteristically impatient injunction, "We've talked for years and years, Bert, and now it's up to you". I also remember his trenchant order to me when I took over the Field Character texts in *BWP* from James Ferguson-Lees, "Put some life in them, Ian".

Max Nicholson's contribution to the Oxford advance in British field ornithology and the associated opportunity for thousands of amateurs to contribute to ornithological and conservation fact was essentially twofold. He was the arch sourcer of funds and the chief scribe and apostle of the new creed stated and necessarily repeated in his *Birds of England* (1926), *How Birds Live* (1927) and a still-flowing stream of reports, books and calls for action to protect an ever more threatened environment at home and abroad. In most of them, the spell of a civilised interface between man and animal is spun and fiercely enjoined.

For 70 years, Max Nicholson consulted on the planet's only important issue – how to sustain its biodiversity – and he attached the manners of a less consumptive and name-calling society to all that he did. When I organised for him (in his years as the Director General of The Nature Conservancy) the last of the Azraq expeditions under the International Biological Program, he was briefly the best of bosses, totally committed and absolutely pointed towards our goals. Particularly in the now global influence of the Worldwide Fund for Nature, of which he was a founding soul, there is still hope, sense prevailing and sanctuaries lasting. His full and remarkable achievements are listed in *Who's Who in Ornithology*. Any more from me would be superfluous.

The original blueprint of Oxford studies and institutions has not been completely followed. Indeed but for another generous act from Witherby, the BTO might have been still-born. He sold his large personal collection of skins to the British Museum and gave the price of £1,400 (over £50,000 today) to the

In the marshes of the freshwater outfall of Azraq Shishan, Jordan, in the 1960s, some reeds were twice as tall as men. In them were Clamourous Reed Warblers (left), while nearby Moustached Warbler (right) and a Marsh Harrier inhabited more open vegetation. Max Nicholson was in his element in such an ecosystem.

BTO as a capital investment. Four years later, in 1937, the BTO agreed to take the ringing scheme off his hands but it never raised sufficient finance to create a sibling institute to co-ordinate amateur studies, effectively becoming as the years went on both trustee for and executor of the increasingly national exercises that stemmed from the Oxford starts. It was the death of a bird-praising statesman in 1933 that initiated the creation of a separate institute at Oxford. Sir Edward, later Viscount Grey of Fallodon, had been Foreign Secretary from 1905 to 1916 and he served in his last years as Chancellor of Oxford University. What James Fisher called his dream-like book *The Charm of Birds* (1927) was a prototype of amateur dedication, the only book on birds that my publican grandfather ever read and still available at least to glance into in most second-hand bookshops. Such was Grey's influence on the British establishment that on his death, his own thought that there should be a national ornithological research centre based at Oxford was taken up as part of his national memorial. Hence the birth in 1938 of the Edward Grey Institute for Field Ornithology (EGI). It is now part of the Zoology Department of the University. Its famous library could certainly do more with greater use or, as Chris Perrins has recently and chillingly remarked, "it will rust along with so much else of our traditions".

Able to digest information with astonishing application, James Fisher served British ornithology many times, being for example the describer of the early invasions of the Collared Dove. Although spurned by many birders, it has made an astonishing westward conquest of Europe since the 1930s.

Finally I should not end this reverie of Oxford without noting that Tucker's college Magdalen was where James Fisher decided to switch from medicine to zoology, to become, after a push by Sir Julian Huxley, the assistant curator of the Zoological Society of London in 1936. During the Second World War he became the chief researcher into the Rook and its effect on crop yields, and after its end, he was one of the foremost conduits for increased public interest in natural history. Intriguingly, Huxley also took a hand in the advance of Fisher's chief Cambridge contemporary, David Lack, influencing his taking of a teaching post at Dartington Hall in Devon and a long summer holiday with the Moreaus in (then) Tanganyika. Eventually after further travels to the USA and Galapagos, which spawned his work on behaviour and ecological speciation, Lack's great powers of observation were concentrated upon the wartime investigation of the mysterious radar echoes called 'angels', alias the birds that initially confused our aerial protection system. In 1945, he took over the directorship of the EGI from Wilfred Alexander and at long last a Gresham's boy cum Magdalene (Cambridge) man took a turn in Oxford's leadership of the new bird study disciplines. Of his seven major books, *The Life of the Robin* (1965) and *Swifts in a Tower* (1956) attracted the widest readership, combining somehow the threads of White's Anglican worship, Grey's ability to charm and his own single-minded concentration. For British birdwatching and field ornithology to lose James Fisher in 1970 and David Lack in 1973 was doubly unfortunate but happily both their families have continued their work.

Other Enthusiasts

The conflict of passions – Abel Chapman, naturalist-gunner with a conservator's heart – Bird Life of the Borders *– Chapman's ornithological universe – Lionel Walter, second Baron Rothschild – his 21st birthday present – the output of his Tring Museum - and its loss to America, due to blackmail – Ernst Hartert's tears – a Cheshire cluster of original talents – T. A. Coward and Archibald Thorburn – two Lapland Buntings – Charles Oldham – Arnold Whitworth Boyd – Charles Frederick Tunnicliffe, first Royal Academician of bird art – the Aberdeen resurgence – the Thomson baronets – two seabird specialists – a soaking and a rejection – a honorary Aberdonian called Bourne – three magic schools*

Recently, the magazine *Birdwatching* has adopted as its by-line the phrase: "Sharing your passion for wild birds". In this context, the third word is meant to convey 'a strong affection for enthusiasm' but in an older sense, passion can also mean 'inflicted suffering'. So in one word, we have the conflict of human emotion and avian danger that arose from the corps of naturalist-field-sportsmen that was active in the same period as the collectors. The corps' most celebrated member was Abel Chapman, born in 1851 to live an incredibly adventurous 78 years. Educated primarily at Rugby School, inspired by F. C. Selous but above all brought up in the traditions of gamebird shooting and wild-fowling, Chapman used the wealth of a family beer and wine business and his attendant leisure to explore particularly the Borders and Northumberland coast, and also Spain and eastern and southern Africa, particularly the savage Sudan. A gifted communicator in word and sketch, he wrote ripping yarns that rival those of Seebohm and other great bird chasers. His natural history was even better than theirs.

I found his first book in a school library and read it from start to finish in one go. With conservation concern not front of mind in the late 1940s, I paid no heed to the then legitimate killings; it was the author's close knowledge of his countryside and its wildlife that led me on page after page. *Bird Life of the Borders* (1889, republished 1990) is another treasure-trove of real, wind in your eyes, exciting and perceptive contacts between hunting Man and wild Bird. I must have read it at least five times and I still marvel at how the addition of a purposeful pursuit sharpens perception. Therefore, do not worry about reconciling the conflict in the ancient sportsman's behaviour, read his book and get 'the best of grounding in field-craft' that it was to me. Even the best of modern guru guides to bird identification dismiss this fundament of successful nature study in a few paragraphs.

In my later readings of Chapman's tales, I was surprised to find that between the purple passages of grouse-shooting and punt-gunning, there were, firstly, references to old debates on ornithological issues, and secondly, obvious springs of conservationist conscience. Among the former were his liking for Canon Tristram's theory that all life originated in the once hot North Polar region and some fascinating speculation on the nature of bird migration coupled with totally accurate separation of the multiple status of some species. Chapman referred to

Abel Chapman's separation of 'southern' Golden Plovers breeding on a Border summit from other 'northern' birds still on spring passage was an early example of status definition.

"double stocks" of Curlews and Golden Plovers, clearly distinguishing his breeding pairs from northbound migrants massing for departure and calling frequently. Among his quarrels with the abusers of wildlife, the most overt group was made up of the collectors who displayed an "insatiable – aye, insane – greed" for "British killed specimens". These men paid "half a guinea" for a young Raven (£35 today) and offered bribes to shepherds and keepers to take their masters' last Peregrines, Hen Harriers and Buzzards. Chapman thought it "regrettable that the fate of the few survivors of these noble aborigines should in some cases be left at the mercy and ignorance" of scoundrel collectors.

Unlike that of Peter Scott, 58 years his junior, Chapman's conversion to conservation was never complete but along the long trail of birds shot for fun and food and game killed for trophies, he supported his local natural history society and researching museums, became the joint lessee of the Coto Donana, discovering its breeding Flamingos, and even played a part in establishing the Kruger National Park in South Africa. For his last three decades, he lived at Houxty on the North Tyne River and according to George Bolam, a friend and fellow ornithologist, this patch became a second Selborne, attracting wildlife and naturalists alike. Of the 134 birds that he recorded there, 110 had been identified from his windows. Is this the record for one house's bird diversity?

To try and divine the ornithological universe of a classic 'naturalist-gunner' of the late 19th century, I searched Chapman's book for his companions in sport and authorities on whom he relied. He mentioned two of his seven junior siblings, nine local persons from families involved in shooting and wild-fowling for at least three generations, two eminent local ornithologists including the most famous Newcastle taxidermist John Hancock, two despised excessive collectors, one the dreadful Charles St. John who persecuted the Osprey, and many hill people like keepers, water bailiffs, shepherds and even postmen. The pure ornithologist in Chapman consulted five authorities on migration including Seebohm and von Middendorf. His general references included the Howard Saunders' edition of Yarrell, Morris, Montagu and Bewick. He also read *The Field*, then a useful source of natural history, paying particular attention to the contributions of H. W. Wheelwright, who wrote under the sobriquet of the 'Old Bushman' and coined the motto of the collector corps. His wildfowling hero was the immortal Colonel Hawker who wrote that sport cum craft's primer. Only one lady gets a mention, the "good wife" who restored him to warmth after his punt 'Boanerges' (the thunderer) had dumped him in the briny.

As geese come down to roost over Holy Island sands, Abel Chapman and a fellow wildfowler manoeuvre his punt *Boanerges* for a raised shot (from Chapman's own drawing).

Having given Oxford due credit for attracting a new engine room of ornithology, I must draw attention to some other places and people that added to the advance of that science and its amateur expression. For example, Tring in Hertfordshire was to house a rival collection to those in London and Cambridge, benefiting from the huge investment of the Rothschild family in a seven-year-old boy's announcement that he was "going to make a museum". The year was 1875. Having completed in his teens his education at Bonn and Cambridge and become a first-class shot and a bird and butterfly enthusiast, the later Lord Walter Rothschild received on his 21st birthday the ultimate ornithological present. It was a brand new building specially designed to house his collections. In 1892, the first Tring Musem was opened to the public; 30,000 visitors a year poured in. Walter had to wait eighteen years to be released from the family chore of vast wealth creation (via N. M. Rothschild and Sons) but in 1910, he was set finally free to concentrate on zoological research and enterprise. The result was astonishing.

Having bossed one curator from the age of 12, Lord Rothschild eventually employed well over 400 collectors and a museum staff sufficient to manage a museum that eventually occupied one and half acres. The latter included a senior ornithological curator, the aforementioned Ernst Hartert, and both he and Rothschild welcomed all fellow enthusiasts of all ages. Particularly pertinent to today's surge in taxonomic interest was their shared view that subspecies were the geographic units of evolving species. They propounded the trinomial nomenclature needed for their definition, in opposition to a conservative majority centred on the British Museum of Natural History in London and including most members of the British Ornithologists' Union. The Rothschild ensemble described 5000 new animal species and published 1200 books and papers on the fauna of seven continents and the islands of three oceans, the flow of work being so great that the museum published its own journal *Novitates Zoologicae* from 1894 to 1939.

Contrary to many of his like, Rothschild clearly enjoyed the company of live animals as much as his drawers of specimens, creating a free-ranging menagerie within Tring Park in a fascinating prequel to the modern safari park. As an undergraduate at Cambridge, he was accompanied by his much loved flock of Brown Kiwis. More far-seeing was his renting of Aldabra Island from 1880 to 1908 in order to save the Giant Tortoise but this prophetic adventure in conservation ended when his large but fixed share of the Rothschild purse became insufficient to cover the cumulative bills of three mistresses. The last of these, the wife of a peer, blackmailed him unmercifully – on pain of informing his dominant German mother – and eventually in 1931, the maker of an amazing museum and storer of so much knowledge was forced to sell nearly all his bird skins to the American Museum of Natural History. Recently returned to Germany, the retired Hartert broke into tears at their last unexpected westward migration. He had always wanted that journey to be the shorter eastern one to London. In an odd quirk of fate in 1972, our national taxonomists in the British Museum of Natural History paid the ghosts of the lost collection the compliment of quitting London and moving the Cromwell Road collection into a new, four-storey 'bird room'

University campuses regularly host unusual sights, but Walter Rothschild's Cambridge flock of Brown Kiwis must have turned many heads.

next to Walter's great library. Called today, the Sub-department of Ornithology or Bird Group, it is the second great ornithological storehouse to bring everyone from British birdwatchers and bird artists to the furthest-stationed foreign ornithologist to Akeman Street in Tring. When I use the resource of the new 'bird room', I always find myself glancing over the wall at the creation of the lord who was never fully accepted in the ancient sanctum of British Ornithologists' Union but nevertheless fostered a forward-looking ornithology, of which today's young systematic Turks may be ignorant but would approve and should applaud. But for the stress of his personal associations and weight of 22 stones, Lord Rothschild might have seen the full flowering of amateur field study that followed the age of specimen display but he died in 1937 at the relatively young age for his social class of 69. Given his remarkable enterprise over six decades, it is rather unjust that it is Hartert through his more direct association with the Witherby regime that is the better remembered.

Identifying two remarkable single cells alias individualists from the second half of the nineteenth century is not difficult. Spotting other multiple cells or clusters of common cause is less easy but in the rest of this chapter, I demonstrate such influences in one inland county, another northern university and its region and three schools. Before my reading for this book, I had not noticed the remarkable gestation of popular ornithology and illustration in Cheshire from the last third of the 19th century. Full of bird-attracting meres often within well-keepered country seats, that county's profile of birds was and still is rich. It is effectively the north-pointing tongue of southern England's diverse community of 164 breeding species and 157 wintering ones, the former total being the highest regional figure within Britain and Ireland. Was it this natural store and its distance from the stuffy capital establishment in London that saw to the emergence in Cheshire of four substantial figures? Namely Thomas Alfred Coward born in 1867, Charles Oldham born a year later, Arnold Whitworth Boyd born in 1885, and Charles Frederick Tunnicliffe born in 1901, the year that Queen Victoria died.

For sheer beauty, Archibald Thorburn's plates remain unequalled.

I address some of Coward's contribution to the popularisation of birdwatching and the understanding of avian phenomena elsewhere but here I want to stress that my generation owes him a huge debt for the wonderful set of three little volumes entitled *The Birds of the British Isles and their Eggs (1920-26)*. Not only was its full prose friendly and informative, but also its illustrations, borrowed from Lord Thomas Lilford's part-work *The Coloured Figures of the Birds of the British Islands* (1885-97), were both remarkably accurate and aesthetically romantic. The latter came mainly from the superb techniques of watercolour and tempera and flake-white practised by the then emerging doyen of Victorian bird artists, the Scot Archibald Thorburn. Gone were the stiff, skin-derived images and incorrect habitats of earlier bird portrayers; there was the stuff to assist real perception of birds as organisms in their correct niches.

Respectively, Coward and Thorburn primed my memory with facts and my mind's eye with trustworthy models. To assist my own attempts at drawing birds, I copied all of the latter's figures into an early sketch book and I can still remember the thrilling moment when, after such input, out came my own first outline of a

Woken from its snooze at a roost, a Dunlin stretches its left wing and leg in preparation for its next move. To appreciate the human perception of bird form and action, there is no better place than the exhibition of the Society of Wildlife Artists, held each September at the Mall Gallery, London.

seemingly live bird. Recovering from the first of many bouts of bronchitis, I was lying propped up in a school sanatorium bed, fiddling with pencil and paper and suddenly somehow there appeared a passable Dunlin in a pose different from any of Thorburn's. That moment of first bird-image creation, since repeated thousands of times, caused a letter home which included a claim that "at long last, I have got the hang of drawing waders". As the year was 1944 and it was not until 1947 that Pat Sellar showed me more Dunlin, this early experience in the miraculous synthesis of perception remains the crux of the identification skills that I have employed over the ensuing 56 years. Look, look, draw and at long last see.

Coward's excellent book also contained many early bird photographs. One featured a mystical creature named Lapland Bunting and it became my most desired bird. It took seven years for the photograph to come to life. On 16th September 1951, after two days of hopeless solo casting over the Fair Isle moors, Maury Meiklejohn took pity on me. He was a Professor of Italian and a witty poet. "Come on, Ian, I'll find you a Lapp Bunt", he said and off we went. "We must look in the grass runs amongst the heather", he enjoined. Spotting a likely line of light-catching seed-heads, he got me alongside him and we pushed up the run. Suddenly two birds flushed almost from our feet and there for real was the owner species of the oft-stared at image. That Sunday had earlier consisted of a westerly gale and the then expected Church attendance but ended with this log entry: "Lapland Buntings, yes, two definite records at last. Just before tea above Gilsetter, two Skylark-like buntings with flatter rattle than Snow were put up and then seen at a range of some feet, not yards, in the long grass". It was another crucial experience not just of a new species but more importantly in understanding how to spot a bird-bearing habitat. Coward, the photographer and a Professor of Italian had together taught me some more ornithological field-craft. Anti-blood sports, secretary to his local RSPB group and level-headed conservationist, Coward wrote four other influential popular books on birds and, always a 'field man', was a model of early purposeful birdwatching.

Two Lapland Buntings inspect their observers from seeding grass above Gilsetter, Fair Isle, on 16th September 1951. Beyond them, a migrant Merlin hunts for Meadow Pipits.

February 17, 1932
Brentwood
Bowdon
Altrincham

Dear Sir

Thank you for your note about the Bittern. Of course the occurrence of Bitterns in Cheshire is exceptional, though almost annual. This is the second this year; the first was shot in Burton. I have seen the bird twice or three times on one mere, I heard it booming on another but I do not advertise this fact too much. Too many are shot and nothing is done to stop it, except by such landowners who like Lord Egerton have given strict instructions that rare birds must not be shot . I must have been in Tatton along that time but did not go to the Mill pool, which I know well, as it was bitterly cold. A few years ago one of the Rostherne Keepers nearly shot one but stopped just in time. There is hardly a water of any size in Cheshire which has not one or two bitterns to its discredit – Rostherne, Tatton, Marbury are three exceptions – birds seen and not killed.

Yours faithfully
T. A. Coward

D. C. Barber, Esq.

Charles Oldham shared many of Coward's skills and interests, becoming an early specialist in field identification, even the top expert in the post First World War period. Unlike Coward, he did join the London circus of ornithological discussion. Horace Alexander met him at the British Ornithologists' Club dinners, and once in the 1920s went with him to the Severn New Grounds (later the site of the Wildfowl Trust). To H.G.'s amazement, Oldham like Coward could not recognise the calls of the common pipits but his visual separation of such similar species was rated "as good as those of any of us younger men". From that ability stemmed the "brief, but in the main very good, field identification (texts)" for the *Practical Handbook* (1912-24). These formed the fundament of British described field characters, so well developed by Bernard Tucker 25 years on. Oldham was also skilled in observing mammals, writing thorough accounts of them when he wrote with Coward *The Vertebrate Fauna of Cheshire and Liverpool Bay* (1910). The third of the Cheshire quartet was Arnold Whitworth Boyd, 18 years Coward's junior but a neighbour and soon his own ornithological heir in the county. Reared on the birds of the meres, he was one of the observers who started the fashion of watching birds at reservoirs, even straying to the Staffordshire gem at Belvide. His two books *The Country Diary of a Cheshire Man* (1946) and *A Country Parish* (1951) show his true allegiance to local natural history but he also served as an editor of *British Birds*, while as true uncle and first field companion he sparked James Fisher's interest in birds.

Making his first visits in the 1920s, Howard Davis blazed the trail to the wild geese of the New Grounds, Slimbridge, Gloucestershire. In the harsh winters of the 1960s, the 'Goose House' field was visited by Eastern Greylag and Red-breasted Geese. Most of their kinds normally winter no nearer than eastern Europe.

A seemingly courteous enquiry to an unknown observer but the leading question on the male's appearance smacks of trap-laying by the local expert.

Last in the influential Cheshire quartet came a remarkable artist, Charles Frederick Tunnicliffe. Living from 1901 to 1979, he took his time in finding the particular appeal of birds and making them his prime passion in drawing and painting. He was 33 when, following his successful etchings in *Tarka the Otter*, another commission from Henry Williamson to illustrate *The Peregrine's Saga* (1934) produced the necessary imperative for full familiarity with a bird. Immediately his gift of a natural flowing line wrapped itself around more and more of them. For me as a young birdwatcher his best book was *Bird Portraiture* (1945). In line with Thorburn's luxurious images, those of Tunnicliffe still looked rather pretty but they had even better structure and shape. Better still, the book was full of helpful hints on handy art materials and the most wonderfully instructive plumage maps of dead birds. Elected to full membership of the Royal Academy in 1954, Tunnicliffe became the leading bird artist of the post Second World War period. From 1947 to 1966, every copy of *Bird Notes*, the title then of the journal of the Royal Society for the Protection of Birds, came with a new Tunnicliffe painting on the cover. For the same charity, his series of Christmas cards started in 1950 and are still being reproduced today. By his death, Tunnicliffe's images had become as much the artistic *lingua franca* of birds as had been those of Bewick almost two centuries earlier. In the end, I personally found his style of painting too chintzy. So it was J. C. Harrison's *Bird Portraits* (1949) that became my second primer and the "peerless eye" of the Swede Bruno Liljefors that filled me with hopeless envy. Even so, Tunnicliffe fully deserved the honours that came in his last years and his etchings stun still.

Another centre of ornithological potential that merits attention in the early 20th century is Aberdeen. Heirs there to the MacGillivray tradition, a family of Thomsons, produced first a professor Sir J. Arthur, who did much to further ornithology as a meaningful study ground within general zoology, and second his son Dr. (later also Sir) A. Landsborough Thomson, who published his first bird note in the first issue of *British Birds*. The son went on to deliver at Aberdeen University the northern of the two starts to bird marking, and several important books, which included another excellent primer for birdwatchers entitled *Bird Migration: A Short Account* (1936). Landsborough Thomson also gave effective leadership and administration to a small flock of institutions beginning with the British Ornithologists' Club and the British Trust for Ornithology during the Second World War, the British Ornithologists' Union and others all the way up to the Trustees of the British Museum of Natural History, of which he was Chairman. How the second officer of the Medical Research Council found the time for all these additional tasks is remarkable but a variant of the work ethic 'if you want something done, give it to a busy man' is an apt phrase for his formula for achievement. Today, the single survivor of the nine young natural historians who went from Aberdeen to the trenches of the First World War is best remember-ed for the voluminous *A New Dictionary of Birds* (1964), of which he was chief editor for the British Ornithologists' Union. As had the earlier *Dictionaries* of Montagu and Newton, this teamwork outclassed even the standard textbooks of ornithology in its wealth of detail and definition. Unusually for his time but not as an Aberdonian used to life in a trading city, his studies at Heidelberg and Vienna led to an early familiarity with European ornithology. It was a teenage fortnight at the pioneer bird-ringing station at Rossitten on the Baltic in 1908 that sparked his lifelong interest in migration and its orientation and the beginning of ringing at Aberdeen in the following year. Sadly I only observed Landsborough Thomson across a crowded room; his was an offstage presence in my search for learning. It seems increasingly strange that today's scene has no elders like him.

One of the British Ornithologists' Union's bequests to the nation was its centenary publication of *A New Dictionary of Birds*.

At Aberdeen, the example of the Thomsons has been followed by a remarkable train of ornithologists and birdwatchers. Born in Yorkshire in 1906, Vero Cooper Wynne-Edwards was a multi-faceted naturalist who pioneered North Atlantic seabird transects during his 15 years of associate professorship in Canada. From 1945, he served for three decades as the Regius Professor of Natural History at Aberdeen University. Although his personal view of animal population limitation jarred with the tradition of Charles Darwin and latterly David Lack, his initiatives

A pack of Red Grouse sail down the approach to Schiehallion, Perthshire. Once enshrined as the only endemic British and Irish species, this favourite gamebird is now relegated to being just a south-western race of the Willow Grouse.

in field ornithology were so well directed and organised that they were taken over by the emerging governmental conservancy. From them and his students came classic work on seabirds and the Red Grouse and its place in moorland ecology. His contacts with more ordinary birdwatchers came mainly from his editorship of *The Scottish Naturalist*. Operating rather in the Witherby mould, he was to me in 1951 a somewhat ethereal elder. I still remember wincing when he excised the nonsenses in my first immature submissions to the journal but he redeemed himself by singling me out for courteous counsel during my first attendance at a Scottish Ornithologists' Club (SOC) event. When eventually he published three wintering Whimbrel found by the Loretto School Ornithological Society at Aberlady Bay, we all knew that there was a benign God. Since Wynne-Edwards kept largely his own company, enjoying nothing more than cross-mountain skiing, his presence in Scottish birdwatching was lofty, but as early as the 1920s, his choice of people to influence was uncanny. It was he who 'press-ganged' Max Nicholson into the Oxford Ornithological Society, transforming the former from "a chronic loner to a team player" and later sharing with him happy cruises after Manx and Balearic Shearwaters in the English Channel and the puzzles set by Highland grouse.

Much closer to modern birdwatching circles was Wynne-Edwards' best student, George Mackenzie Dunnet. Born in Caithness in 1928, he led another typically itinerant Scottish life, following the southern ocean counterparts to his beloved Orkney Fulmars as far as Macquarie Island south of New Zealand. It was after his stunning lecture on that place that I first met him, marking him immediately as another true gentleman of field ornithology but not spotting until I researched this chapter that we probably shared some close genes due to the small 'k' in our maternal surname. George's chief memorial is the Culterty Field Station on the River Ythan, opened by him on his return to Aberdeen in 1958. When its work was first published, I was amazed by the intricacy of the supporting observations on, for example, the feeding ecology of Redshanks. Suddenly it came to us that the ways of life for birds could be fully fathomed; mere identification and numeration were only the 'starting blocks' of birdwatching. Student and scientist traffic to and from Culterty spanned the world and George's skills, not least in marshalling co-operative effort, caused Bill Bourne and the other initiators of the Seabird Group to pick him as its first chairman. In his last 25 years, George became as had Joe Eggling before him, the leading Scottish conservationist, tackling one crisis after another. He led the successful arguments that saved the Loch of Strathbeg of many geese (and Crimond of the 23rd psalm tune) from being over-run by a pipeline terminal. Requests for his help came from all directions and although supposedly retired, he still worked flat out in the cause of conservation until his premature death in 1995.

In the 1950s and 1970s, my involvement in two fairly hopeless attempts to maintain a specialised Scottish herring industry saw me cross the Ythan at least twice a week. I only visited Culterty once but two other Aberdeenshire cameos may suffice to sketch in the rest of its birdwatching universe. On a January day in 1959, I took a longer look at the estuary in the hope of a good bird. What showed was a Hebrides-type cloudburst. Squelching back to Newburgh, my business suit creaseless, I was rescued by Elizabeth Garden, who allowed me not only to drown her car but also to strip off in front of her glowing kitchen range. No cup of tea came closer to elixir than hers and our talk was all of how we and Scottish birdwatching had fared since our shared tutorials on Fair Isle. Betty Garden was a friend of Valerie Thom, who became Scotland's third 'Good Lady' of ornithology, writing in time for the SOC's Golden Jubilee the excellent *Birds in Scotland* (1986). Not content with that, Valerie showed her devotion to Scotland's *locus classicus avium* in *Fair Isle: An Island Saga* (1989). Unlike me, Valerie rarely wrote in the first person but her heart spoke out in the poetic conclusion to the preface of the first of her books. Somehow the manners of today's birdwatchers do not allow such limpid, untrammelled expressions of belief, as witness the events in my second cameo of Aberdeenshire behaviour. In the mid-1970s, I became aware of a band of 'young Turks' at the university, much addicted to migrant watching at Girdleness. With their publication of a *North-east Scotland Bird Report*, I thought to have found a ready repository for my Fraserburgh sea watches and other records. Off they went only to occasion peremptory requests for full descriptions of Black-throated Divers and like uncommon species. I bridled, suggesting that as, for example, I had seen my first Black-throat off Raasay in 1947, had competed with a pair for trout on Loch Achall in 1948 and had also shared in the Cambridge Bird Club's elucidation of their winter appearances in 1956, I could be let off further chore. Back came an ever brusquer retort from Mark Beaman who opined that, as the recent chairman of the Rarities Committee, I should not be avoiding any discipline from any ornithological caucus. I bent no such knee but did muse on the passing of kindly cups of tea!

In the last two decades, I have seen wretched little of Aberdeenshire, and the city and university have lived on mainly in the singular personality of Bill Bourne. Although born in Bedford in 1930, Bill is by now a honorary Aberdonian, earning his ornithological re-domicile in four bouts of inspired research, first for the EGI

Fifty years ago, Redshanks were common and irritated birdwatchers immensely by flushing other waders with their noisy alarms. Nowadays it is yet another species in need of conservation.

A pair of Black-throated Divers grace the highland summer on a Scottish sea loch. There are no identification problems with birds in breeding dress.

as a field assistant in the radar study of migration, and then for the university as ordinary, senior and finally honorary research fellow in the study of his beloved seabirds. Between these bouts and others of fitful doctoring, Bill has reached both ends of the Atlantic and Pacific Oceans and many of their seabird colonies. His exploits have become the rare stuff of living legend and his battle honours have been earned from at least six major confrontations with the world's despoilers. Some of Bill's doings are addressed elsewhere but here I must add that to all in his attentive audience, his most engaging habit is his self-confessed "metaphorical throwing (of) stones at stained glass windows". Actually, although those words are his own, they imply far too much vandalism. The reality is that with his deep affection for birds and our entire ornithological history, particularly its human characters, Bill is their foremost champion (and obituarist). Furthermore, if you do become a subject for his criticism, then be prepared not for a stone but one of Thor's thunderbolts, let alone an Exocet. Bill Bourne is not always right but thank God, he makes you sting and think again. Cerebral pack-drill does no harm. Apparently the rest of Aberdeen's once vibrant birdwatchers sound lower notes today and so I move on to my last three examples of what other ornithological helixes have produced at sub-university level.

My own last school, Loretto, was a rather homely, rugger-lunatic institution in Musselburgh until it acquired in 1951 the painted ceilings of Pinkie House. Yet it reared a small but urgently precocious brood of teenage birdwatchers in the late 1940s and 1950s. With some guidance from SOC sages but mostly by aping early observatory practice on the Forth foreshore, we wrote a full 'patchwork' report in 1951. In a unique post-scholastic coincidence, P. J. (Pat) Sellar and I went our ways to meet again as editors of *BWP* in the 1980s. We both feel gratitude for still-precious letters received from Dr. Baxter and Miss Rintoul and all the other natural history encouragements of our learning time, which culminated for Pat in Aberdeen and myself at Cambridge. Sadly the trail of the Loretto School Ornithological Society petered out but both Greshams at Holt in Norfolk and Leighton Park near Reading in Berkshire continue to foster broad nature study. From the former and its birdwatching master cum early bird cinematographer R. P. (Dick) Bagnall-Oakley came birdwatchers and ornithologists as engaging as Maury Meiklejohn, David Lack and David Hunt. Dick himself could also have doubled as Stewart Granger; my Mum would look faint when they met in Norfolk society. From the latter and the long-living keeper of the Quaker ornithological tradition, J. Duncan Wood, came James Cadbury, Jeremy Sorensen, Humphrey Dobinson and Robert Gillmor. Without the respective efforts of the last quartet, the research programme of the RSPB, its peerless reserve at Minsmere, the avian story of the 'ice-age' winter of 1962-63 (and the Inland Observation Point Scheme), and the Society of Wildlife Artists (and the excellent school of British bird artists) would all have been less complete or indeed absent.

At this point, I see that this chapter has been a display of exemplary exploits occasioned by favourable circumstances. No doubt at many other stations, other selections could have been made from our rich past. Hopefully 50 years on, a roll-call of similar contrasting honours will fill another tablet.

Embarking for Fair Isle, Grutness, Shetland, 22nd September 1954. From left, Joe Eggling, Guy Mountfort, James Ferguson-Lees, Bill Conn, author. WJW.

Exposed passengers on the Good Shepherd, 22nd September 1954, From left, Bill Conn, Guy Mountfort, James Ferguson-Lees (with pipe and duffel-coat, not otherwise universal mackintosh), Joe Eggling. WJW.

The Good Shepherd at anchor and the original observatory, North Haven, Fair Isle, September 1951.

The Good Shepherd hauled up (in order not to blow away), North Haven, September 1951. Above the boat the grass sward of Buness.

The original observatory on Fair Isle taken across South Haven, September 1951. The huts were inherited from The Royal Navy's time on the isle in World War II. Note Heligoland trap (in front of right-hand hut) constructed for the observatory's official opening on 28th August 1948.

The Haa and southern limit of crop fields, Fair Isle, September 1951. Note the island's first Heligoland Trap between cottage and dyke, built in 1945.

Fair Isle's ornithological conviviality, mid-September 1951. From left, standing, Holger Holgersen, Esther Williamson, author, pretty girl, Tom Yeoman, Peggy Condliffe, pretty girl, Kenneth Williamson (first director of the observatory); from left, sitting on the wall, Maury Meiklejohn, Hervor Williamson, pretty girl, George Waterston. WJW.

Fair Isle's ornithological conviviality, late September 1954. From left, author, James Ferguson-Lees (recently appointed executive editor of *British Birds*, Joe Eggling (Chief Scot in the then Nature Conservancy), Isobel Thom, Guy Mountfort, Valerie Thom, Horace Alexander (with broken arm) Bill Conn. WJW.

'Fieldy', alias George Stout of Field, with pannier of driftwood, Fair Isle, mid-1950s. 'Fieldy' referred to the then ultimate rarity, Pechora Pipit, simply as "one of thaim pipits" (per Bill Conn) and was the longest survivor of the islanders that assisted the shotgunners of the collecting era. JC.

Under Ward Hill, Fair Isle, September 1954. From left, Jackson Wallace, Valerie Thom (later assistant warden and author of *Fair Isle: an island saga*, 1989) and Bill Conn (also assistant warden and identifier of Britain's second ever, first live Paddyfield Warbler on 16th September 1953). DIMW.

Setting an 8' by 3' clapnet to catch an unidentified terrestrial passerine, Buness, Fair Isle, 15th September 1951.

A bagged Tawny Pipit, Buness, Fair Isle, 15th September 1951. Its shepherding into the 24 square feet of catch area took two hours. From left, George Waterston (then Laird of Fair Isle), three pretty girls, author, Peggy Condliffe, Tom Yeoman (inventor of the first portable net) and Holger Holgersen (of Stavanger Museum, Norway). WJW.

Nicotine celebration at Sumburgh Airport, Shetland, 30th September 1954. Guy Mountfort and author, happy with their memories of Britain's first Citrine Wagtail and the imminent launch of the famous 'Peterson' *Field Guide*. Guy's main gift to birdwatchers was his wide apostleship of field identification. WJW.

The Founts of
Purposeful Birdwatching

*The longest innings – perils of field identification – for one Booted Warbler, read two –
the useful taxon – a pocket book – the Witherby* Handbook's *commonwealth of
observations and disciplines – its ancestry – its Dutch paintings – its lasting value –
other most-read books and journals in the late 20th century –
the size of private libraries – changing reading behaviour*

By now, in your reading of this book, you may have had enough of ancient history. It was my worst subject at school, all boring Kings of England and dates, not half as much fun as its retelling in Shakespeare's wordplay or the stories of William Wallace and Robert the Bruce. Noting my continuing low marks, Dad said, "Don't worry; it'll come to you" and it has. Now everywhere I look in the ancestry of my hobby of purposeful and joyous birdwatching, I find extraordinary people and amazing bird spectacles. My fascination with our ornithological past grows and grows.

The longest individual innings of young birdwatcher and adult ornithologist in one person that has so far reached the record books belongs to Horace (H.G.) Alexander. His first score was a singing Chiffchaff on 25th March 1897; it became his first diary entry 24 days later and although he lost his upper hearing in his early sixties, his eyes stayed alert until his final confinement in September 1989. He had carried his ornithological bat not for the *Seventy Years of Birdwatching*, about which he wrote in 1974, but for over ninety. Given the increasing life-span of our species, Horace's record may not hold for much longer but it will remain unique in its witness of birdwatching's progress from pleasurable hobby to purposeful amateur science or, to be accurate, the forming of the enduring enjoyable compound of those two pursuits.

In one of his 1961 letters to me discussing the separation of once hideously difficult, uni-coloured warblers – 'little brown jobs' to some – Horace counselled me rather sharply on "the ever present perils of bird identification". Three years later in my 1964 paper on the identification of the genus *Hippolais*, I unintentionally took him to task for a seemingly loose attribution of pale outer tail feathers to the Booted Warbler. He was hurt but characteristically forgave my impudence before signing his letter. Horace's ghost was, however, to have the last smile. In the Mays of 1998 and 2000, I was privileged to assist Lars Svensson and Andrew Lassey with the observation and netting of both western and eastern forms of the Booted Warbler in Kazakhstan. Lo and behold the latter race, now increasingly split as a separate species called Sykes's Warbler, proved to have more white on the outer edge and tips of its outermost tail feathers than the former, still called the Booted Warbler. The penny did not finally drop until I began my preparations for this chapter. Horace's "useful but undependable feature" had not only existed but clearly was relevant to the distinction of the two forms. So why had he rated it

"Last winter I had a small thrush in view on a telegraph-wire near Poole Harbour for several minutes, with broad pale wing-bar. When it flew down into some swampy woodland, I saw a little white on the tips of the tail. Is there any such bird? Well, in the end I found there is, namely the female Siberian Thrush. And so I have entered it in my list. I am not inviting anyone else to believe it! – but only to note that those do seem to be diagnostic features of a species that probably occurs in Britain from time to time. It isn't an African winterer, is it?" Horace Alexander, *in litt.* (1961)

A shorter account of what was "almost certainly" Britain's second Siberian Thrush, also discussed in *Seventy Years of Birdwatching* (1974). The habit of keeping a private list has softened the blows dealt by the increasingly zealous bureaucracy for many observers.

"undependable"? Because throughout most of India, the pesky little sods mingle in winter. He had almost certainly been seeing first one, then the other, and although he firmly believed that some subspecies could be distinguished in the field, he had not quite got to the complete visual perception of the two different taxa. (Incidentally, the word taxon, plural taxa, is terribly useful when you get fed up with modern taxonomic diarrhoea, alias the excessive splitting or sub-division of former species. It has come to mean a discernible group of virtually identical organisms and its use is essentially non-contentious.) This last anecdote illustrates the amazing development of field identification, even field taxonomy at the end of the last century and nobody would have approved of it more than Horace.

I must now write the rest of its early story, taking up first the wonderful legacy of the Witherby *Handbook*. Although I have rightly given Bernard Tucker most of the credit for its Field Characters texts, it is important to recognise the full promise in their actual, longer title: Field-Characters and General Habits. This clear attempt not to repeat yet again a collection of plumage colours and contours but instead to describe the whole being of a species was not actually new. Indeed my first bird bible, *A Bird Book for the Pocket* (1927) by the independently-minded Edmund Sandars, had presented very astute notes on Flight and, charmingly, Manners (alias behaviour) but the *Handbook* essays went much further. Drawn from the publications or notes of at least 349 named observers published over at least 279 years, the portraits allowed readers to picture 424 species and 96 subspecies more fully than ever before. Tucker's canvass of information was almost worldwide, with his major sources being 40 references from North and South America and 225 from Eurasia and India. Strangely, there were only three contributions from Ireland and just one from the Antipodes. Interestingly, the social classes of the British observers showed much change from those of Gilbert White and Francis Morris. The nobility and clerics mustered only seven mentions and there was only one of a native tribe, Seebohm's Samoyeds, but in their places had appeared two legions of true experts and eager adventurous birdwatchers. The new ecumenism of field ornithology begun at Oxford had become fully visible.

The main stores used for American species had been stocked mainly by A. C. Bent, Dr. R. C. Murphy, C. E. Alford, E. H. Forbush, W. Rowan and the young Roger Tory Peterson. The chief cupboards for East European and Asiatic species had been filled by J. A. and J. F. Naumann, H. Whistler and O. Heinroth. The most frequent contributions of British observers, both at home and from the Empire, came from J. G. Millais, G. W. Phillips, C. B. Ticehurst, F. C. R. Jourdain, V. C. Wynne-Edwards, J. M. Dewar, T. A. Coward, H. F. Witherby, H. G. Alexander, R. Meinertzhagen and Collingwood Ingram. The profile of most studied species was still skewed toward sporting targets like ducks and gamebirds but seabirds and waders had already acquired a full press. Uncommon species and particularly small vagrants usually got much shorter shrift. Nevertheless, the general product was a totally invigorating and truly international distillation of avian beings. It all added up to a remarkable conveyance of knowledge and perception and this led immediately to increased ability in field identifications and, importantly, much greater acceptance of the sight records of unshot birds. The burst of observer

It can take decades to perceive the true appearance of birds. Compare the near-cariacature (top) of a Booted Warbler from brief Jordanian experience in 1963 with the full 'distillations' of other Booted Warblers (middle) and Sykes's Warblers (bottom) made after extensive Kazakh studies in 1998 and 2000. Today long lenses and digital cameras add even more security of recorded image.

"Can you imagine what it was like for a 17-year-old when he first set eyes on full sets of Dresser's Birds of Europe and the Witherby Handbook in Peggy Meiklejohn's living room in 1949? She became something of a hero too. She and Richard Richardson. He used to go round to her home "Arcady" every Tuesday night for his weekly bath – the only time one would see him not wearing his army beret!" Mike Rogers, *in litt.* (1999)

Early observations on ducks formed a major plank toward the understanding of avian behaviour. They were lyrically portrayed by John Guille Millais, the fourth son of a famous President of the Royal Academy. Depicted here are a pair of Goosanders which first nested in Scotland in 1871 and have now colonised much of upland England and Wales.

"I didn't have any birdwatching books. The first bird book that I ever read was Lockley's Shearwaters. *Then with the money from a school natural history prize, I bought the Witherby* Handbook ..." The first two literary stepping stones of Professor Christopher Perrins, the final editor of *BWP* and *BWPC, in litt.* (2000)

confidence that followed soon bred a new challenge, called succinctly 'beating the *Handbook*'. Proofs of new characters and previously unseen behaviour flooded into the pages of *British Birds*.

The Witherby *Handbook* did not only improve field identification. The full development of the content framework first used in *The Handlist* and *Practical Handbook* gave us a wonderfully rhythmic discipline for the study of any species. Regular reading of its text sections amounted to a degree course in systematics, nomenclature, ecology, field appearance, general behaviour, voice, display and later breeding behaviour, food, status and distribution in Britain, Ireland and abroad, morphology, moult, measurements, structure, soft parts and allied taxa. Effectively, frequent study produced graduate field ornithologists, unexamined but keenly attuned to all the attributes of their subjects. As Witherby wrote in the preface to the fifth volume published in 1941, a year of no great optimism about the outcome of the Second World War, "we are at the beginning of … a new era in the study of birds in which the main emphasis will be on the *living* bird and its problems, and … the congenial task of ornithologists in such a study will never be finished, for it is inexhaustible."

Moreover, Witherby decided that the *Handbook* must supply "a much-felt want … and ardent desire … figures of (British) birds in their various stages of plumage". In this "long … cherished, if somewhat faint hope", he had an enormous stroke of luck. Three hundred and ninety-one plates of relevant figures had already been painted for Dr. E. D. van Oort's "magnificent work" on *The Birds of the Netherlands* by the then "well-known bird artist of Leiden … Mr. M. A. Koekkoek". Following a generous permission to extend the copyright and further work by the artist to make some of the plates "suit our birds", Witherby only had to fill in 94 gaps – the purely British indigenes and vagrants – and for these, he commissioned George E. Lodge, assisted by H. Gronvold, Roland Green, Philip Rickman, J. C. Harrison and J. W. Frohawk. Their efforts, together with the repetition of the text figures of the *Practical Handbook* drawings of the nestlings, photographs of nest down, pellets, Willow and Marsh Tit and a nestling Cuckoo ejecting an egg, poured bird images and plumage maps all through the *Handbook*. The "accomplishment (was) a very great satisfaction to the editor". Who said that the dry, tweedy elder could not emote? The wealth of illustrations and his own meticulous plumage descriptions removed at one fell swoop the barriers to ageing, sexing and even sub-specifying birds in the field. Always remembering the flying start given by Koekkoek, the achievement of the *Handbook*'s 499 illustrations was of enormous benefit to every reader. Even the future wonders of photographic guides had been foreseen.

I have almost destroyed my Witherby *Handbook*. All five volumes are spineless; many plate pages are loose; scribbled notes and comments despoil the text; hideous early ticks disfigure the full list of bird species and subspecies at the end of Volume V. Yet it remains the friendliest and most treasured reference work in my library and many, many others. Second-hand sets can be found for about £70. Do not hesitate to buy one. Truly, it is a treasure-trove, still constantly consulted throughout the ornithological world.

I have put the best British mid-20th century book first and it might seem that my praising of it will put all other founts of knowledge in the shade. This is not the case, however, and I have two sets of witnesses to prove the point. The first group consists of the 44 birdwatchers and field ornithologists aged between 24 and 95 who answered my questions about the books and journals that had had the greatest influence on their learning and attitudes. Altogether they came up with 130 titles. The most featured were the Witherby *Handbook* and Phil Hollom's succinct reprise of it *The Popular Handbook of British Birds* (1952-88), listed 25 times by all decade classes. Next came T. A. Coward's *The Birds of the British Isles and their Eggs* (1920-1926), mentioned 18 times. Coward's friendly, compact bible which glowed with plates by Archibald Thorburn had been the foremost conveyance of bird knowledge for the over 65 year olds, being listed almost twice as often by them as the Witherby *Handbook*. Coward's pre-eminence in elder circles must have come partly from its earlier publication (by 12 years) but as Bill Bourne has pointed out, the author had an experienced journalist's knack of making his writing accessible and assimilable. After the above triptych, no other general introduction was noted more than five times but F .B. Kirkman and Rev. F. C. R. Jourdain's *British Birds* (1934) with Allen Seaby's wonderful illustrations led the tail. As primers of the hobby, the cheap and widely distributed S. Vere Benson's *Observers Book of Birds* (1937, updated by Rob Hume in 1988) got 12 listings in a clear reflection of its world record sale (for a bird book) of three million copies. James Fisher's *Watching Birds* (1940) and Edmund Sandar's rarer but interestingly maverick *A Bird Book for the Pocket* (1927) achieved respectively only four and three mentions. I thought it strange that no one in the first group of witnesses recalled the engaging cigarette and other similar cards that featured birds (and have since become valuable collectables) or the similarly involving *I-Spy Birds* booklet of the *News Chronicle*. Perhaps they were too ephemeral or aimed a class too low, as Roger McGough would recognise.

The original covers of the Witherby *Handbook* advertised the first ever series of plates that showed every plumage. George Lodge's grouse adorned Volume V. The work was serious but unlike BWP easy to read!

Filling in your I-SPY booklet with each new bird got you points. For 1,000, you became a BIRD SPOTTER (2nd Class); for 1,500, you went up to the rank of RED-SKIN (1st Class). It was fun!

"It has to be 'Peterson' – my first edition fell into pieces. Then Witherby's Handbook *which I bought (27/6d) volume by volume over 3 birthdays and 2 Christmases with book tokens and cash gifts.* British Birds *because it set the gold standard for ever."* David Parkin, *in litt.* (2000)

The three books most helpful to a young Geordie, later Professor of Genetics and past chairman of the BOU Records Committee.

"Max Nicholson, whose How Birds Live *was my first pointer to field ornithology; Edward Wilson (of the Antarctic) as both person and naturalist; Dr Sydney Long who was a great uncle and founder of the Norfolk and Norwich Naturalists' Trust and passed on to me the Witherby* Handbook, *each volume as published; and importantly Norman Moore, a school friend from whom I learned a great deal on excursions both in term times and holidays."* David Snow, *in litt.* (1999)

The mentors and heros of one of Britain's most accomplished ornithologists, largely responsible fo the more reader-friendly concise edition of the *Birds of the Western Palearctic* .

"The News Chronicle's I Spy Birds, *which I still have with my first entries in 1954."* Moss Taylor, *in litt.* (2000)

The start point of an eleven year old who went on to be co-author of a county avifauna and a 'A' ringer for 40 years.

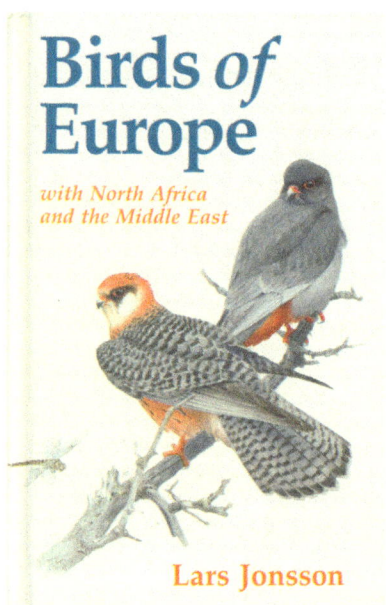

Lars Jonsson is acknowledged to have the greatest gift for bird portraiture in the late 20th century. His birds live; his feather maps prove the power of the human eye to fix detail. This is the cover of Europe's now-favourite field guide.

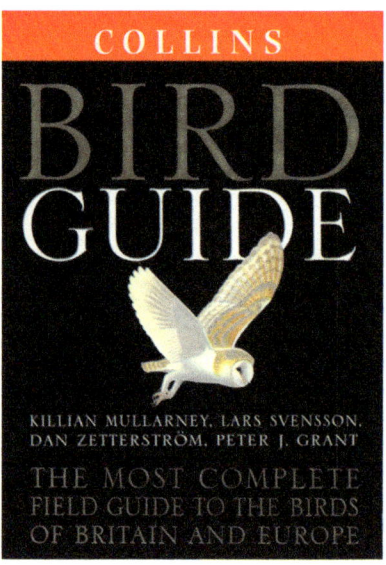

Not content with measuring *n* thousands of skins, Lars Svensson joined forces with Peter Grant, Killian Mullarney and Dan Zetterstrom and together they poured the 'New Approach' into another excellent field guide.

At the 12th of every month, the topical energetic journal of the Birding Information Service comes through letterboxes. If on occasion rather demand-driven, it is nevertheless a sparkling mix of travellers' tales and identification lore.

"I first met Richard Richardson at Wilstone Reservoir near Tring, on 27th August 1939. He was 17, had no binoculars but patently great eyes ... as assistant editor of The Countryman *magazine, I was the first to publish his pictures ... The plans for* The Pocket Guide to British Birds *were first laid in Norwich in 1947 ... Collins were very helpful, even paying for field trips to France and Norway ... it finally came out in 1953 and I was distinctly put out by the arrival of Peterson only one year later and went on to do the flowers guide with Maclintock."* Richard Fitter, pers. comm. (1999)

Although so rapidly overtaken by the archetype of the genre, the two Richards' guide was rated "an excellent work ... especially helpful to beginners" by Peter Tate in his *Birds, Men and Books*

From the 1950s, the early primers were rapidly displaced by the modern genre of field guides of which Richard Fitter was the British instigator. *The Pocket Guide to British Birds* (1952) was full of delightful figures by Richard Richardson but given its non-systematic sequence, it failed to compete with *A Field Guide to the Birds of Britain and Europe* (1954, updated to 1993). The latter is known as 'Peterson', after its senior author, and was listed 15 times as a fundament of knowledge, being most cherished by 50 year olds. Fitter's second guide, *Birds of Britain and Europe with North Africa and the Middle East* (1972, updated to 1995), written with John Parslow and illustrated by Hermann Heinzel, outflanked 'Peterson' in species span. Although only mentioned four times by my respondents, it became an essential companion to extra-European travels. Remarkably the two modern 'guru guides' – Lars Jonsson's wondrous *Birds of Europe with North Africa and the Middle East* (1992), a painstaking regurgitation of his earlier five delights entitled *Birds in the Wild* (1976-80), and the *Collins Bird Guide* (1999), Lars Svensson's dream of a 'perfect field guide' given discipline by Peter Grant and well-worked illustrations by Killian Mullarney and Dan Zetterstrom – got only three mentions each and only from the 20 and 30 year olds. For my 44 respondents, the pre-eminent journal was and still is Witherby's *British Birds* (1907 to present). It has been the most influential topical medium for all age classes; its 32 loyal readerships outreached any like title or indeed book. A long way from *British Birds* come the BOU's seriously scientific *Ibis*, listed by five of the over 60 year olds, and the rare-bird-driven *Birding World* (1987 onwards), mentioned by four of the 40 and less year olds.

of British Ornithologists' Union,
Natural History Museum, Tring, Herts. HP23 6AP

Sacred; old and threatened? The symbol of the BOU has been revered in three centuries but most birdwatchers are more interested in a live Glossy Ibis. In the peak of the collecting epoch, it was a much more frequent vagrant than it is now. In the 1900s alone, 138 were recorded in Britain and Ireland; at least 87 were shot.

Another earlier sample of the favourite literature of birdwatchers and ornithologists was taken by Andy Richford at the conference on 'Birds in Words and Art', held by the British Ornithologists' Union and Society of Wildlife Artists at Dartington Hall, Devon in September 1996. Because its attendees were mostly professional or lifelong amateur birdwatchers alias addictive ornithological bibliophiles, their opinions on the British library of bird books are not strictly comparable to those already given here but 51 males aged on average 59 years answered a questionnaire. They all subscribed to journals and magazines, reading on average six topical offerings, and 44 had established their own libraries, containing on average 506 books amassed over 40 years. (At today's benchmark price of about £40 for a good new or replaced second-hand bird book, the value of the 44 libraries would begin at £880,000.) Since the bibliophiles were clearly compulsive book purchasers still buying up to 50 new titles per annum, and also critics of their ornithological value, their responses presented more studied preferences. The positions of the leading titles and authors matched fairly closely those in my sample but a full analysis of 244 titles of the most-read authors produced a distinctly different order of armchair enjoyment or thought influence. James Fisher (19 mentions) achieved a clear lead over David Lack (14 mentions) and the artist-scribe Charles Tunnicliffe shared third place with the always lucid David Snow (10 mentions each). After this quartet, readership quickly fragmented with no other author achieving more than six listings but three relatively ancient writers Coward, Hudson and Seebohm were still ahead of modern luminaries like Janet Kear and Ian Newton. Also striking were the changes in the lists of 'top ten' favourites with the passage of time since publication. Of the 84 titles, the leading 45 (with 150 mentions) had all been published before 1975. Although the later, younger publications were subject to more frequent use, the implication of more 'reference' to and less 'reading' of newer books is strengthened by the fact that the

"It sometimes strikes me that identification has for many people become an end in itself rather than a means to another end. This has led to a search for ever more difficult identification problems to solve (e.g. endless scrutiny of immigrant gulls) and unrealistic solutions (i.e. to identify this species, it is necessary to observe the patterning of the innermost median covert – does anyone really identify birds this way?) I know some gurus of bird identification (e.g. Keith Vinicombe) agree with me on this." Rupert Higgins, *in litt.* (2000)

"Many – if not most – of the most readable pre-war books were written by exploring collectors. And the impact of Peter Scott's two pre-war books was enormous. Even to non-shooters such as myself, he communicated the thrill of the chase, and he somehow contrived to establish birdwatching as an activity acceptable at the highest social level. Lord Grey of Fallodon had started this earlier." Dougal Andrew, *in litt.* (1999)

incidence of texts in full prose format has fallen (61% of books published before 1975, only 40% after that year), this trend being undoubtedly caused by the proliferation of tense identification guides. What a 'read-bite' strategy will do for future students of ornithology is debatable but certainly it will do little to support the tradition of literary enjoyment that flowed so freely from the older books. (The opinions of journal junkies, CD-ROM users and video voyeurs on such issues would make interesting reading.) The Dartington Hallites were also asked for their other tastes in true literature. Novels, classics and biographies (17 mentions), general reference works (seven mentions) and poetry and plays (five mentions) led the rather short selection given by the ornithological obsessives.

As will be seen from the circulation figures of currently popular books and journals given later, these two snapshots of the most influential or enjoyed British writings about birds cannot be taken to be accurate research but they are likely to have identified the most gushing springs of ornithological knowledge tapped in the second half of the 20th century. Among the most recent British titles likely to dominate the next half century, the two panels gave sufficient comment to indicate continuing success for the Jonsson guide, Peter Harrison's two books on seabird identification and Peter Hayman's amazingly meticulous portraits in *Shorebirds* (1986) and long use of *The Birds of the Western Palearctic* (1977-98) and the series of BTO and IWC atlases. Where cometh more inspring narrative, they knew not.

"Some of the excitement and spirit has been lost … Among all the biometrics, statistics, tertial fringes and tonnage of books and journals, it is sometimes easy to forget the spectacle of a pre-dawn winter duck flight, or the raw excitement of a skua passage over a dark spring sea, or the magic of Hobbies hawking insects on a warm summer evening, or the expectation of an autumn migrant hunt."
Peter Combridge, *in litt.* (2000)

Regret expressed at the loss of narrative that enthused over and interpreted the spectacles provided by birds.

One Whimbrel that lifted off from the Broomfleet Channel, Humber Estuary, on 17th September 1978 had a dull, spotted rump, indicative of an East Siberian or American bird (target subspecies for keen and patient hunters). Peter Hayman in *Shorebirds* was the first and only artist who bothered to illustrate them.

The Winding Road of National Review

The happy-go-lucky post war years – the champion list-keeper –
new bird and wet lunch – the birth of the 10 Rare Men – their good start –
a small grey but model book not repeated – a bewilderment of utterances –
the vexations of modern rarity review – the conflicts of identification approaches –
the origin of 'jizz' – the views of the major players – a future resolution of final criteria

In spite of the amazing harvest of the collector-ornithology and coastal lights scheme, the realisation that Britain and Ireland could display the greatest diversity of birds in Europe was slow to dawn. Until the mid-1950s, there was no sign of the modern thrust for and hype of rarities but he or she was a rare birdwatcher who did not want to build a lifelist. Even those addicted early to purposeful study or general interpretation of birds were prone to the occasional migrant hunt or chase of a local rarity. Given the wealth of distribution data published in the Witherby era, observers could quickly learn enough of vagrant history and geography to recognise the potential of annual pilgrimages to Cley, Dungeness and the other known fleshpots. Richard Fitter, one of the first doyens of widely enjoyed, intelligently presented birdwatching, freely admits to indulging in "a form of twitching" in the 1930s but his pre-war trophies and those of others were usually self-found, not passed far on and certainly never queued for. Further, the still high degree of estate security and keeping meant that many pathways to birds were still private or seasonally interrupted. As late as the early 1950s, the Royal prerogative of Snipe shooting (of up to 1,600 birds) on the north Norfolk marshes prevented any free ranging.

Equally significant to the former calmness of birdwatching was the lack of competition in list length, with the private goals for such quotients being modest. I well remember Dougal Andrew suggesting as late as the early 1960s that all those who managed to scrape together a British list of 250 species should be admitted to membership of a lofty 'K-club' and be allowed to purr in satisfaction. For two years in the late 1950s, there was an unofficial listing competition between England and Scotland administered by the then Bedford office of *British Birds*. Although full or part Scots by blood, James Ferguson-Lees and I played for England in 1958 and managed between us to assemble 198 species. To our chagrin, the true Scottish pair 'stroked' by Maury Meiklejohn got to 202 and took the imaginary ornithological Calcutta Cup by four. It had been great fun but the fixture did not take hold unlike the modern Bird Races, which can assemble over 150 species in a day and do so much to raise funds. Many other post-war birdwatchers used geographical listing more purposefully in their own record files, none more so than Bruce Campbell, Honorary Secretary of the BTO and Britain's most accomplished ever birds' nester (but not egg collector). His matrix of county lists was amazing, building to significantly comparable data, but I never heard him

"Actually there were 4 or 5 'mild twitchers' who sallied out from the London Natural History Society before the 1939-45 war – Dick Homes, John Wightman, Geoffrey Paulson, D. A. T. Morgan come to mind – but the targets were of no great rarity, e.g. Purple Sandpipers at Hastings ...
I remember too C. A. Norris chasing something at Slimbridge in March 1939 and Phil Hollom, demobbed from the RAF, definitely keen for a Lesser Whitefront there on 24th December 1945". Richard Fitter, *pers. comm.* (1999)

To most sporting gentlemen with shotguns, a flushed Snipe was just that but Seebohm probably knew – and today's birders certainly know – that it could be a Great Snipe or even an American Snipe or even better a Pintailed Snipe. Most old records of Great Snipe came accidentally from sport, as the industrious Keith Naylor's first catalogue of rarity records (to 1957) makes clear.

trumpet his national list. The specious status of mere high cost collection did not exist and truly the rarity fortune of others was greeted by applause, just a shred of envy and a mental note to hunt harder for one's own. With no instant communication of bird news, one or two rarities a year kept one happy and common birds were just so, everywhere.

The happy-go-lucky bird hunts of the 1950s can be illustrated by the sheer coincidence required for my best man Bill Conn and me to collect a late share in Britain's first Wilson's Phalarope at Rosyth, Fife, on 16 September 1954. Done with National Service, set for university, we had planned to spend most of that month exploring the fleshpots of eastern England and the Borders before setting steam from Leith for ports north to Lerwick and eventually Fair Isle. Along the way we ticked off in the first 13 days General Wainwright, the commander of duck ringing at Abberton Reservoir, Derek Goodwin at The Bird Room, the wardens of Gibraltar Point and Monk's House Bird observatories and at most ten other itinerant birdwatchers. Not one of them told us of a rarity and our best birds were 220 Black Terns, pushed into the bottom of The Wash.

What changed our bird quality so dramatically was a promise of luncheon with Dougal Andrew, our shared Edinburgh mentor, and the necessary phone-call on the morning of the 16th to ascertain which restaurant was to listen to our likely hoots of mirth. It was to be the Café Royal at the East End of Princes Street but Dougal suggested a dog-leg approach via "the Wilson's Phalarope, an American bird reported to be at Rosyth", my log entry of that night sneered. It continued the next day: "Bill's Mum insisted on properly filled tummies but we were across the Queensferry by 11:00, to begin our search of the reed-belted pools in alternate blinding sun and rain. After a frustrating twenty minutes and a complete soaking (of tweed jackets and moleskin trousers), we left the smaller of the pools and pushed through the reeds and wortsward for the big one. There it was, shining away like a small Greenshank. Up it got; white rump and uniform wings, plus yellow legs, established its identity without doubt. A first for Europe. All hail, Frank Hamilton and Keith MacGregor (the original patch workers of Aberlady Bay) for finding it and Dougal for naming it from Peterson's American field guide." So back we went, dripping ever more slowly, to take lunch with the cream of Edinburgh business society. My green wellies, peeking out from now drying and shrinking trousers, produced on the face of the head waiter a look that no twitcher could ever imagine. Ah, the days of purely accidental rarity collection but purposefully stylish celebration, they were splendid.

Four years later in 1958 the relative innocence of the early rarity hunts, local chases and slow enjoyments was lost for ever, when the scientific products of such activities, rarity claims and records, became subject to a new national discipline and inadvertently a dramatically advertised and thus inflated consumer value. Suddenly the years of local, quiet but not necessarily soft judgements and the God-like rulings of a Witherby or a Tucker were to give way to the acceptances and rejections of an apostolic committee of ten experts, all enshrined in an annual tablet of avian richnesses. The epoch of the *British Birds* Rarities Committee (BBRC), immediately christened 'the ten Rare Men', and their annual Rarity

Original sketch and later painting of Britain's first accepted Wilson's Phalarope, which occupied a marsh at Rosyth, Fife, from 11th September to 5th October 1954. Its backcloth included Snipe, gulls and ships of the Royal Navy; its origin was Canada.

Report continues still after 46 years. In my final role of occasional fierce adversary, I find it difficult to write about the modern conduct of the Committee's affairs with impartiality but I have never counselled desertion from its disciplines, only their more logical improvement. The case for its institution was well considered, particularly by its chief architect James Ferguson-Lees and its first chairman Phil Hollom. Its purposes were also clearly described in a written constitution, published in *British Birds*. Importantly, these included pro-active duties in the promotion of new standards of submission and the dissemination of new criteria. In the early years, there were some misgivings and even some temporary local dissents. For example, the co-operation of the Ornithological Section of the London Natural History Society turned on a single vote, while Bill Bourne and the early sea-watchers fumed at their lack of direct representation. Nevertheless, by the end of its first decade, the Committee and the strenuous efforts of its first Honorary Secretary, the meticulous Geoffrey Pyman, had securely established the national discipline. Its annual report, complemented by a steady stream of ground-breaking identification papers kicked off by Ken Williamson's 1960 revelation of the marsh terns out of breeding plumage, had also demonstrated that it did not only judge, it also served. Indeed, its overall performance was good enough for the Committee's

"Clearly we must have an "election" shortly but given the inconclusiveness of I. J. F-L's initiatives regarding Cheshire, I should like to ask D. G. B. to stay on pro tem. What do you think? I spent much time in October and November with the Liverpool tribe led by R. J. R. and got nowhere, except to the bottom of several whisky bottles." Myself, when Chairman of the British Birds Rarities Committee, to Ray Smith, Hon. Secretary of same body, *in litt.* (1973)

The quasi-democracy of the rarity review system is worn most thin by regional politics. The initials belong to James Ferguson-Lees, Graham Bell and John Raines.

structure and methods to be taken as a model system first by other west European countries and eventually by almost any other nation with organised birdwatching. Today, the international vetting system has produced two further tiers of review bureaucracy, the Association of European Rarities Committee and the West Palearctic List Committee. For a new bird for the region, the descriptions of the actual observer(s) could now have to match the perceptions of the species lodged in 30 to 40 other minds. We have come a long way from a single letter to the editor of *British Birds* and his final decision on its content. Intriguingly, the comparative efficiency and accuracy of the two situations are being effectively monitored in the current retrospective review of the 1950-57 rarities by a sub-committee of the Rarities Committee. As a member of this unit, my impression is that the incidence of records that may fall to the (?) final cut will be far less than that forecast by today's zealots.

The next major act of the *British Birds* team was to set the record straight on the puzzling concentration of ancient rarities in the Hastings area of Sussex and Kent. This is described in another chapter but here it should be noted that Nicholson and Ferguson-Lees' eventual rejections of all the 'Hastings Rarities' were framed as recommendations, presumably out of deference to the other authorities that had been involved in the welter of publications that the records had provoked. The fact that the British Ornithologists' Union was not directly involved in the investigation stems from the long prior separation of review duties which had left the Records Committee of the BOU with responsibility only for admitting the first fully accepted example of a new species to the British list, fitfully answering national taxonomic questions via *ad hoc* sub-committees and once in a blue moon producing a new checklist for Britain and, until 1985, Ireland. The last edition of the BOU checklist had come out in 1952 and had let all the suspect records stand. Given the BOU's acceptance of the *British Birds* re-assessments and its relaxation in 1956 of prior attitudes against boat-assisted passage across the Atlantic, the way to a major revision of the status of birds in Britain and Ireland was open. The task was laid mainly on the shoulders of the BOU Records Committee. Once again, the chief architect of the report's final

In 1940, George Lodge's plate of the juvenile White-winged Black Tern (top right standing bird) had shown its diagnostic dark saddle but it required fresh observations in 1959 to convince birdwatchers that such a mark was a field character.

format was the far-seeing James Ferguson-Lees but the real core of intelligence came from the founder of the St. Agnes Bird Observatory, John Parslow, who had spent 1966 investigating as never before the status of British and Irish breeding birds for the Nature Conservancy. Two other crucial features of the BOU's work were a special distribution survey in 1964-65 and the adoption in 1970 of a tier of four categories designed to make the list a more dynamic instrument of avian classification. Altogether, ten professional and 12 amateur ornithologists, drawing on information supplied by 135 survey respondents, contributed to a small, grey, but still shining, model of fact-trove. Entitled *The Status of Birds in Britain and Ireland* (1971), it was one of the first books that I bought on my return from Nigeria and I have treasured it ever since for its compact distillation of systematics, relative abundance, status class and seasonal and geographical distribution. Common and rare birds were put together in a wonderfully succinct way; the British and Irish avifauna was at last made visible as a structured avian community in a world context. There was some human pain, particularly over nomenclature and the sequence of species within families, and the later stages of the work were marred by resignations by a few amateur members of the committee. Thirty years on, however, the overall achievement remains singularly efficient, the only question arising being about the sad lack of any revision but for the mere peripheral additions of new species. A recent series of politically correct handlists, occasioned by the sad loss of solidarity with Ireland and responsive only to the now-divided geography of ticking, has not amounted to a proper sequel. It may seem that the equally helpful Atlases of the British Trust for Ornithology and county societies, and the new wealth of national handbooks written for Scotland, Wales and Ireland between 1986 and 1994, make a new *Status* unnecessary. Nevertheless it feels odd that nowadays the BOU does far more to describe the bird communities of the once British colonies and other foreign parts than those of its own home nations.

In the last quarter of the 20th century, the vacuum of official utterances has been filled by many voices and organs. Where once the Witherby *Handbook* or the BOU *Status* compendium provided a clear focus, there are rarity reviews from at least four report channels, separate reports on rare breeding birds and scarce migrants, annual reports for all the nations but England, conservation reviews, a stream of new county avifaunas and atlases and a steady flow of books and papers on individual species and families. The plurality of authorship is totally healthy but it does not always deliver an easy synthesis of British birdwatching and field ornithology, particularly to the minds of the all-important beginner generations. The greatest vexations lie inevitably in the now competitive and repetitive statements on rarities. The BBRC and the BOU Records Committee struggle to defend their roles, the former employing as already indicated an ever more zealous approach to record review and the latter co-opting foreign taxonomists to contain the hydra-headed developments in dynamic taxonomy, now radiated into three concepts of speciation and incorporating genetic sampling techniques beyond the ken of most observers. While respectively retaining the right of official pronouncement on the annual crop of rarities and the precise limitation and status classification of the British avifauna, the two committees have lost much popular

A female Lesser Kestrel floats across a wide sky. Recently the Rarities Committee rejected the five post-1957 records of that species for October and November. Yet of the ten old dated specimens, five were first seen or shot in those same months and three more occurred in winter or surprisingly early in spring. A loss of logic in a rush of zeal?

"I disapprove the prospective increase (of perhaps several hundred percent) in the total number of species of birds, arising from the DNA upgrading to species level of many subspecies (e.g. four or so species of chiffchaff instead of one). Many of these new "species" will not be identifiable by ordinary people interested in birds (nor by experts), and will cause despair among those birdwatchers who like to be able to name what they are watching. Despair may be followed by feelings of futility – and then turn-off. The hobby/science will suffer. Reassessments, re-classifications of subspecies should not be labelled species. And think of the turmoil in race-relations circles." Phil Hollom, *in litt.* (1999)

A forthright critique of the modern taxonomic diarrhoea from a senior field ornithologist in the mould of Horace Alexander.

support. Furthermore, their beliefs that the new voices and constituencies of *Birding World* and *Rare Birds* would have no lasting influence on their realms are clearly mistaken. The topical press which thrives on the virtual market economy of rarities, particularly as they are increasingly split into more tickable taxa, has not just swamped any cautious advance with a confusing flood of challenges but has also publicised as never before whole new agendas of debate, even arguing almost for democratic rather than scientific votes to force their conclusions. There never were only ten 'Rare Men' but today there are literally thousands with urgent opinions and no little skill or knowledge. Underlying all this fascinating turbulence of human behaviour is the unresolved divide in attitudes towards the safe identification of all birds. This was briefly debated in both committees in the late 1970s, particularly with a view to separating the calm proving of criteria from the sometimes heated judgements on actual records, but no consensus for change was assembled. As already noted, the conflict between what Roger Peterson called the traditional 'Holistic Approach' and the new 'Micro Method' or 'New Approach', as it was renamed in Britain, has become increasingly apparent. The former system is normally credited in full to Peterson, who described it as "almost more of an art than a science" and enshrined it from 1934 onwards in his famous series of field guides. All of these concentrated on defining the essential differences in the field marks of birds. He also referred to the process as the "boiling down" or "simplification" of their general appearance and actions into a distinctive compound image or 'jizz', the last term apparently stemming from an acronym used by fighter pilots, GISS or General Impression of Size and Shape.

The truth is that while Peterson was undoubtedly the greatest exponent of the art, he was not the first author to describe birds in simple but engaging language alongside lyrical illustrations. Thomas Alfred Coward was, in his three-volume gem *The Birds of the British Isles and their Eggs* (1920-26), in a contemporaneous essay in the (then) *Manchester Guardian* (spotted by Bourne, 1985) and in his *Bird Haunts and Nature Memories* (1922). Particularly in the last two, Coward discussed the recognition of birds by a synthesis of their forms and appearances and, as noted by Ken Osborne in 1999 (*Birding World* 12:82), the supplement to the Oxford English Dictionary places the first printed appearance of the term 'jizz' in the third reference. The word was not coined but reported by Coward as having been used by a "West Coast Irishman" to encapsulate the instant process by which he could "name … usually correctly … the wild creatures which dwelt on or visited his rocks and shores", the man familiar with them to a point where their naming took only "a glance". In spite of quite frequent debate on the origin of 'jizz', there has never been an Irish claim to it and certainly its three consonants do not indicate a Gaelic root. My bet is that it is a corruption of 'guise', a word falling into disuse except within its opposite 'disguise' but meaning 'external appearance in general', apt in an identification context, and also 'semblance, pretence and customary behaviour or manner', all again close to the issue. Intriguingly, 'guise' does not stem from Old English but from Old French and, via the related Archaic 'wise' also meaning 'manner', German and Old Norse (*Collins Dictionary of the English Language* 1979). Thus as a term for the essence of perception, the root of 'jizz' may be as old as the Vikings to whom western Ireland was no strange coastline.

There can be hardly one senior birdwatcher in Britain who did not absorb as schoolboys in the 1920s, 1930s and 1940s the original clues given by Coward to ready field identification. These formed an excellent mental preparation for its more detailed treatment in the Witherby *Handbook*, and his perception of the essential approach clearly antedated both Peterson's syntheses and Bernard Tucker's lucid texts by at least 15 years.

The instrument that ushered in the 'Micro Method' was the Questar, a veritable howitzer of a telescope invented in America in the 1970s. Look at a nettle through it, as I once did in Siberia in June 1980, and you recoiled from the sight of every last stinging hair; look at a bird through any similar glass and field observers who knew their avian topography could indeed compete with the inspections by eye of specimens in a museum tray. All of a sudden, it seemed, there was a levelling of the study field; no longer did the odds favour the museum bench. In Britain, the inspection of the minutiae of plumage and structure became the particular objective and province of the late Peter Grant assisted by Killian Mullarney, Ireland's most gifted bird artist. Although during its development some unnecessary time was spent up the cul de sacs of size illusion and moult vagary, their painstaking treatise on *The New Approach to Identification* appeared first as a series of articles in *Birding World* and finally as a pamphlet in 1989. It was quickly accepted by younger birders as the new almost biblical text for age-specific plumage determinations and certain identifications. It also acted as a useful glossary of revised descriptive terms. To those who had really learnt the Witherby series of handbooks, the content was less fresh, being as it was little more than a reinterpretation of the disciplines of plumage and other features first worked out for *The Practical Handbook*. To understand the rest of avian personalities, however, one looked in vain for any new guidance on the search for the more vital and just as species-defining clues in action, voice and behaviour. The gaps in these had actually been analysed in some detail by myself during the *BWP* process but most still yawn.. Thus in the wake of the *New Approach*, the strictures on narrative holistic descriptions increased. Although Grant and others did call for the continued use of the older method or, better, the combination of the two, the issues still vex.

Roger Tory Peterson's view on the new failsafe discipline was never directly solicited but I got it by the unhappy accident of the excessive revisions that, in his opinion, I made to the original *Field Guide to the Birds of Britain and Europe* (1954, rewritten in 1992). Roger let Guy Mountfort know that it had never been his intent to obscure the cruces of diagnoses with excessive detail or wearisome caveat. My attempt to make his classic book more competitive with its later rivals had not pleased the greatest distiller of field marks or characters of all time. At the 2000 Bird Fair, I repeated his transatlantic rocket to an audience which included Killian Mullarney. The surviving author of the *New Approach* said that he saw the point and went away to think.

The conflict is not irresolvable. It simply requires the balanced use of both approaches and a fresh attempt to understand how seemingly identical species do themselves tell each other apart. The brief given to would-be contributors to the

The cover of Peter Grant and Killian Mullarney's *New Approach*, with the figure making the point that with a modern telescope and a knowledge of plumage topography, a careful observer could compete with a museum worker.

In the 1950s, the Crossbills of the Rothiemurchus Forest, Strathspey, were assumed to be wholly and truly Scottish. Nowadays after recent interruptions, both Common and Parrot Crossbills co-habit in the same woods.

One hundred and seventy years on, aided by ever more accurate recording analysis, such early aural perceptions are being proved and used in the new taxonomy.

Field Character texts for *BWP* (Wallace 1983) could still be the plan for the next stage of our advanced perceptions. The final answers, which might well be simpler than most observers currently imagine, are likely still to be provided by gifted individual observers, such as Martin Garner and David Quinn, who finally showed a whole nation how truly different some large gulls were from others, and Magnus Robb, who has used his musicology to penetrate the mysteries of crossbill radiation through the arboreal medium of different conifer species and their cones. Much remains to be captured by open eyes, unwaxed ears and un-indoctrinated minds. A fascinating, final fling towards fully sensitive field identification beckons. I note again that young observers need not fear for employment. Only the infectious traumas of the reviewing committees will require some careful medication. Otherwise the most adventurous perceivers of field characters will only continue their current tendency to "bird off piste", Anthony McGeehan's lovely analogy for leaving the officially beaten track!

In blessed contrast to the affections and disaffections with modern identification practice and uncommon bird reporting, the achievements of the conservator-commentators on the fortunes of common birds have become increasingly unified and politically persuasive. Developing the sound research first put in place to turn back the threat of toxic chemicals in the early 1960s, the *BTO* and *RSPB* have shown increasing confidence in their commentaries on both avian and general wildlife issues. Their work deserves a full chapter of its own, which follows.

The mountain of Kuhi Taftan was the perfect antidote to the Baluchistan coast.

Right: Richard Porter bends down for his camera and Lindon Cornwallis reaches for his notebook. DIMW.

Left: newly met with Plain Leaf Warbler and Hume's Lesser Whitethroat, the author pauses to give thanks. Above his bonnet, the smoke plume of the mountain's volcanic summit trails east.

An Irano-British expedition exploring Baluchistan, Iran. All expedition convoys stop for raptors. DIMW.

Sudden floods constantly damage roads in arid land. Here Derek Scott supervises a temporary repair. DIMW.

Probably the worst camp site in the world at Gwatre. Hot, humid hell. Even if you went for a swim, you met lethal seasnake! DIMW.

Ornithological Shysters

The phenomenal magnetism of the Hastings area to rare birds – its reinterpretation as a major fraud – contemporaneous and later views of this act – the prescience of vagrancies and some undoubted recurrences – a fresh test of provenance – an earlier cluster of similarly tainted records – the ultimate hunter–gatherer, Col. Richard Meinertzhagen, C.B.E., D.S.O. – his footsteps followed – a singular meeting – his last military act – his fall from ornithological grace – the continuing enigma – a last salute – petty crimes of the modern era

Having instituted in 1958 a national and quasi-democratic discipline on rarity records, the *British Birds* editorial board proceeded to tackle one of the strangest features of rare bird history in Britain. This was the astonishing concentration of exceptional records in west Kent and east Sussex in the first two decades of the 20th century. The phenomenon had been evident piecemeal on the pages of various journals since the early 1890s but its full distillation in the Witherby *Handbook* in 1938-41 had raised many eyebrows even higher. How could 28 of the 219 taxa occasional or rare in Britain have occurred *only* within 20 miles of Hastings?

With much care and the help of John Nelder as statistician, Max Nicholson and James Ferguson-Lees investigated 1,015 occurrences, involving 1,360 birds of 168 species and subspecies. One of their first findings showed again the extraordinary magnetism of the Hastings area between 1903 and 1916. Thirty-two of the 49 taxa added to the British list in that period had come from it. Yet between 1919 and 1941, only 2 of the next 24 new taxa had found their way there. The disparity of the former 65% and latter 8% share of discoveries glared. The outcome of eight years of sleuthing into the increasingly suspect handling of rare birds, backed by analyses of both geographic and annual patterns, was the excision of 16 species and 13 sub-species from the British list and the retrospective rejection in total of 542 specimen and 53 sight records. As had palaeontologists with Piltdown Man and botanists would with the Rhum affair (Sabbagh 1999), British field ornithologists had firstly choked on and secondly diagnosed and disgorged a major fraud in the specimen marketplace. The findings were trailed in Phil Hollom's prompt discard of all the Hastings Rarities from *The Popular Handbook of Rarer British Birds* (1960) and the full chapter and verse was given in the voluminous August 1962 issue of *British Birds,* beginning with an editorial entitled *Setting The Record Straight.*

As far as the ornithological establishment is concerned, none of the immediate heated debate over, nor any of the later observations made on the chief suspect, George Bristow, a taxidermist and gun-maker with a shop in St. Leonards-on-Sea, and the flood of rarities that exited from it, are likely to change the result. The most overt response to the excisions came from Jeffrey Harrison in his *Bristow and the Hastings Rarities Affair* (1968) but his rather parochial arguments drew no blood from the sleuths and no re-acceptances from the national committees. More troubling to my mind was the fact that because the Dungeness rarity records made contemporaneously to the review did not fully reflect that area's actual early promise as demonstrated for example by H. G. Alexander, the reaper blades of Nelder's statistical analysis could have been set too low. This concern recurred in fuller measure when, in March 1972, Phil Hollom passed to me notes on the detailed comments of Dr. N. F. Ticehurst on 102 of the rejected records. The fourth editor of the Witherby *Handbook* and author of *A History of the Birds of Kent* (1909) had been correctly billed by Nicholson and Ferguson-Lees as an expert scrutineer. Even so, their rider that his earlier supervision of Bristow's role, as the depute of Witherby, would have brought a speedier end to the deceptions sat uncomfortably with the balance of Ticehurst's own remarks. As early as 1951 in correspondence with David Bannerman, he had been unwilling to condemn the major players and once again he offered proofs of Bristow being nowhere near ten suspect specimens and receiving no more than a stuffing fee on seven other occasions. "Where was the extraordinary gain?" was his repeated question. Even more positive were the supports given by Ticehurst for at least 45 of the records, these including not only confirmation by reliable witnesses but also impressive links with associated occurrences and close-dated weather vectors.

Conscious of the need to answer such lingering questions on their work, Nicholson and Ferguson-Lees have a further statement in draft but the senior author became nervous of adding further allegations to those already published. One example of how the plot has thickened is the role of Michael Nicoll, a keen observer and collector trusted by the reviewing ornithologists of the period, all the way up to Ticehurst and Witherby, but involved himself in 50 of the records. No fewer than 43 of these were discussed by Ticehurst who defended them all vehemently, regarding any suggestion of collusion between Nicoll and Bristow as ludicrous. Unnervingly, however, it must now be noted that Nicoll was a particularly close friend of Colonel Richard Meinertzhagen, the latter now proved to be both thief and fraudster and suspected of turning a blind eye to the Hastings affair. Does such liaison colour even more ill the awkward discovery that an Eastern Black-eared Wheatear, shot by Nicoll at Pett on 9th September 1905 and then locally skinned, became somehow a traded specimen? It had been prepared "in the unmistakably oriental manner", this suspect preparation being diagnosed by the expert taxidermists Pratt and Sons of Brighton. Nicoll has been given the benefit of the doubt in a seeming case of later substitution but it is hard not to feel yet another retrospective shiver about collector-driven ornithology. In human terms, the Hastings affair rumbles confusingly on.

Meanwhile, to the interest of many and the amusement of some, its subject taxa continue to reclaim their places on the British list through undoubted modern records. Of the 16 excised species, only two remain absent. The 13 subspecies are proving more stubborn but claims have been made for four. One of these, the Semipalmated Plover, has been promoted to species and its modern records are fully accepted. As many have noted, Bristow, or some ornithologically minded person in his circle, had been amazingly prescient of avian vagrancies. Indeed, the acute choices of the Hastings rarities remain as intriguing as their circumstances were suspicious, made as they were when British understanding of migration vectors

On the 15th October 1991, there was a daintier bird among several Wheatears at Flamborough Head. To one observer, it was an Eastern Black-eared Wheatear, a subspecies no less likely to occur than the Pied Wheatear but strangely under-recorded in modern times. No other observer thought of shooting it, as a collector would have done, and it remained a puzzle.

Impatient with discussions on the identity of a lark on Fair Isle on 6th and 7th October 1952, Richard Meinertzhagen unfolded a hidden shotgun and killed it. He presented the corpse of a certain Short-toed Lark at the evening log on the 7th to the utter shock of Ken Williamson, the observatory's director. It was the last vagrant to be collected on Scotland's 'Heligoland'.

was minimal. They have led one recent reviewer of rarities, Phil Palmer, in his splendidly energetic romp *First for Britain and Ireland 1600–1999* (2000), to suggest once again that many of the records were probably genuine. Accordingly, I went back to the raw data and applied a fresh test, namely the cross-matching of the monthly occurrence patterns of *first* the 493 Hastings rarities with some possibility of wild occurrence in 1892-1930, *second* the 90 with retrospectively little or no possibility of wild occurrences in 1892-1930, and *third* 385 relatively modern extreme vagrants providing less than ten accepted records in 1958-85. The three patterns do not match. The virtually impossible taxa appeared most frequently in winter, with the combined December to February share a staggering 44.4% and smacking heavily of the artefact of filling a normally slack specimen season with attractive goods. Furthermore, the total of March to June records was 80% greater than the August to November tally. The possible taxa were less frequent in winter but the combined December to February share was still high at 12.8%, compared with 7.5% for the recent extreme vagrants, while once again the former's spring records exceeded their autumn ones by 60%. The winter and spring preponderance of the Hastings rarities is truly incredible. Additionally, in the light of both prior and post rarity patterns, many of the associations of taxa and their origins do not reflect the realities of trans-Atlantic and trans-European vagrancy. For example, classically predominant species such as Lesser Yellowlegs, Short-toed Lark and Ortolan Bunting were either rare in or absent from the Hastings records. No sense, no science; just every reason to see the over-valued rarity as a dangerous thing in any era.

This conclusion may also apply to an even earlier but happily much smaller cluster of records from Yorkshire. The so-called 'Tadcaster rarities' share with those of Hastings a similar set of odd places and dates and a human cast again including unnamed obtainers, a suspect taxidermist and an acquisitive baronet collector. The records include the tattered hen from a highly improbable pair of Orphean Warblers said to have been breeding in a plantation near Wetherby in July 1848. An investigation into it and the other two first British acceptances that remain extant from the cluster was begun by the BOU Records Committee in 1999 but no conclusion has yet been reached. One other case has, however, been opened and shut. It is the retrospective ornithological drumming-out of Colonel Richard Meinertzhagen, D.S.O., O.B.E.

The world needs mentors and heroes and nowhere more so than in its ornithological fields, where such characters are becoming rare. Along the fortunate stations of my birdwatching course, many appeared to teach and inspire me but one alone totally gripped me. It was a man of far horizons and military derring-do and, to me, the ultimate hunter–gatherer among the several generations of field ornithologists and collectors which were spawned by the British Imperial spirit and who thought that almost the entire world was their hunting ground. For fully three decades, his writings presented me with a sharp thrusting interpretation of birds that I found irresistible. His name was Richard Meinertzhagen and I owe my almost entirely vicarious experience of him to one of my other early tutors, Dougal Andrew. "Remember to buy Meinertzhagen's new book on the birds of

Arabia," he advised with an eye to our mutual edification and eventual gain, "it is reckoned to be a good investment." Off I trotted to Thin's bookshop in Edinburgh and there it was; a few pages turned, I had to have it. Thus on 11th January 1955 began my enthralment by the once most famous, now most infamous soldier-ornithologist of the 20th century. During my National Service, I had followed in his footsteps in Kenya. Nearly 50 years had separated our tours of duty but their causes were identical, what were then called tribal rebellions.

For Meinertzhagen, it was the haughtily beautiful Nandi and characteristically he had brooked no nonsense when threatened by the Laiban and his spear-fidgeting court. He shot the main man dead. For me and the other hopelessly juvenile subalterns, commanded initially by incredibly ill-fitted field officers (effectively retired to Kenya by the War Office), it was the lowly, superstitious, land-exhausted Kikuyu. The Mau Mau rebellion was a hideous mess and but for the spell of the country, its still-prolific game and wonderful birds, I might have come even closer than I did to the atrocious behaviour of other officers, who felt that actions similar to Meinertzhagen's imperious gesture and the French in Algeria were still sanctioned. Accustomed to the insecurity of a declining Britain that had already given the Indian sub-continent back to its indigenes, I found the later publication of Meinertzhagen's *Kenya Diary 1902-1906* (1957) a warp not just of time but also of manners. Entrancing but truly fossilized in terms of its personal and general politic, only the animals shone out from it as eternal truths. Uncannily, however, during my one brief conversation with Meinertzhagen, it was about Kenyan politics that we exchanged a few words.

Once in the late 1950s, I was a guest of Eric and Thelma Simms at a British Ornithologists' Club dinner at the Rembrandt Hotel in Kensington. The evening's celebration was of the emerging school of young British bird artists and we ate against a panoply of canvasses and frames. Keith Shackleton's wonderful pattern of flying Crowned Cranes was the stopper for me until, at the end of our table I spotted a gaunt elder, attended by a majestic Lammergeier. It was almost alive, glowing out from its rare medium of sumptuous tapestry hung in a majestic frame mounted on wheels. The magnificent artefact had the air of an ancient engine of war. My attention returned to the discussion in which voices from the then-fashionable Bond Street natural history galleries were much to the fore. The elder seated at the end of our table said not one word but there was no doubt about who were the Alpha males of the whole august assembly. It was he and his vulture familiar; they made a riveting pair, brooding over us like figures from the Old Testament. With no knowledge of the man's field characters, I had to ask Eric who he was.

"Meinertzhagen," came the whispered answer and next the urgent question, "Do you want to meet him?" "Oh, God, yes," I managed as every bird-soaked synapse in my being discharged. "Come on, then", enjoined Eric but as we rose, so did Meinertzhagen to stride away from the hubbub of which patently he (and his faint hearing) had had enough. We caught up with him just as he reached the top of the marble steps that led down to the street. Happily he turned in recognition of Eric's cheery call of "Colonel". I stared at him; gaunt indeed, yet fiercely erect,

The Lammergier tapestry sewn by Richard Meinertzhagen (from his *Pirates and Predators*).

he was awesome. Three parts of him are indelibly imprinted on my retinal memory. Incongruously in those days of prim dress, his trousers were distinctly 'drain-pipeish' and his 'brothel-creeper' shoes had the deepest imaginable soles. Unnerved, I met his gaze and was instantly transfixed by his piercing eyes. (Forty years on, I recalled for Mark Cocker that I also detected the fires of hell flickering therein but that could have been a retrospective embellishment.) Introductions exchanged, a handshake allowed, I got out a few words on the outcome of the Mau Mau rebellion, the inevitable granting of independence to Kenya. "It should never have happened," snapped the one-time Captain in the King's African Rifles who had served such short shrift upon the Nandi Laiban. Unnerved by the strength of feeling in his retort, I fumbled my further response and he was gone, down the steps, an almost spectral figure as he faded away into the London night. Meinertzhagen lived on for another two decades but our paths never crossed again. Eventually, however, I found in David Bannerman's obituary for him an explanation of his tetchy retort. Meinertzhagen had unerringly forecast the outcome of the then Colonial Office muddle in Kenya and the still-continuing shambles of the Levant. Due partly to his espousal of Zionism, his prophecies had not found favour with the regional politicos and unable to influence their policies, he had moved on. Unwittingly, I had touched some scar tissue.

Knowingly, I followed in the Colonel's later footsteps in Syria and Jordan and came close to them again in the seeringly humid, broken land of Baluchistan. Once again, his writings formed excellent guides to habitat and bird. Indeed the diary entries for his journey to and past the Azraq oasis were an uncannily precise Baedeker for the first of Guy Mountfort's expeditions there in 1963. Only in a possible exaggeration of the Spectacled Warbler's past status was there any sign of flawed perception. Tragically the cameo that he painted of the Azraq ecosystem as "a perfect paradise for birds", and over which we so enthused in the 1960s in the then real hope of seeing it preserved as a National Park, has been smashed. Its subject place, no longer isolated in the Hammada desert, has been exterminated by the compound of lorry (and tank) highway and almost total water-table exhaustion. Meinertzhagen's paradise had served also as a cold but safe healing ground to his fellow officer in the Turkish–Arab campaign of the First World War, the equally enigmatic T. E. Lawrence, but it is no more. In 1999, another chance meeting with a native of Azraq Druz, Qusay Ahmed, brought further news. Qusay had been only one year old when we first looked down upon the oasis but, in his later life as the Tourism Officer of the Royal Jordanian Society for Nature Conservation, he had witnessed King Hussein's frustration with its destruction. With the continuing buffeting of the limited Jordan economy, however, he had felt unable to deny his people its water. Meinertzhagen would have recognised and disapproved the loss of imperial or kingly writ.

Meinertzhagen's military-political career was outstanding. There are few absolute impossibilities but the subverting from the late British Imperial establishment of the double award of C.B.E. and D.S.O. is one. Unreasoning bravery and quick intellect were the twin engines of one part of his self-diagnosed double life. Too old to fight in the Second World War, his hand was nevertheless on the tiller of one of the small boats that recovered the British Expeditionary Force from Dunkerque. At 62, he still did not flinch from an enemy that, like so many of his own and close generations of British soldiers, he had thought to have defeated several times and in several places before. No wonder he drew praise from Field Marshal and Prime Minister alike. Yet the ornithologist in Meinertzhagen has proved to have feet of clay and meanness of mind. Incomprehensively to his reviewers, even the unfortunate Mark Cocker who set out to laud him but had in the end to bring him down, and infuriatingly to me, a student disciple for 30 years, he was a thief of specimens, an irrational and savage critic, a plagiarist of rivals and a false witness to records. The proofs are certain, no more so than that provided by the specimen forenses undertaken by Alan Knox, and one is left impotent in any attempt at gainsay. I felt the last keenly at the BOU Conference on 'Birds in Words and Art' held at Dartington Hall, Devon, in September 1996. There, still with contrary notes of reverence in his voice, Mark Cocker exhumed my hero and besmirched him. I ignored the implications of a somehow tortured pysche put down to an unsympathetic mother and his maltreatment at prep school, culpable fantasising over a nymph-like niece, and even the chillingly mysterious loss of his second wife to the almost incomprehensible discharge from a revolver in her gun-familiar hands. The first two vagaries were the common lot of legions of public school boys and the third had been adjudged an accident. I was nevertheless forced to accept that the ornithological sins were real. Only Max Nicholson and I said anything in his defence. My Colonel's colours were in the mud.

Oscar Wilde diagnosed the killing of a love by excessive obsession with it as the ultimate absurdity in relationships. It seems that a similar departure from accepted mores has to be Colonel Meinertzhagen's ornithological epitaph. His calumny and physical plagiarism formed, as Barbara and Richard Mearns have aptly put it, "a mighty peculiar and unnecessary course of action". It amounts to treason against his own self in that part of his double life that he professed most to value and enjoyed even in the face of more dangerous companion duty, as Lawrence observed. I cannot see as his motive the desire for enhanced scientific status. In his personal life, he had faced triumph and torment time and again. Meinertzhagen lost not just a wife who was a full soul-mate in ornithology, but also a beloved brother, his eldest son and a brother-in-law, the last again of similar interests and like addiction to wilderness adventure. He was also wounded several times, once severely on the Western Front. The chances of any conceit surviving that assault course of personal trials seem remote. In any case, many of his writings verge on the description of, if not confession to his worst traits. What I intuitively sense is an overwhelming impatience with the lack of evidence for favourite theories and with the opinions of lesser mortals who could not or would not follow his drifts. Lawrence noted that in war Meinertzhagen's "hot immoral hatred of the enemy expressed itself as readily in trickery as in violence" and given what has passed in recent years, I find the occasional baleful asides in his books increasingly telling. At times, he comes close to condemning the evolution of *Homo sapiens* but, more importantly in this context, his forceful personal creed of science is clearly stated in the Introduction to his *Birds of Arabia* (1954):

"We had Meinertzhagen up here for a few days. He was in excellent form and was talking about chartering the Little Polar Bear hunting boat in Spitzbergen next summer. At age 77 or thereabouts, he remains quite irrespressible." Dougal Andrew, *in litt.* (1957)

"The dreamers of the day are dangerous men, for they may act out their dream with open eyes, to make it possible." T. E. Lawrence, thinking of Meinertzhagen?

111

"If I were to record views in full agreement with modern practice I should be writing nothing new; there would be neither advance nor progress; I should be stuck in the orthodox groove. I have endeavoured to climb out of that groove and suggest new ideas which, I believe, will take root among the younger generation of ornithologists. I would sooner express definite erroneous opinions on debatable subjects than compromise with an indefinite statement or avoid the subject altogether. The truth often arises from error."

Had he added "or even false construct" to the last sentence, there would be far less of an ornithological enigma to make us sad, scornful or angry. Given his less emotional commitment to, and much more scholarly research into, the Colonel, Mark Cocker feels that this is too simple an equation for his behaviour. Perhaps one day a young ornithologist more distant from Meinertzhagen's treason will find the time to sort the soiled chaff from the clean wheat in his work. For example, the warbler skins prepared by "my man (actually Trooper) Powell" are indeed the most beautiful in the Bird Room at Tring; the dark Desert Lark, named after his wife Annie as *annae*, does indeed inhabit the black basalt around Azraq where he had first described it in 1923, and he did indeed sew the tapestry of the Lammergeier, a bird that incidentally he saw as an inspiring image but a poor coward of a predator. Furthermore, in spite of his own deprecation of them, his majestic *Birds of Arabia* and his wonderful gallop through the avian *Pirates and Predators* (1959) are too important to be ignored. To deny the wealth of perception in their pages to later birdwatchers and ornithologists is a needlessly excessive sentence on innocent students and guilty author alike.

In the basalt flow of the north Jordan desert, cryptic forms of the Mourning Wheatear (top left) and the Desert Lark (left centre) occur. In the sandy wadis alongside, Dunn's Lark (bottom right) was discovered in 1965. Richard Meinetzhagen named the dark Desert Larks *annae* after his wife.

In order to gain other opinions on Meinertzhagen from those who had known better or studied him with less affection, I sent the draft of the above comments to Dougal Andrew, Bill Bourne, Mark Cocker, Mike Rogers and Keith Shackleton. Their responses were as divergent as my own disordered views. On the positive side, all remembered that his personal strength and attractiveness had been exceptional, even putting stars in the eyes of another African ornithologist Reg Moreau's wife Winnie, and his enjoyable ornithological dinners and other hospitality were recalled with relish. "I swallowed his tall stories hook, line and sinker", admitted Dougal, "as I drank his generous measures of vintage port distributed from a large gun-metal flask". Keith witnessed the real happiness of his last years spent in the long-desired company of Theresa Clay. On the less positive but still amusing side, Bill and Keith also noted the general tendency of Meinertzhagen's generation to indulge in playful fibbing. His own sallies included not just the oft-related enforced witnessing of a birth in a cramped train carriage, shared with an equally embarrassed King of Saxony, but also his shooting of a German Archduke in Kenya, recalled as his first human trophy of that rank (and the completion of a sporting 'pair' with the Nandi Laiban?) and with the tale topped "with quiet satisfaction" in the final comment: "I still have his picnic basket as a memento!" Keith mentioned in particular the kite-flying request in *The Times*, given his recently acquired understanding of the supposed value of rhino-horn as an aphrodisiac, for someone to value an attic box full of dusty, long-tumescent trophies. Both Dougal and Keith wonder if such acts were not determinedly wicked tests of the credulity of his listeners. The Devil take those who believed them? Conversely, Dougal noted the aberration, compared to the secretiveness of most military heroes, of Meinertzhagen's flaunting of his D.S.O. On the entirely negative side, Mark produced the taped and unequivocal attesting by Pamela Rasmussen and Nigel Collar to his multiple thefts of specimens and associated label frauds and resultant claims. Finally, there remain the dreadful question marks over his wife's tragic death and his subsequent mental breakdown. The choices of answer are several. Any others than that the former was a true, however unlikely accident and the latter the product of overwhelming grief are appalling.

Overmuch contemplation leads to indecision. There is clearly no saving Meinertzhagen's ornithological reputation, needlessly self-destroyed, but his books remain one of my chief and lasting inspirations. So I come back to the, as yet, unsoiled parts of Meinertzhagen: near peerless member of the British officer-class first commissioned in 1899 to the snobbish but foolishly brave 60th Regiment of Foot which marched at 90 paces to the minute, scholar of the cradle of civilisation and troubled romantic. I still hope that when he reached the other side in 1967, the buglers were in dress uniform and blew their proudest notes. *Kwahiri, bwana nkuba*.

Compared to the sins promoted by the collector marketplace or committed by an extraordinary but flawed soldier, there has been relatively little to occupy the fraud squads of British ornithology in the last four decades. As collecting ceased, so did the coveting of trophies and their conceited display. As museum expeditions and purchases decreased, specimen-based ornithology yielded mostly to the new study methods of observation and trapping. The profits at all levels of the prior trade in skins evaporated, except for curio value, and a new ethos of identification took hold, as I have already described. The intention and hope was for commitment to the new disciplines of field notes and checked diagnosis and trust in the observers

"He must have been very well-liked in his day (he apparently gave splendid parties) for people to defend him so long-nobody who really knew him has ever let him down." Bill Bourne, *in litt.* (2001)

`*"It was normal practice for Meinertzhagen's generation to tell fishing stories, and bad form for their contemporaries to question them, and of course he had to tell the best ones. It was also impolite for ladies to peach on them – and Theresa Clay was a real honey, which hardly seems to have applied to Annie Jackson, who was a blue-stocking tomboy."* Bill Bourne, *in litt.* (2001)

"Have I said that it seems to me truly sad that ornithology has undergone a split into academics and twitchers? The first are now cocksure, ignorant and narrow-minded, and the second spend all their time slandering each other. The type of the second in my youth was D. D. Harber, who never said a nice word about anyone. The type of the first was David Lack, who was awfully supercilious about ordinary birdwatchers." Bill Bourne, *in litt.* (2001)

The impressions that human beings create depend much on the prior attitudes of the impressed. I shall never forget the evident 'birder's desire' in David Lack's voice when in 1965 he interrupted my use of the EGI library with a whispered request for the whereabouts and field characters of eagles on the Austro-Hungarian border and the thanks that followed his later acquisition there of two new species.

The suppression of rarities is abhorred by the twitching fraternity. The most 'infamous' episode featured the Tengmalm's Owl which was allowed three weeks of peace in the Warren of Spurn Point, Yorkshire, in March 1983.

who met them fully. In large measure, these new traits in recording behaviour did appear and have stuck, monitored not just by the tiers of local and national review committees but also, at the end of the second millennium, by the new fierce moralists of the twitching fraternity, particularly the most competitive listers. Inevitably, however, a tiny minority of observers have been suspected of over-reaching perception or determined falsification of records. The former fault has been particularly associated with the relict observers such as myself who still favour the holistic approach to diagnosis. In my experience, these people are actually fully aware of the fine line between real and illusory difference and consider the minutiae of the New Approach as less important than the search for new behavioural characters. A few fresh springs of observation upon such signals have begun to appear in the topical journals, and young observers need not fear for employment in the further discoveries that will hopefully follow.

The situation with over-claimed unsubstantiated sightings, now known universally as 'string', deliberately fraudulent submissions and, for the competitive listers, the almost heinous sin of non-disclosure or suppression of rarities will be much more difficult to set to rights. Since the mid-1980s, an increasing zealous attitude has been shown by both the committees and the twitcher-moralists. Although often at odds on the taxonomy and status of individual vagrants, they appear nevertheless to be increasingly united in running books on the behaviour of observers, particularly the adventurous loners and claimers of solo records, and in prescribing full suites of characters as the *sinae qua non* for acceptable claims of so-called difficult species. These trends are heartily disliked by some experienced observers, not least by the several prior members of committees who have themselves deserted from the disciplines. Equally, sheer fatigue with the full system of reviews seems to be responsible for the loss of many records from the national record (Steven Gantlett and Lee Evans, pers. comm.). In the meantime, almost interminable restrospective reviews of so-called problem species continue to increase the sanitation of the British List and each year's rarities. Whereas at one time such actions would have commanded total acceptance from all interested birdwatchers, they no longer do so and the Rarities Committee now accepts that its universe is restricted to the rarity finders/submittors, who now submit over 1,000 claims annually. The divide in allegiance between the last camp and the full twitcher corps is potentially serious; once again, the demand for rarities is outrunning the approved supply. If not the shades of Hastings, two dogs of future war?

"No single observer should find more than two or three Scottish rarities a year!"
A remark of a modern record assessor who was clearly not a hunter, hearsay from Allan Vittery, *in litt.* (2000)

Just the sort of prejudice that troubles the observer who hunts alone and flat-out!

To my astonishment, late on 25th October 1997, a Red-flanked Bluetail presented itself on the fence of the Straight Lonnin, Holy Island. Dutifully put on the overnight Birdline, it left that night. In my submission to the Rarities Committee, I noted that the marvel had been a case of "fourth time fortunate" in my 47 years of rarity hunting in Britain. The seeming impossibility of such a contact rate saw to its rejection. Well, at least the local Robin knew that it was different!

The Genie of a New Handbook

The ageing of the Witherby Handbook *– its international competitors in the 1950s and 1960s – Oxford to the pump – a dark horse to the fore – an escalating target – the revolving cast – medals won but not awarded – crisis and resolution – editorial and artistic productivity – life after death*

By the 1960s, the Witherby *Handbook* was two decades old. Its stimulus was still enormous, particularly upon the 2,000 or so young fact-contributing observers that had entered its post-war British universe and the scores, if not hundreds, of similar European students that had added it to their national references. Except for the journals *British Birds* and *Ibis*, however, there was no obvious repository for the multiplying products of new ornithological research.

Abroad, the writing of new ornithologies was not so laggardly. Amazingly in view of the wartime devastation of their western regions, the Soviet Union's state ornithologists found the energy to produce, through the joint editorship of G. P. Dementiev and N. A. Gladkov, a new work on the *Birds of the Soviet Union*. Its five volumes appeared between 1951 and 1954 and although its accessibility was hampered by the double difference of language and script, its importance was made clear in lengthy reviews. Five years later, Ernst Hartert's systematic catalogue *Die Vogel der Palaarktischen Fauna* (1903-22) was finally superseded by the arrival of the first volume of Charles Vaurie's classic *The Birds of the Palearctic Fauna: Passeriformes* (1959). Beautifully created by H. F. and G. Witherby Ltd, still honouring their tradition of ornithological publishing, this essentially American initiative brought the zoogeography of so-called British birds into full continental relief. The collapse in avian parochialism was further hastened by the launch in 1962 of the *Handbook of North American Birds* of which the chief editor was R. S.Palmer, the appearance in 1965 of Vaurie's second volume on Palearctic non-passerines, and the initiation from 1966 of a new central European compendium, the *Handbuch der Vogel Mitteleuropas*, edited by K. M. Bauer and U. N. Glutz von Blotzheim. Faced with this increasing breadth of content and canvas, the heirs to the Witherby tradition saw that the years of resting on his *Handbook*'s laurels were over. Furthermore, with the restoration of wide ornithological exploration, the mere revision of a British and Irish handbook would not meet the new much wider span of zoogeographic vision. This target had already been hit with instant success by *A Field Guide to the Birds of Britain and Europe*, a collaboration between Roger Tory Peterson, Guy Mountfort and Phillip Hollom, which was first publish-ed in 1954 and has since been translated into a dozen tongues.

According to Max Nicholson, the early consultations on a major new British initiative did not prosper. Witherby (the firm), naturally the first choice as publishers, felt unable to finance any large undertaking and there was "a general flinching from commitment" throughout the early 1960s. Finally, in 1966, the ice was broken. Having persuaded the Oxford University Press (OUP) to take on its production, Max Nicholson publicised the scheme of the book at the International

Often tame, familiar to more people than birdwatchers, the Moorhen belongs to the gallinules, a worldwide tribe of rails. It shows a marked ability to prospect food sources in nocturnal flights. For an understanding of such behaviour, ownership of a major handbook is essential.

Ornithological Congress held at Oxford in July. Offers of co-operation came instantly from the leading ornithologists of at least seven European nations. A small flock of advisory and associate editors was nominated around a core of seven prospective writing editors in Britain and Holland. There was, however, a major competitor for one of the latters' times. This was the International Biological Program, a path-finding initiative in the protection of the world's habitats, and Max Nicholson as Director General of the then Nature Conservancy had to give this priority. Who was to be chief editor and engine driver?

The answer was an ornithological 'dark horse', self-taught in the science but with a mounting run of delivered achievements to show for his personal commitment to birds, then suffering from the first major crisis of environmental pollution to affect Western Europe. The baton of *The Birds of Western Palearctic* was passed to Stanley Cramp, already (since 1963) senior editor of *British Birds*. Able to take early retirement from H.M. Customs and Excise, Stanley became the first full-time employee of the management company West Palearctic Birds Ltd in 1969. His garret offices at 71 Grays Inn Road in London were to be the epicentre of *British Birds* and *BWP* for the next 15 years. For the latter huge task, the aim was to harvest full texts and illustrations of the ever-expanding number of west Palearctic species, originally thought to be about 740 but eventually panning out at 857. Their order and nomenclature came from a new list of Holarctic birds specially prepared by Professor Karel H. Voous. His early input was indicative of the international co-operation that would occur throughout the entire production. As with the Witherby *Handbook* illustrations, it continued to feature a particularly close association with Dutch ornithologists.

To the eyes of the ordinary British birdwatcher, the pregnancy of the great new handbook was more of a close confinement – an early indication perhaps of its managers' sensitivity to what would be its frequently close sail into the squalls of rampant cost inflation and editorial flux – but one day after my return from Nigeria in June 1971, I stumbled upon it. Calling at the then Bedford office of *British Birds*, I found James Ferguson-Lees crouched over a typewriter, unusually secretive and flailing away with two fingers at list after list of bird names. Spying my spying, he eased back in his chair and identified the papers as the plan of the new seven-volume work. The subject avifauna would indeed be that of the whole Western Palearctic. The late inclusion of European Russia in its geographic cover would also ensure that the gap between the new American and Soviet ornithologies would be more than filled. I remember blanching at the thought of the huge task but went away excited by the prospect of a new mightier successor to the Witherby *Handbook*.

With the passage of a whole decade between the first formal announcement and the first typescript delivery to *OUP*, the *dramatis personae* for the first volume differed from those forecast in many positions and the working casts for the next six tomes which became actually nine also revolved. In the end over its 24 years of volunteered or paid employment, the whole exercise tested a chief editor, three senior editors, 22 other editors and its backers severely. The last camp included the RSPB who offered a specially jacketed edition to its members. Within the corps of writing editors, only eight – Ken Simmons, Phil Hollom, Max Nicholson, Karel Voous, Joan Hall-Craggs, Pat Sellar, Kees Roselaar and myself – were involved in every volume but sterling efforts came from another ten – Robert Hudson, Malcolm Ogilvie, Peter Olney, Jan Wattel, Euan Dunn, Michael Wilson, David Snow, Dorothy Vincent, Brian Hillcoat and William Seale. These editors had not only to distil the essence of the international literature on their subjects but also to sift though two other constant streams of information. Measured as ranges of input over the nine volumes, additional data came from 56 to 85 correspondents speaking from 41 to 50 countries or regions and from 199 to 385 individual contributors of facts on 45 to 143 species. In total, the additional contributions from all correspondents came close to 1,600 and featured a particularly high interest in waders, gulls and warblers. I kept particular tally of the discussions of field characters that were received from individual observers. These had been a telling feature in Bernard Tucker's texts throughout the Witherby *Handbook* but sadly the flow for *BWP* was not as constant, with over 20 contributors to each of the first three volumes but only six on average by the last three. At the time, it occurred to me that the mounting tiers of review committees and the squelch of a new breed of identification gurus were effectively stifling new perceptions in individual observers. This worry has not left me but I do remember with delight Stanley's disdainful snort at the British Birds Rarities Committee's belated suggestion that it should approve my work.

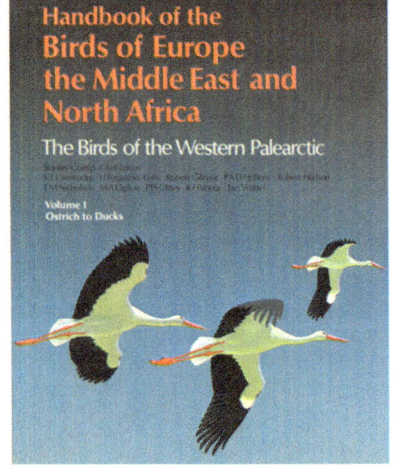

The cover of the first volume of *The Handbook of the Birds of Europe, the Middle East and North Africa,* also entitled *The Birds of the Western Palearctic* and (thank God) usually referred to as just BWP, featured Robert Gillmor's inspiring White Storks.

Closely linked to the expanded treatment of field characters, plumages and biometrics was the creation of an even more ambitious set of illustrations than those in the Witherby *Handbook*. The content of the major colour plates was briefed by myself and Kees Roselaar and despatched as detailed figure and plumage schedules to Robert Gillmor. He then marshalled the artists best suited to each volume's species groups and, adding his customary charm to the rather slight scale of fees, secured from four European and 18 British artists the intricate portraits. Over the two decades, draft figures were criticised and skins despatched from museums to studios and back again. For the nine volumes, a total of 573 plates flowed in to Gray's Inn Road and, after an office move, to Tring. Very few failed to receive immediate approval but for me the most loving perceptions were Robert Gillmor's series of stunning covers, Peter Hayman's beautifully constructed tubenoses, Peter Scott's living geese, Hakin Delin's ethereal owls and Viggo Ree's lovely chats. Quite how all the telling drawings of display and other behaviour were assembled, I never discovered but once again Robert Gillmor enthused a small flock of often younger artists. Thus along with the photographs of eggs so precisely matched by R. J. Connor and A. C. Parker, the second Herculean task was achieved.

Still coming up with great images, Robert Gillmor put dancing Snow Buntings on the cover of *BWP's* ninth and last volume.

Returning to the full-time workers, two truly deserved medals. As whipper-in for seven volumes and executive editor for the last four, Duncan Brooks played a crucial part, managing the interface with the other editors and often filling in their gaps. No less important was the typing and secretarial service of Ruth Wootton, who in the early years before word processing typed every word two and three times. As a paired performance in work endurance and tact, theirs was a remarkable feat. Ruth and Duncan also had to cope with Stanley Cramp's decline into ill health. Although always warm to fellow loners and *BWP's* many correspondents, Stanley remained amazingly distant from his two greatest aides sitting only a few footsteps away and too often lit yet another cigarette when an exchanged joke (or curse) would have done all three more good. Of course, few of the peripheral editors were party to the additional strains of financial management that became focused upon Stanley, the other directors of West Palearctic Birds Ltd and the increasingly exasperated Oxford University Press. Nevertheless, it became for me an increasingly uncomfortable experience to deliver texts to the office and sense its lack of union.

Originally the *BWP* schedule had called for the volumes to appear every 18 months but this rhythm was never achieved. The first four took ten years, not-withstanding the appointment of two more full-time editors who effectively became residents of the library of The Edward Grey Institute for Field Ornithology at Oxford. Even with their help, the compilation of the passerine texts became more and more protracted and an impasse loomed. Worse still, from 1985, Stanley Cramp's increasing incapacity prevented any chance of acceleration. I last saw Stanley at a 1937 Bird Club dinner early in 1987 and was appalled by his unusually withdrawn and almost speechless state. Almost gone was my chief London mentor, my eager field companion in places as far apart as Dungeness and Baluchistan and the sharer of frequent mirth, one intoxicated fit of which was so loud that we were threatened with eviction from a Chingford steak bar. Hit finally by a stroke and pneumonia, Stanley died on 20th October 1987. Everybody concerned with *BWP* was immensely saddened but the overriding imperative was to right a listing, if not sinking ship. Who would be the *deus ex machina*?

The fortunate solutions to the crisis were the inter-personal skills and long experience of the warm, balanced merchant banker cum ornithologist Sir William Wilkinson. He stepped into the chair of the management company and, with rare resolution no doubt honed from his time as chairman of the Nature Conservancy Council, he sorted out the long list of problems, soothed the feathers of OUP to the point of obtaining the seeming impossibility of further patience (both temporal and financial) and was also instrumental in getting firstly Dr. David Snow and secondly Professor Christopher Perrins to take the final watches at the senior editorial tiller. Following the relocation of the executive office at Tring, there was one particularly interesting day which was billed as a general review of progress but turned out to be little short of a thunderous 'group rocket', with several thin-lipped suits from OUP seated behind the square of editors. As ever, the most incisive remark came from Max Nicholson, "If I had thought 20 years ago it would have taken us so long, we should never have started." I was, however, more intrigued by the quiet but spirited defence of their apparent slowness put up by the Oxford editors, by then grown to six. Some time later, I enquired of Euan Dunn, "Just between you and me, why did the later volumes take so long?" With his customary wry smile, Euan replied, "There was so much to research that somehow we let the genie out of the bottle and there never was any chance of getting him back in." "Stanley even came on a rare foray to see us", he continued, "and tried his best to engender haste but we really did want to try and get it all in." "It all" was of course the European (particularly East European) explosion of amateur and professional ornithology that had been written during the *BWP* epoch. The true measure of the final avian curricula had been quite unforeseen in the early 1960s or even the 1970s. In this respect, the actual productivity of the editors merits a belated salute. In terms of the average number of pages devoted to the science of each species, the quotient of delivered information rose steadily from 4.9 in Volume 1 to a peak of 12.3 in Volume 7 and only fell back markedly to 5.1 in Volume 9, due to the high incidence therein of American vagrants not accorded full treatment. Thus although some reviewers carped on about the reduction of plates in the later volumes (of fewer species), the essential purpose of *BWP* was in fact enhanced steadily throughout its long birth. I consider it invidious to review the reviews of *BWP* content for the realms of the other editors but given the modern intensity of interest in field identification, I will admit to mixed delight and horror at having my Field Characters texts rated respectively by Colin Bradshaw and Simon Harrap as "eloquent and detailed" and "pretentious waffle". Who ever pleases all?

 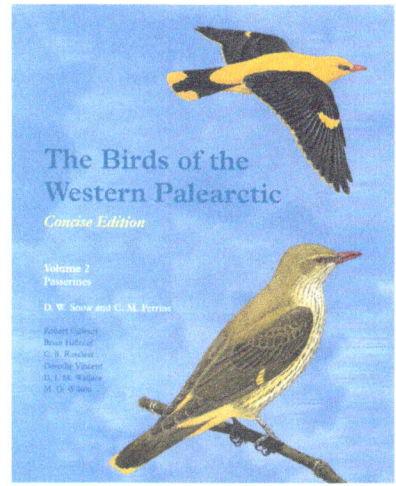

The covers of the two volumes of the *Concise Edition* of *BWP* were designer cobbles but the revised integral plates remain the most disciplined series of European bird illustrations ever created.

Given all the vicissitudes of the near quarter-century of its actual preparation, let alone the 31 years since its first conception and the near exhaustion of the final team, it needed the announcement of a closing celebration at an Oxford college to confirm that we had actually reached the end of our labours. Promptly re-christened "The *BWP* wake" by the younger editors, it passed off without mementoes (other than my awards of small paintings) but there was a good deal of resolving conviviality and, correctly, a special vote of thanks and presentation to Ruth Wootton. The real surprise was, however, yet another enthusiastic speech by Max Nicholson who announced the possibility of putting the entire work onto CD-ROM and finding other ways to insure an afterlife for *BWP*. Accordingly, although most of the final corps of 18 editors went their separate ways, seven found themselves revolving again on the *BWP* roundabout as a CD-ROM version, a Concise Edition and a new journal urgently entitled *BWP Update* were all planned and executed. Inevitably the original work's condensation involved a wince-making amount of 'cut and paste' but David Snow wielded the knife with considerable finesse. The main result was a much more attractive and accessible two-volume work published in 1998. It treated not 857 but a staggering 936 species and featured a revised presentation of 594 plates. The much-improved maps and associated accounts of distribution and population, re-harvested from correspondence and the growing number of atlases issued in 63 countries and regions, came from the

The first cover for *BWP Update*, alias *The Journal of Birds of the Western Palearctic*, gave Peter Scott's family of Pink-footed Geese another outing.

ever-faithful Dorothy Vincent and Mike Wilson. The final breadth on the shelf of *BWP* and *BWPC* came to 22 inches but once again the reprised achievement was tinged with sadness. Sir William Wilkinson, who had done so much to bring the main work to a successful end and had seen the Concise Edition well on its way, died in April 1996.

In spite of the efforts of first Malcolm Ogilvie and now David Parkin as editor, *BWP Update* has been erratic in its appearance and its high cost has certainly prevented any large following by the general readers of the antecedent 11 volumes. The main problem appears to be the eternal trait of some authors to promise but not deliver. Hopefully, the journal will survive but with all its less serious siblings facing currently static or reducing subscriptions, it may well require another long act of faith and finance from OUP to achieve the final objective of giving *BWP* a useful future to complement its often tense but nevertheless passionate and scholarly past.

As replacement 'tail-end Charlie' in both my editorial and artistic roles and with the lightest research task, I will always regard my part in *BWP* as a most fortunate privilege. For my colleagues, all I can add is a triumphant *Vivent scribae et pingerendi BWPensis*. Who indeed will ever do it all again? In the view of Chris Perrins, "No one!"

At the BWP 'wake' in 1994, most of the team gathered to celebrate the end of what had been for some a 28 year sentence to hard labour! Phil Hollom (second left) beams back at the book's last boss Sir William Wilkinson (far right).

The late Chris Mead looks down at the author as they contemplated the end of an era (or compared the costs of horse-riding daughters).

The Benefits of Addictive Counting

*The conflicts of common versus rare bird study – an accusation of treason –
the 1960s commitment to bird censusing – its development over four decades –
its chief products – the crucial cameo of the Common Bird Census – its succession by
the Breeding Birds Survey – the challenge of bird atlasing and its four answers –
their chief architects – a vagrant marsh tern – a register of good places for birds –
some stresses and strains – a prophet's view of the state of the nation's birds – his
optimism, others' pessimism – the last year of CBC – a precedent to the atlases –
Alexander More, first national Surveyor*

Arguments over the worth of common versus rare bird study go back at least 40 years. I particularly recall three episodes. The first was a radio debate put out by the BBC on the Home Service (today's Radio 4) in the mid- or late 1950s. High on the all bird charge induced by the Cambridge Bird Club, its chief priest Ian Nisbet and early European adventures, I listened with increasing incredulity as the proposer of the merits of common bird study, the Reverend P. H. T. Hartley, and his witnesses pointed out mercilessly the scientific insignificance of rarities and won the vote. When in the same period, James Fisher, Peter Scott and Roger Peterson all admitted their own embarrassment over enjoying rarities and also put the case for prolonged, dutiful study of common species and communities, it was clearly time to reconsider my personal motives in birdwatching. In fact, my early logs were brimful with counts of all sorts of birds but it was not until I began my first concentrated patchwork in Regent's Park, London in 1959, using the daily census technique learnt on Fair Isle and maps, that I experienced the springs of the lasting fascination of total bird study. When in the mid-1960s the awesome grasp of ecosystems in the minds of complete naturalists like Max Nicholson and

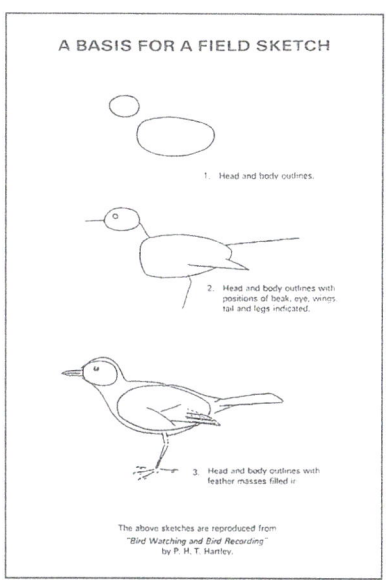

A BASIS FOR A FIELD SKETCH

1. Head and body outlines.

2. Head and body outlines with positions of beak, eye, wings, tail and legs indicated.

3. Head and body outlines with feather masses filled in.

The above sketches are reproduced from "Bird Watching and Bird Recording" by P. H. T. Hartley.

Real help for beginners used to come from kindly elders like the Reverend P. H. T. Hartley. Here is his basis for a field sketch. Today many observe that similar acts of teaching are insufficient.

In the latter half of the 20th century, Pochards have bred annually in Regent's Park, London, producing up to 25 broods. Yet for all of Britain in 2000, only 545 pairs could be logged by the Rare Breeding Birds Panel. The remarkable colonisations of the park have continued. It now hosts over 20 pairs of Grey Herons.

Duncan Poore was one of the lessons delivered by my participation in the expeditions to Jordan, I did not give up the dreams of rarities nor the occasional chase after them. I did, however, join those who were committing new energies and much more time to their local birds. At one meeting of the 1937 Bird Club in the 1965/1966 winter, the after-dinner discourse took up again the relative merits of the two birding attitudes and their follower types. Then rather carried away by the holy grails of the International Biological Program, I passionately urged total commitment to the nascent monitoring schemes in Britain. "Traitor", exclaimed someone and turning my head, I saw that the accusation had come from of all people, Bob Spencer. I had thought of him as one of the BTO's leaders wedded to duty but patently even his Quaker soul called for the refreshment of exceptional finds. After the laughter had subsided, we agreed that the only sensible course was to go on enjoying all birds, while increasing our care of the vulnerable ones.

Sensible care does, however, require many more facts than passionate outbursts. Without them, the national defence suddenly required for the birds of agricultural land and vulnerable coasts in the pollution crises of the late 1950s and early 1960s would have been much weaker. At least seven national organisations were soon concentrating on conservation but with its origin in such early studies as the 'economic importance' of the Rook, it was above all the BTO who stood up and got its amateur observer corps counting and the birds counted. The scientific imperative was to provide statistically sound, lobby-worthy annual indices of bird fortunes. The Nest Record Scheme, the Wildfowl Counts and the periodic censuses of individual species had already histories of up to 30 years, but talk of eggs and nestlings, Arctic-breeding geese, Herons and Great Crested Grebes did little to expose the mounting perils for all wildlife in a poisoned or reduced food chain on land and carelessly discharged oil at sea. So began a flurry of new schemes to monitor the breeding populations of Britain's main bird communities. With hindsight, these can be divided into lengthy annual or period marathons requiring truly dedicated slogging and disciplined sprints needing briefer but no less total concentration. Of the former type, the most important have been the Common Bird Census begun in 1962, the three surveys of seabird colonies in 1969/70, 1985/87 and 1999/2002, the Waterways Birds Survey begun in 1974, the Constant (Ringing) Effort Sites Scheme begun in 1983 and the Wetlands Bird Survey begun in 1992. The first of these became the fundament of farmland and woodland monitoring and its acceptance by bodies all the way up to H.M. Government was a tribute to the early work of the staff of the BTO Population Section. This unit had been founded by Kenneth Williamson who had migrated south from Fair Isle and St. Kilda to moult first into the BTO's Migration Research Officer and second into its formulator of modern population surveys. Owing to the wanders in my career, I took no part in any of them. By 1959, I had in any case become addicted to my own, more-adaptable, all-year method of patchwork. Its practice also made me worry that the Common Birds Census presented more a cameo of English birds than a full picture of the British avifauna. Very testing and time-consuming, the CBC never attracted more than 350 adherents and this noble contingent declined over its 38 years of life to 130.

Once the province of both professional and amateur contribution, *Bird Study* – the BTO's journal – cannot now lack for a fully scientific and statistical discipline. Of the 39 authors of the 12 papers in the November 2001 issue, only three were not working for research or conservation bodies. The bird news was, however, still exciting, containing for example a new census of 150,000 pairs of Manx Shearwaters from three Welsh islands.

Facing the need to supplement and extend their monitoring, the BTO and the other conservation bodies invented other less onerous but still significant surveys. Of these relative sprints, the most important are the Garden Bird Feeding Survey begun in 1969/70, the Rare Breeding Birds Panel in 1972 and, in an end of millennium rush, the Breeding Birds Survey from 1994, the Garden Birdwatch in 1995, the (Irish) Countryside Bird Survey in 1998 and the Winter Farmland Bird Survey in 1999/2000. It was with immense relief that I found one of the randomly chosen one kilometre squares for BBS and WFBS to be on the edge of my current Staffordshire patch. Four decades of guilt ended; at long last, my own and the national count imperatives could be jointly served. Of the new initiatives, the BBS is the crux of the much-needed nationwide farmland index. Simpler than the CBC, lacking its spatial mapping component, finished in four to five hours over three days dated in early and late spring, it has rapidly attracted soon addicted observers to nearly 2,500 km squares. The eventual goal is 3,590 squares and a complete farmland sample of the United Kingdom. Analysing the data from the annually walked and counted distance of about 10,000 kilometres is a simpler task than that required for the CBC and the context of each observer's input is quickly demonstrable. Thus, while in any one visit my east Staffordshire square SK 12/29 has never held more than 33 breeding or attempting species, it is comforting to know that this figure represents a third of the county's diversity of breeding birds. Actually, over the eight springs of my participation so far, I have accumulated in the square 73 species, including the delicious surprises of Yellow-legged Gulls gleaning silage cuts and a Whimbrel heading north. When will the first rarity appear? The BBS makes additional use of observer effort to monitor

A Caspian Gull (flying, second from left) and a Yellow-legged Gull (standing, fifth from right) join a party of British Lesser Black-backed Gulls in gleaning a silage cut. Before the 1990s, the first two taxa were ignored by most birdwatchers. Now they light up many an otherwise dull day or count!

habitat use and change in great detail and to record all visible mammals. Ireland's equivalent CBS was launched in 1998 and so the hope rises of complete geographic cover in the five ornithological nations. The contrast between the patient commitment of the potential 3500 observers serving this prime conservation study and the fast-driving obsession of a similar number of twitchers constantly chasing single rarities constitutes the widest divide in modern birding behaviour.

Could any form of purposeful birdwatching attract an even larger observer force than the BTO regulars? The answer has been affirmative. The most adventurous challenge ever, that of mapping the birds of the 3858 (or is it 3862?) 10-kilometre squares that contain some piece of British or Irish land, enthused a universe of about 10,000, perhaps initially 15,000 observers. They took up the habit of atlas work readily between 1968 and 1991 and so allowed the BTO and its associate institutions to produce two breeding bird and one wintering bird atlases. I recall an early discussion of atlas feasibility by the BTO's Scientific Advisory Committee (SAC) in 1963. Its place on the day's agenda had undoubtedly stemmed from the publication in 1962 of the *Atlas of the British Flora*, which had used for the first time the National Grid of 10-kilometre squares to plot the distribution of plants, but on the day only a minority of the SAC believed that birdwatchers could be persuaded to follow the botanists' example. In fact, there had been attempts at mapping the local distributions of birds as early as 1912 but it was not until 1947 that the use of the 10km National Grid appeared in a national work. This was Frank Fraser Darling's magic distillation of *Natural History in the Highlands and Islands*, still perhaps the best of the famous New Naturalist series published by Collins. Tucked in its last pages are 15 maps of mammal and bird plots and eleven of these had dots which represented "the intersection of two 10km lines … nearest to each place … inhabited by the species in the breeding season". The drawer of the maps was none other than James Fisher. In 1950 and 1952, the innovative C. A. Norris led the West Midland Bird Club into the use of a 25km system for renewed regional plotting. By 1965, it seems, the BTO became persuaded of atlas feasibility and 1966 saw not only the independent publication of raptor grids by Ian Prestt and A. A. Bell but also field tests of mapping practice by BTO staff. In 1967 their promise and the construction of a disseminable observer brief allowed the huge exercise to be approved. In the spring and early summers of 1968 and 1972, over 150 Regional Organisers, about 1500 main observers and between 10,000 and 15,000 other contributors searched all over the surfaces of Britain and Ireland and brought in a record six-figure databank, which exceeded all expectations. Contemporaneously, other BTO stalwarts raised £30,000 to support many of the search and production costs. Ten years after the first trials, *The Atlas of Breeding Birds in Britain and Ireland* (1976) appeared as a joint publication of the BTO and the Irish Wildbird Conservancy. The plaudits were loud and long, particularly for the chairman of the Steering Group, James Ferguson-Lees, and the chief organiser cum editor, Tim Sharrock. The textual complements to the maps had, however, been particularly hard won, causing a period of visible stress at the BTO's then headquarters in Beech Grove in Tring. As ever, the staff rallied to the final knitting but I was especially saddened to learn much later from

David Glue that Ken Williamson's early death in 1977 had been hastened by his share of the overload. The amateur ethos of British birdwatching lost one of its greatest apostles and those of us who had been directly inspired by him, on Fair Isle, in lectures, at conferences and in adult education classes, still miss him. His dedicated prospecting of population measurement methods earned Ken the British Ornithologists' Union Medal but he never quite lost his fancy for rare vagrants. "The most tantalising I ever saw", he told me, "was a Whiskered Tern in the Belfast docks in 1941. There I was, rooted in the middle rank of my platoon waiting to board ship for the Faeroes, and there it was, passing to and fro at a few yards. All I could do to follow it was to move my eyes!"

The next major initiative of the BTO was to create a register of ornithological sites. This was completed between 1973 and 1978, with about 4,000 areas described and their birds listed and seasonally quantified. Although birdwatchers had found the task less enjoyable than the *Atlas*, a second important set of data was nevertheless delivered to the Nature Conservancy Council and the RSPB. The book that followed in 1982 was written by Robert J. Fuller and entitled *Bird Habitats of Britain*. Anyone wanting to place their observations into the ecosystems of our lands should read it.

Given his early days in Greenwich Park and his large share in the study of a cleaner Thames and the eventually shipless Surrey docks, Peter Grant was no stranger to purposeful counting and it was he who first raised, in 1977, the prospect of sending the 10,000 observers out again in winter. The BTO approved the initiative in the summer of 1980, and the three winters from 1981/82 to 1983/84 saw over 100 regional organisers re-marshalling the efforts of the field workers. The Irish came on board in June 1982 and with Peter Lack as national organiser and his mother Elizabeth as principal editor, *The Atlas of Wintering Birds in Britain*

Kenneth Williamson's most awkward rarity was a Whiskered Tern in Belfast Docks. Stood in three ranks about to board ship for war service in the Faeroes, he was unable to get to full grips with it.

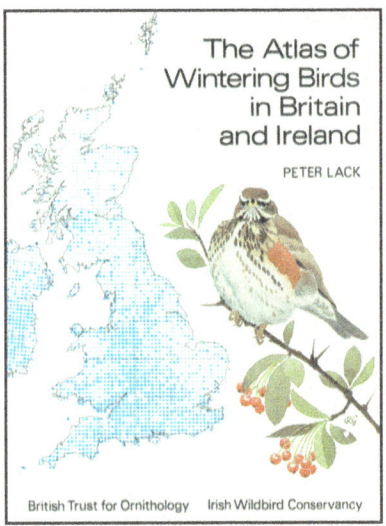

The cover design of the 'Winter Atlas' set Robert Gillmor's Redwing alongside the blue dots of its winter distribution in Britain and Ireland.

and Ireland was published in 1986. Its maps dotted in chilly blue made a fascinating comparison with the warm red ones of the breeding atlas and our understanding of the British and Irish avifauna took another bound forward. Some of the winter population sizes, for example those for the thrushes, sat uneasily with other studies but once again applause rang out, while the conservation lobby gained another crucial reference. Yet again, they also served who raised funds; only three people were responsible for half the £65,000 that went in the kitty from voluntary contributions.

1985 saw the start of real work on the *Atlas of European Breeding Birds*, effectively a synthesis of the now almost universal mapping of birds that had taken hold as far east as the Urals. First mooted in 1971, the production was organised by The European Bird Census Council and executed by a team of six. Two of the members were BTO staff, namely Mike Blair and Simon Gillings, and the former was joint editor. Truly monumental, its publication in 1997 fixed in time and space the European distribution of 495 species in over 4400 50-kilometre squares and their texts displayed population estimates and trends, with acute comments on adverse factors. At the other end of the atlas scale, more and more counties published or committed to their own avian Baedekers using the finer tetrad grid of two-kilometre squares. Many of these also contain population sizes and it became apparent in 1986 that a fresh effort was required to add such measures to a breeding atlas and to catch up with the dynamism of bird distribution in Britain and Ireland, an increasingly obvious phenomenon. Thus in the four springs and early summers of 1988 to 1991, the New Atlas Working Group used the now-customary charm of national organisers including a first for Scotland, over 180 regional organisers and "several thousand of the Old and New Guards" of British and Irish birds to harvest a fresh set of breeding bird data. For the first time, however, there was some serious vexation in observers minds over the assessment of bird densities. The principal method of a two-hour-only tetrad count left many observers with what they thought were low scores but importantly the frequency of occurrence quotient for commoner birds and the actual counts of scarcer ones had been designed to be repeatable. The essential objective was to show how density varied geographically within individual species. The goal of an overall matrix of comparable densities between species remains to be achieved. As some loyalty was lost from the earlier Atlas corps in its third campaign, any more complicated task will need to be better sold and packaged. Notwithstanding these issues, David Gibbons, James Reid and Robert Chapman and not least the third-time-round publisher Trevor Poyser brought in another remarkable volume of maps and texts. The lessons in the variably coloured abundance maps were immediately fascinating but too many of the trend calculations showed depressing declines. Somehow the applause for the third Atlas sounded a bit muted but this reaction did not stem from disappointment with the team or the now-formidable array of corporate sponsors. As for the birding available along the course, it was always engaging. I remember one magic moment as I tried to sort out just how many Wood Warblers were shivering ahead of me. One step nearer them and up got what remains the only fledgling Woodcock for my patch.

Given what was fast becoming the huge waves of information in the atlases and the ever-increasing conservation reviews, the ordinary birdwatcher, birder and twitcher could be forgiven for the occasional blink and head scratch. Who would summarise simply the overall state of Britain's and Ireland's avifauna? Entered 'Moses', alias Chris Mead, another of the BTO stalwarts from 1961 to 1995, bird feeder extraordinary, ringer of countless enterprises, triple medal-winner, and always the briskest of all the ornithological voices in any medium. Although seemingly retired, Chris decided in 1998 to take a serious look at the history, current state and prospect of every British and Irish breeding bird as the millennium changed over from second to third. In eighteen months flat, he produced not another tome but a winningly succinct and friendly little book entitled *The State of the Nation's Birds* (2000). It filled, at last, the 30-year gap in easy assimilation that had yawned since the BOU's best effort described in the preceding chapter. In only 276 pages of actual text, he highlighted the history of breeding population censuses, surveys and atlases, discussed six major habitat zones, climate, bad and good human attitudes to birds, and then launched into graphic accounts of and personal judgements upon the fortunes of 182 non-passerine and 98 passerine species. Chris's personal postscript contained the conclusion that of the 280 species that were breeding or had bred or might conceivably breed, 204 could be allocated good or bad prospects. To those of us feeling real fear for farmland species, the balance of his judgements was surprising. One hundred and eighteen species were seen as 'set fair' or better.

It is difficult to compare Chris's taking of the overall avian pulse with that not stated but lurking throughout the *New Atlas*, but undoubtedly he was overall more optimistic than the earlier authors that had commented on their individual subjects' trends. They had seen only 77 species with stable or improving fortunes. Where the two sets agreed most was in their flagging of the plights of several vulnerable families. Six gamebirds, four rails, 12 waders, five chats and three buntings are in shock due to the chaos of the agri-environment and continuing pollution, and five gulls, four terns and three auks are seriously troubled by the continued over-fishing of virtually every age class of fish. Never mind your parakeets (or rather watch out for their pest potential), enjoy your Mandarins but let us not forget that the core conservation challenge of sustaining the widespread and diverse mix of healthy common bird populations of the mid-20th century and before remains to be answered. The reserve managers have undoubtedly earned their keep and justified the huge generosity of donors to such bodies as the RSPB, but I would add to Chris's verdict that we "could have done better" the reminder that we "should have woken up earlier". One of the dangerous by-products of the tragic foot and mouth outbreak of 2001 has been the interruption of most bird monitoring in farmland. The BTO and IWC and the charities and other agencies that have supported them, most notably the RSPB, JNCC and latterly DoE and WWF (UK), cannot relax their vigilance.

Rather uncannily on 23rd July 2001, the day on which I finished the first draft of this chapter, I received from John Marchant, chief mentor of the CBC, a press release covering its virtual demise in 2000 and the take-over by BBS of the breeding

The State of the Nation's Birds by Chris Mead should be on every birdwatcher's bookshelf.

population index. For once, the bird news from the Mohican rump of 130 plots were mostly good, with 14 significant increases to show for a "conspicuously better" season than in 1999. Four featured birds of conservation concern, namely Tawny Owl, Dunnock, Blackbird and Song Thrush. Conversely, the seed-eaters of fields remained at risk. In its last year, the total of CBC surveys had passed 10,000 and it is expected that in its dying fall, the last CBC map will plot the two millionth territory of its target, the singing male bird. I sense that another remarkable text will tell us all more of the spatiality of bird territories. As can be seen throughout the above summary of its most productive epoch, the BTO wastes not the efforts of its labour force nor makes it want for the interpretation of its findings. Small wonder that today over 30,000 birdwatchers of all hues contribute to the full array of the BTO's crucial databank for conservation action in the UK.

The counting cum mapping of birds may seem to be a relatively recent product of the tandem effort of professional and amateur birdwatchers. In the sense of the increasingly accurate small-scale atlases, it is, but at the outset of bird distribution surveys, there had been a remarkable precedent of ornithologist following botanist. The former was Alexander Goodman More (1830-95), a Londoner educated at Rugby and Cambridge. In his 30s, he became aware that an amateur botanist Hewett Cottrrell Watson (1804-81) had devised a nationwide system of provincial down to vice-county subdivisions in order to plot more precisely the distribution of British plants. When Watson published the main part of his survey in *Cybele Britannica* (1847-59), More set out to provide an ornithological equivalent. Corresponding with naturalists and sportsmen in Watson's 38 sub-provinces, he elicited breeding bird distribution from most of England and parts of Wales and Scotland and published three part-papers in *Ibis* in 1865. More's statement on *"On the distribution of Birds in Great Britain during the Nesting Season"* made an immensely valuable contribution to the ornithology of the mid-19th century. More regretted its incompleteness but his solo initiative in co-operative investigation antedated the wider-flung coastal lights exercise in migration study by at least 13 years.

Two years after the publication of his papers More took himself to Ireland and his ability to the Natural History Department of Dublin. Having done his bit for Britain, he then contributed to the golden age of Irish ornithological discovery by stimulating younger men like R. M. Barrington and co-operating with R. J. Ussher in the research and preparation that led to the latter's chief authorship of the classic *Birds of Ireland* (1900).

Above: even birdwatching marriages began at Cambridge. Anne-Marie, future Mrs Christopher Smout, punts him and Margit Nillson down the River Cam on a summer's day.

Left: members of the Cambridge Bird Club in the mid-50s. Holbeach, 4th February 1956. Chris Smout (in pale duffel coat), Ian Nisbet (with balaclava) and Bill Bourne (hatless but fleece-gloved) scan the saltings of the Wash. JC.

Left: a good field meeting makes for good morale support. The Cambridge Bird Club was invited to Horsey Mere by the Buxton family on 3rd June 1956. Standing on the left scanning for harriers Ian Nisbet; sitting to his right waiting for the next crossing Clive Minton. JC.

"It must be all of 40 years since we went to East Anglia and were stopped by the police after the IRA had raided the Army depot at Blandford Forum." Eric Simms, *in litt.* (1999)

Drives through the night were rare in the 1950s and our ornithological foray was seriously questioned.

To the Ouse Washes on 3rd March 1957, Eric Simms brought an early sound recorder to record Bewick's Swans. JC.

At Wisbech Sewage Farm on 13th November 1955, John Larmouth (left) and Clive Minton (right) compare a White-rumped Sandpiper with a Dunlin and check that the former's ring number is logged correctly. The sandpiper was the first of its species to be trapped in Britain. JC.

The seawatch point in the marram dunes at Holme, Norfolk. From left, Colin Kirtland, Tony Vine and Graham Easy. JI.

The camp at Walsey Hills, autumn 1957. From left, Bob Emmett, Marianne Medhurst, Shelia Izzard, Jan, John Izzard. Behind the coast the road leads on to the start of the East Bank, the Cley marshes and the wooded bluff of Cley village. HPM.

Peter Colston and (right) Richard Richardson, the 'guardian spirit', of the East Bank. Note Richard's shining black biker's leggings, early 'trainers', sweater from Fair Isle (his other home) and Norfolk Terrier. His bins were the small ones from Ross. REE.

In 1958 a 'secret guard' based in the New Forest was established, dedicated to the protection of the area's breeding Honey Buzzards. Bob Emmett on a 650 AJS. The longest-serving member of the guard – able to brew fresh tea up a tree.

'Lilo camp' in the New Forest, summer 1958. Four watchers, one sleeping wife; nearest to camera John Parslow. Note the Broadhurst Clarkson four-draw telescopes beside John and special kit panniers designed by Bob Emmett and attached to the Matchless 350. REE.

The guards' love of raptors took them to Sweden, summer, early 1960s. From left, Gedge Westerhof, Len Mummery, Eddie Wiseman, creating raft.

A tryst in Matley Wood on 16th May 1959. From right, Mike Goodman, Richard Richardson and Ron Johns, with Howard Medhurst, Bob Emmett, Brian Newport and Marianne Medhurst.

Searching for Red Kites in Wales. Bob Emmett scans, Peter Colston mends the car.

Howard Medhurst puzzling over an *Aquila* eagle seen over Aden. Note mighty binoculars, string of rings, ringing pliers, Meinertzhagen's *Birds of Arabia*, an eagle drawing by Peter Hayman, a field guide and Kershaw 10x40 binoculars.

The Diversity of
Modern Birdwatchers

*From joke butt to huge lobby – the twitchers and rarity collectors – their annual
budget – the fact contributors – the main journal readerships – the potential universe
of serious birdwatchers – the decline in youthful support – specialist groups –
three men, two wives and a dog*

As a teenage birdwatcher in the late 1940s, I felt mortally offended when my
choice of tribal hobby became the butt of music hall comedians, not yet descended
from the war's macho values, but even their jokes did not prepare me for the real
stopper whispered by the interview advisor at my May 1952 War Office Selection
Board: "Don't tell the members about your hobby, lad; they might think you're
queer". What they thought when I did my obligatory piece of "intellectual presentation" – on the hermaphrodite nature of the oyster – went unrecorded in Britain's
military history. Miraculously the whole episode was salvaged by the discovery
that Eric Ennion's son, Hugh, was at the same board, also after a National Service
commission. Stifling in hot khaki barathea, we had both been drawn to the cool
trees and meadows of the then bird-thronged tributary of the River Test at Barton
Stacey. Zeroing simultaneously on to a pair of nesting Grey Wagtails, we bonded
immediately as only birdwatchers can, military careers instantly forgotten. I left
Hampshire not only with a commission but also with an invitation to Monk's
House Bird Observatory and Field Centre. My next pass and three buses got me
there. I was enduringly captivated by the warmth of Hugh's father, indubitably
the bird artist with the most graced eye for avian character ever, and also the
kitchen window from which Britain's second Pallas's Warbler had been seen on
the previous 13th October. The rest of the magic house held a courseful of students
of both sexes as keen as mustard on the Northumberland coast and the sparkling

In military Hampshire in May 1952, a pair of
Grey Wagtails gave avian solace to two
young birdwatchers intent on convincing the
War Office Selection Board that they should
command other National Servicemen.

Farne Islands. Monk's House was the perfect antidote to the Hampshire trial of faith and became for a precious time the best mainland learning venue available to young birdwatchers in the post-war revival of natural history.

Fifty years on, birdwatchers are largely free of ridicule, due in no small measure to the increasingly wonderful presentation on television of the recreative joy in birds. We have been particularly fortunate to live in a nation that has founded yet another great tradition in the portrayal of the natural world's habitats and animals. The BBC Natural History Unit at Bristol and Anglia TV's Survival team at Norwich gave the lead in the brilliant images, and an interesting variety of avian interpreters, from converted wildfowler Sir Peter Scott, planet ambassador Sir David Attenborough to former Goodie and real birder Bill Oddie have added the entrancing and compelling messages. Now there is hardly a day of the week when television windows do not open onto precious havens of wild places and their variety of life are not offered directly to over 21 million homes containing 59 million people. Within this total human universe, about 5.5 million are reported to watch birds occasionally. Certainly 600,000 households support the Royal Society for the Protection of Birds, providing a bird conservation constituency of over 1,000,000 individual adult members, a number greater than the lists of all the political parties put together. The development of the televised interface with nature could be even more dramatic. Briefly, in 2001, the birds of the new RSPB Belfast Harbour reserve were the first ever to go on-line in Britain, once again thanks to BBC technology and the camera siting and creative interpretation of Anthony McGeehan and John Scovell. Ever more compelling encounters with birds beckon and hopefully these will be followed by another increase in birdwatchers and conservation supporters.

Newshounds are odd dogs, however, and too often the modern media still pick out the rarity collectors, alias twitchers, as seemingly the only newsworthy

"It is a common place that novelty exercises such an attraction that it frequently diverts itself a measure of attention out of all proportion to the true value of the subject or object. In science the field of every new discovery forthwith becomes the focal point round which attention centres, to the detriment of other fields more important but less glamorous." H. W. Parker, past Keeper of Zoology, British Museum of Natural History

The first two sentences of the Preface to the *Checklist of Palearctic and Indian Mammals* 1758 to 1946 (Ellerman & Morrison-Scott 1951). The expressed sentiments could be equally applied to the modern obsession with so-called rare birds.

For young minds to catch fire, the spark of interpretation is all. Graham Wynne rated the posters of the RSPB's Belfast Harbour reserve as the Society's best.

When a Yellow-throated Vireo crossed the Atlantic and found cover and food at Kenidjack, Cornwall, on 20th September 1990, it provoked a collecting investment of £75,000 from those who twitched it in the following week.

Denzil Harber was the second secretary of the British Birds Rarities Committee. His atavistic manner became a serious issue in the early public relations of the reviewing bureaucracy, removed later by the tact of Chris Swaine and Ray Smith, both schoolmasters, not embittered as Harber was by a narrow life.

class of birdwatcher, alias birder. Tabloid pictures of twitchers putting Tesco car parks under siege, TV documentaries of them fidgeting in queues for the last hireable plane or boat, awards of multiple speeding tickets to drivers on hectic through-the-night, trans-Britain rarity rallies, and, most recently, the tense dispute over the ownership of the biggest-ever annual list have all been broadcast as behaviours somehow typical of all birdwatchers. Occasionally the half tongue-in-cheek of the announcer shows but the confusion in birdwatcher types and purposes is heartily disliked by nearly all the surviving creators and exponents of proper purposeful birdwatching and field ornithology. The reality is that the competitive twitching scene is inhabited by not more than a thousand obsessives and less than 150 of them immediately drop everything, including wives, when a new target is signalled. In their essentially "follow-the-finder and, please God, let his or her bird still be there for me" sport, they spend thousands of pounds a year in almost ceaseless travel after the ultimate prize of the longest British list and the grail of 400 species in Britain in 12 months. In pursuit of the latter, Lee Evans' 1996 achievement of 383 is so far the undisputed highest. The balance of the 3,000, exceptionally with a long-staying target perhaps 6,000, observers who go to collect rarities, particularly when they are species new to them, are far less obsessive, eager when challenged to argue that they do also contribute to more purposeful birdwatching. There are signs that twitching is past its zenith. The 2001 peak attendance for the Nottinghamshire Little Swift was only 3,000, but the latest 'Hastings Rarity' to reappear, a Black Lark which wandered from Kazakhstan to Anglesey in June 2003, attracted 4,000 eager viewers. Meanwhile the recent pursuits of rare subspecies are continuing to produce rather inflated fervour. The latter taxa are 'banked' just in case someone splits them into tickable species later on!

In 1990, I estimated that the eight-day September twitch of Britain's first Yellow-throated Vireo from America involved the expenditure of £75,000 (£102,500 today), mostly on fossil fuel. Any conservation body would have whooped with joy at such a week's cash flow. In all senses, the modern inflation of rarity value is astonishing. In recent years, the British rarity chase could well have a total annual travel budget of over £2.25 million to move a minimum capital investment in wheels, optical and communication equipment of £7.5 million that I estimate to belong to the 2,500 most avid rarity collectors. The leading twitchers have pagers strapped to their wrists, cellular phones, CB radios, timetables bulging in their pockets and never less than £300 cash, let alone multiple credit cards in their wallets. When the peeps sound, they go, eager for another avian scalp and hopefully a march-stealing on or the dethroning of Ron Johns, still the 'King of Twitchers'. His British list of over 550 species has been harvested from 40 years of frenetic effort and is approved by Lee Evans, the self-appointed monitor of twitcher list purity. Engagingly ready to acknowledge the over-consumption of rarities, Ron has recently announced his retirement from the tournament but the word is that his habit dies hard. From the late 1970s, it has also become highly infectious, becoming the chief addiction of young entrants to birdwatching. With today's instant identification lore, they leap over the knowledge gaps that older generations took decades to cross and mistakenly ignore the fuller, lifelong sustenance in the rounder birdcraft still on offer from RSPB Members' Groups, bird clubs, patchworkers and the now sadly neglected and struggling bird observatories. Accordingly, young twitchers can nowadays reach a British list of 460 in three years but then their interest just burns out. There is still freedom of choice in human recreation but mere collection is no recipe for a lifetime of joyful bird-watching.

After the above brigade, there is the far more important corps of another 7,500 to 10,000 fact-contributing observers of broader interests, personal studies, full conservation conscience and infrequent twitches. They provide the main part of the membership of the BTO which has in the 1990s grown significantly to an all time peak in 2001 of 12,500. This is a sure sign that when challenged with worthwhile work, birdwatchers not only respond but can actually enjoy the scientific disciplines required for the research results essential to conservation logic and attentive politicians. Given the overall drift away from bird club or local society membership in the last two decades, estimated from *The Birdwatchers Yearbook* figures to be a loss of 6% from the peak of about 48,000, the success of the BTO in growing its force of field workers is totally commendable. It has been no accident but the product of improved management and energetic publicity, a sure sign that no ornithological doldrum is currently threatened. Importantly, BTO members and especially its 2,000 or so 'hard core' surveyors can see that their efforts are well used in the conservation battles. It bears repeating, however, that many are stung by the contrasts between the instant spends on rarity collection or other ephemera and the always-short funds available for crucial research, the restoration of the bird observatories or even the proper manning of The Bird Group of the Natural History Museum, to name another imperilled national resource.

It follows from the above that not all members of county societies and local clubs submit records. In 1980 with the help of 27 county recorders, the total British and Irish universe of fact-contributing observers was estimated to have grown from 2,070 in 1950 to about 7,850 in 1980 (Wallace 1981). I had despaired of seeing an update of this quotient of bird recording but in the nick of time for this chapter, a paper in *British Birds* on the UK status of the Hawfinch (Langston, Gregory and Adams 2002) presented a welcome graph illustrating the growth in record-submitting observers in 38 English and Welsh counties from 1975 to 2000. The re-interpolation of the maximal universe is dogged by changes in both political and ornithological constituencies but applying the 50% growth in the 38 counties over the last quarter-century to the mid-point of my 1970 and 1980 estimates, the total current number of regular fact-contributors alias purposeful birdwatchers in the traditional local recording framework comes out at 9,975.

The likelihood of 10,000 or so purposeful birdwatchers and the BTO's recent resurgence contrast distinctly with the depressingly shrunk commitment to the master journal and central voice of *British Birds*. A subscription to it once designated membership of an all-action cell of informed and informing birdwatchers and ornithologists. In recent decades, this grew from an earlier nadir of only 2,900 in the early 1970s to a peak of nearly 11,000 in the late 1980s, only to plunge down to a maximum circulation of 5,250 in 2000. The recent case history of *British*

"The apparent infatuation with rarity identification is a modern phenomenon. I have met people who could (they said so anyway) distinguish the various rare Phylloscopus and other warblers but could not tell a Stock Dove from a Feral Pigeon and (their being both 'common') had no wish to." Derek Goodwin, *in litt.* (1999)

A telling comment from the author's most helpful museum mentor.

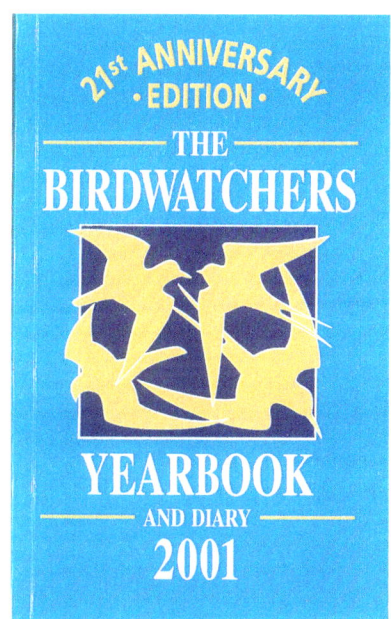

The amazing service of *The Birdwatcher's Yearbook* now comes from David and Hilary Cromack. This is no better reference to the scene of British birdwatching and field ornithology – and no more caring service providers.

"Few major clubs existed between the Wars and many came into existence during the 1970s. I suspect that, in some instances, the latter births were due to energetic post-war generations reaching the ages when they could oust members of the old guards who still tried to cling on to their original castles." Mike Rogers, *in litt.* (1999)

The cover of the *British Birds* now carries the obligatory glossy photograph. This issue featured the most involved and protracted tests ever of the characters of a rarity. The bird remained what it always was, a Slender-billed Curlew or Asia's rarest wader.

Two more faces of popular birdwatching magazines, Emap Active's ever-improving *Birdwatching* – the best monthly for beginners and those who want a pleasing hobby – and Dominic Mitchell's *Birdwatch* – a half-way house to the more official or rarity-heavy journals.

"There is a niche for a really high quality birding/popular ornithological magazine in Britain. I'd like to see BB *fulfil this role, but don't think it's even close at the moment."*
Jane Reid, *in litt.* (1999)

A comment typical of those that indicate a niche for a new authoritative but readable journal to match the early style of *BB*.

"Not only was Bernard an enthusiastic and kind man, but his famous BB *notes demonstrated that as an amateur you could still add something to ornithological knowledge, and that even apparently minor observations might be of wider interest. We need more people like him."*
Peter Combridge, *in litt.* (2000)

Somehow everybody's birdwatching uncle, particularly in the 1960s and 1970s, Bernard King became the first and so far only centurion of notes in *British Birds*.

Birds is a sharp reminder that even a semi-institutional journal with a total life history of over 160 years must also compete with the more topical birdwatching press with 20 years of some success, as a valued mail-order product. Hopefully the energetic revivalism of the reconstituted editorial board and the ornithologically streetwise Roger Riddington's appointment as editor will regain for *British Birds* the greater support that it still deserves.

Because of the competitiveness of the more youthful media, my enquiries after hard circulation figures have been partly ignored but after the hard won success of Emap in sustaining the constantly improving magazine *Birdwatching* at up to 28,000 (32,000 at best) bought issues a mouth, the hearsay order of paid purchases is the engaging Dominic Mitchell's *Birdwatch* at around 6,000 (8,000 at peak), the ever thrusting Bird Information Service's *Birding World* at 4,000 (5,800 at peak), the imported, increasingly excellent *Dutch Birding* at 850 and Lee Evans' officially spurned but industrious *Rare Birds* also at about 850 (1,000 at peak). Clearly all of the last four have nibbled at, if not bitten hard into the former allegiance to (and single choice of) *British Birds*, but I can only guess that the overlaps in subscription bring the net total loyalty to birding journals down to about 35,000. The overlaps with the general natural history titles like the Wildlife Trusts' *Natural World*, the BBC's *Wildlife* and the more recent *British Wildlife* and the myriad local bird bulletins are simply not known.

What can be calculated, however, is the likely total universe of people who could move up from just enjoying birds to more energetic, mindful study of them. Determined to improve their marketing, the RSPB researched the characteristics of their one million members and found that 8% were committed not just to the

The always-enthralling spectacles of wildfowl in a wide landscape have underwritten the loyalty of their counters for half a century. This gathering of Snipe (foreground), Whooper Swans, Greylag (upper most) and Greenland White-fronted Geese can be seen at the Insh Lands, Co. Londonderry.

'warm, fuzzy' support so beloved of the American wildlife 'non-profits' alias charities but also to the greater personal fulfilment in purposeful observation. That quotient applied to the RSPB's 600,000 households allows a minimum of 48,000 keen observers or, applied to the 1.25 million adult and young members, an absolute maximum of 100,000 potentially productive birdwatchers. The lower estimate is uncannily but perhaps coincidentally close to the past peak membership of the bird clubs. The higher figure suggests that, provided more can be done to sustain or re-engage the interest of the computer-obsessed young, the continued growth of the traditional hobby and its now contingent conservation studies could be assured. For the moment, however, there is real evidence of insufficient recruitment to the crucial observer classes that provided, for example, the ceiling of 350 former Common Bird Census workers and the 2,000 ringers of the BTO and the observatories. This syndrome of an ageing community of skilled volunteers is also discernible elsewhere in Western Europe.

The RSPB, with its Wildlife Explorers and Phoenix Club designed to attract and hold the interest of children, and the Wildfowl and Wetland Trust (WWT), with its excellent collections and adoption schemes affording real emotional attachment, try hard to deliver the educational platform needed for a secure lifelong hobby. The WWT has achieved a membership of 100,000 and close on 175,000 youngsters do show allegiance to it and the RSPB. Nevertheless it seems that nowadays an early grasp of Mark Cocker's formulaic love triangle for a lifetime of dynamic birding – the natural poetry in the union of beautiful place, wild bird and hunting observer – streams only fitfully from the bird clubs and local societies. Like the national organisations, most report serious problems in attracting and sustaining younger members (under 40). This trend is also mirrored in the near extinction of school natural history and birdwatching societies. David Ballance has confirmed that the ornithological crucibles of our generation are nearly all cracked.

"In 1958, a rabble of disaffected youths broke away from the local natural history society to form the Tyneside Bird Club: Brian Little, Brian Galloway, Michael Bell, David Howey, and me were founder members. We met in the YMCA in Newcastle and produced a monthly bulletin. The room was about 12 feet square and we didn't have any visual aids. One week someone brought some slides and we had to pass them round and hold them to the light. Magic!" David Parkin, *in litt.* (2000)

Some observers do not just join; they form new ornithological units, characterised by enthusiasm and new achievements. Witness also the Fife Bird Club and its new initiative to unite Scottish birders.

"Most of us at Cambridge had received encouragement from school natural history societies. Very few now exist, and of these hardly one publishes anything. Yet Marlborough, Winchester, Rugby published reports back into the 1860s. Those that do survive (Bootham, Gresham's, Sedbergh e.g.) have mostly become specialist–scientific rather than sporting–birding. A pity, a great loss of youthful talent. Hence no undergraduate birders. I gather from Mike Wilson that Oxford has gone the same way, which it was already set on in the 1950s." David Ballance, *in litt.* (2000)

If you are looking for an experience of Mark Cocker's triangular formula of magic bird, majestic habitat and human wonder, go to the Cairngorm Mountains and Strathspey – and take your time. Depicted here are Snow Buntings (bottom right), Ptarmigan (bottom left), Hen Harrier (centre) and Golden Eagle (top left).

Other groupings of adult birdwatchers are visible in the more-specialist charities such as the Hawk and Owl and Pheasant Trusts. At least 20,000 full members now belong to this group but again their overlap with other cadres is not known. This fog also descends on the British-initiated but more internationally oriented bird clubs like the African Bird Club (with 1,280 members), the Neotropical Bird Club (1,200), the Ornithological Society of the Middle East, now also encompassing west-central Asia (960), and the Oriental Bird Club (750). Nearly half of their members are not British, reflecting their international character and their universal attempts to assist native observers, but of the rest, few will not have been covered by earlier figures. In spite of the reduced sizes of our military arms, the Royal Navy Birdwatching Society, the Army Ornithological Society and the Royal Air Force Ornithological Society are each holding on to between 200 and 300 members. From Easter 2001, however, their journals or reports were combined in a new journal entitled *Osprey*.

In contrast to the emergence particularly since the 1960s of the above energetic bodies all vitally concerned with fresh discoveries and the protection of vulnerable species across four continents and sub-continents, the recent decline in support for the most venerable institution of British ornithology is marked. The British Ornithologists' Union retains only around 1,700 members, with nearly half of its rump composed of other nationalities. Although its recent series of joint conferences has been successful, the BOU's much-reduced attachment to British ornithology appears to be no recipe for local loyalty. Furthermore, even submissions to its truly august journal *Ibis* have become subject to subvention by the internet and general ecology journals. The BOU did, however, hold its nerve in the recent taxonomic debate and its 2002 guidelines on speciation have informed birders of the criteria by which its Taxonomic Sub-Committee will be making its future decisions.

Last but not least in the parade of British birdwatching units are the afficionados of special causes and extraordinary birds. These include the Seabird Group, the North Sea Bird Club, the Severn Gull Project, the SK58 Birders (winners of *BB*'s award for the best small-group local report), John Callion and his fellow Stonechat enthusiasts, and many others. With memberships numbered in only low hundreds, scores and tens, these cells are the havens of driven birdwatchers and secret crusaders, enjoying their hobby and contributing to ornithology and conservation more than most. In one of them, the secretive, shadow-flitting guardians of the New Forest's Honey Buzzards, there is even a link to the twitchers, for their early ranks were led by a motor-biking squad of three men, two wives and a dog. It was this unit that originated the once occasional adventure, now manic sport of rarity hunt and collection. So, hats off for Howard and Marianne Medhurst, the late John and Sheila Izzard, Bob Emmett and Jan (a warm bitch), who did what all sensible birdwatchers do, combine fun and duty in their passion for birds. Their watch over some of Britain's most precious raptors is in its 47th year and one of its early tales is a gem. Totally conscious of the need to put *Pernis apivorus* (the Honey Buzzard) first and keep its once last relict community as safe as possible, they defended its forest lair against nearly all comers. To the New Forest, I was allowed two utterly magical visits and distant views but Horace Alexander, in pursuit of an additional feast of five Red-footed Falcons, was not so lucky. Although brought to the 'pony' watchers' base camp by a beseeching acolyte, his access to any rare

From 1993, the SK58 *Birders* have published a wonderfully enthusiastic and detailed annual report on the birds of their 10 km square on the South Yorkshire and Nottinghamshire border. The 1994 cover showed the four Dotterels – "easily the birds of the year".

Wondrously strange, as much kite- as buzzard-like, a Honey Buzzard cleaves the canopy of the New Forest, Hampshire. Bob Emmett, one of the original twitchers, has proved his devotion to the eater of 'porridge' (wasp larvae) every year for nearly half a century.

"When I was President of the BOU, I tried hard to reduce the stuffiness and involve more young people and, happily, had a Council who were very prepared to give time and thought as to how to do this. Our meetings are now always held with other Societies and quite a lot of our money goes to sponsoring students in order that they can attend. So the age of participants has gone down, and the interest is great – most meetings are well attended these days." Janet Kear, *in litt.* (2000)

The most enjoyable and recreative conference that I ever attended was the joint BOU/Society of Wildlife Artists presentation of "Birds in Words and Art" held in September 1996.

raptor was nevertheless denied. As far as the guardians were concerned, "Not even H. G. (f. word deleted) Wells" would have been allowed to day-trip into their pregnant trees. Although Ron Johns has told me that he and Richard Richardson witnessed this incident, it has since been denied by the senior veteran. Does this show how socio-ornithological crack is as often wished as actually spoken or written?

There is no doubt, however, over the first-ever fit of actual twitching. Truly it was displayed by Howard Medhurst who, after 110 miles from London to Cley on a bike that lacked the Izzards' tummy- and back-warming hound, was wind-chilled and in sore need of a nicotine hit. With shaking hands, he lit up and lo, his shivering, alias twitching, subsided.

The 'original twitchers' in north Norfolk. From left, Marianne Medhurst, Jan (the dog), Howard Medhurst, Bob Emmett and Sheila Izzard. Note Kershaw 10x40 binoculars on Howard. JI.

Mount Kenya dominated the eastern 'White Highlands'. On 21st December 1952 and 17th May 1953 with various guides, its endemic species were found during 'one-day slogs' that reached over 4,000 metres, but never the peaks. DIMW.

An experiment at skin collection with a 22mm rifle was happily not repeated. Nauyaki, December 1952.

The 1960s expedition to the Azraq Oasis, Jordan. With perhaps a King to meet, the 1963 party traveled in suits. Happily arrived at the Philadel-phia Hotel, Amman, were (from left) Guy Mountfort, James Ferguson-Lees, Sdeuard Bisserot (mammalogist), Eric Hosking (with twin Zeiss Contarex cameras), author, George Shannon, unknown and Jan Gillett (botanist).

The best driver of the three expeditions was Corporal Mahmoud, able to drive a Landrover over basalt blocks. He got the author and James Ferguson-Lees to every last desert niche. EH.

However long the day, every one of each day's records was logged before sleep. Here the author tries to map the Shishan marsh while James Ferguson-Lees enters counts in the pre-printed species registers. EH.

Volunteering excellent guidance to the expedition, Azraq Shisham's hunter Fayk Wazani and his Austrian wife Lily were more helpful than anyone from the Tourism Authority.

James Ferguson-Lees looking over the tamarisk swamp at Azraq Shishan. DIMW.

James Ferguson-Lees counting waders on the qa edge. DIMW.

Checking the race of a Reed Warbler by reference to Ken Williamson's guide to its genus. DIMW.

Bob Spencer and John Ash ringing birds at a temporary station consisting of the tail door of a Landrover.

A Hoopoe Lark objects to James's inspection of its nest in a silt wadi. DIMW.

James triumphant after the discovery of a Purple Heron's nest. DIMW.

The Flamborough Ornithological Group first visited Kazhakstan with Ibisbill Tours. From left, Algirdas Knystautus (founder), Anna and Barry Giles, Andrius Kunigauskas. DIMW.

Lunch at a trapping site in the Sor Bulak lakes. From left, Andrew Lassey, Algirdas Knystautus, Mike Pearson, Lars Svensson, Terry Dolan. DIMW.

'Nirvana' alias the Tien Shan at the Almaty Great Gorge. JMP.

The Chu River camp and the bus that did not die. The demands of the expeditions sent one in four vehicles to 'hospital'! JMP.

Lunch in the saxual trees of the Saryesik Atyrau Desert. DIMW.

The Makings and Modes of Purpose

The characters of purposeful birdwatchers – from Who's Who in Ornithology –
*from a questionnaire – their origins and influences – their heroes – their allegiances –
their fascinations – their views on the best and worst of birdwatching –
and its organisation – a final cameo from Devon*

"The only time I ever really birdwatched 'seriously' in the UK was while I was an undergraduate, at Dungeness Bird Observatory. One day or weekend we had something like Lesser Grey and Woodchat Shrike, Gull-billed Tern, Avocet, Bluethroat and a rare wader, if I remember rightly. One of our party was Horace Alexander who said it was by far the most exciting field trip he had ever made in the UK. So I thought, well, I'll probably never better it and therefore I might as well give up. So I did." Professor Christopher Perrins, *in litt.* (2000)

Not all professional ornithologists stay the birdwatching course.

Among birdwatchers who took on the study of individual species was E. R. (John) Parrinder. He monitored the colonisation of England by the Little Ringed Plover from 1938, in which year the first pair bred at Tring, Hertfordshire. In the background here are Sand Martins, a favourite subject of Bob Spencer and other studious ringers.

My assay of the diversity of British birdwatching and ornithological types was framed largely in numbered camps. Given the earlier cameos drawn of past observer universes as far back as the late 17th century, I now attempt to paint a comparable picture of the leading players of the last 50 years. One source of some colour and many facts is the 1,100 entries in *Who's Who in Ornithology* (1997), compiled by the industrious John Pemberton from information solicited from 1993. From these, I analysed the first and last hundred to produce a 9% sample, nearly twice as big as that normally required for reliable quality control. My other source of facts and many colourful comments is the sum of the responses to a questionnaire of my own construction returned by 44 birdwatchers and ornithologists known to me or suggested by Mike Rogers, the daily scribe of the British Birds Rarities Committee for the last 26 years. The commonest denominators of all 244 subjects and respondents were their lifetime enjoyment of birds, their purposeful study of them and their wide involvement with peers. So always remembering John Pemberton's telling comment that 'modesty never made a good *Who's Who*' (or questionnaire response), here follows an approximation of the main personal characters and attitudes of leading observers in Britain at the end of the 20th century.

The *Who's Who* analysis indicated that the recognised, experienced, fact or service contributing birdwatcher cum field ornithologist in Britain at the end of the 20th century was typically a 51-year-old man. He outnumbered his (listed) distaff peers by 27 to one and displayed three foremost ornithological traits: the development of special interests, particularly in species or families (89%), the communication of these to others, particularly by writing (83%) and the holding of offices, particularly in local club or society (67%). The established birdwatcher's wide ornithological interests show all through the *Who's Who* entries, but, in contrast to the almost blanket support for the RSPB and BTO, the noted memberships of the relatively new, extra-limital clubs and societies are still uncommon with the Oriental Bird Club (10%) just ahead of the African Bird Club and the Ornithological Society for the Middle East (9% each), followed by the West African Ornithological Society (4%) and the Neotropical Bird Club (3.5%). Nevertheless all of these do, however, attract greater membership than the Hawk and Owl Trust (3%) and Wader Study Group (2.5%). Other allegiances are caviar to the general. Within the range of special interests, the commonest were fieldwork on favourite birds or families (46%), population cum conservation studies (42%), migration (18%), identification and taxonomy (16%), photography (12%), ringing (claimed at 7%, although permits were actually held by 23% of the sample) and bird distribution with general zoogeography (6%). Within the first special interest, the most seductive bird subjects were seabirds (8%), waders (4.5%) and raptors (3.5%).

In the rest of his life, the established birdwatcher cum field ornithologist was well educated, usually holding at least one degree (69%). He was rarely a business man (only 8%), earning his living mainly from the practice of science or conservation (37%), the professions and the Civil Service (22.5%), teaching (14%) and communication or the arts (10%). His once deemed prominent position in clerical ranks had collapsed to one priest and two community servers (1.5%). (Of the landed gentry or inheritors of shotguns, there was in the Pemberton sample no sign.) He was usually married or living with a partner (74%). His other interests were noticeably varied and often multiple, with the commonest companions to or reliefs from ornithology being music (20.5%, from classics to jazz), sport (18%), other branches of natural history (18%), walking (14%), world travel and exploration (9.5%), food and drink (9%), gardening (8.5%), photography (8%) and collecting or reading books (7%). As many wives and partners may witness, professed dedication to the family did not feature as highly as any of the above (only 4.5%). His rarer pastimes included fishing and island going (2% each), foreign languages (1.5%) and the church (1%). Once-common hobbies from the early and mid-20th century such as railways, specimen collecting, taxidermy and model aircraft were mentioned only once each. Party politics fared no better!

Given the mainly factual character of the *Who's Who* entries, it was difficult to divine therein the flutters of ornithological hearts or the concerns of accompanying souls. The 44 respondents to my own questionnaire were, however, free to offer both facts and feelings. Their average age was 62 years, 11 more than in the *Who's Who* sample, but otherwise they shared completely the essential characteristics of the committed, experienced birdwatcher or field ornithologist. Sadly, again only two were ladies but neither held back. Asked what had been the germs of their interest in birds, 22 gave thanks for the introductions to and interpretations of birds made by parents, other relatives, teachers and friends. Fascinatingly, eight of the 22 affected a belief in an innate affection for or reaction to all animals, with eight also specifying birds as their start point and three coming quickly to them because mammals and insects proved too elusive. The next most frequent entry point was the once-widespread childhood habit of birds' nesting and egg collecting. Fifteen had initially indulged in the behaviour but its incidence decreased steadily with the reducing age of the respondents; only one of the ten 20 to 40 year olds had taken any eggs. With seven mentions each, initially entrancing experiences (given by images as separate as a museum display, the turquoise eggs of the Song Thrush and an inspiring holiday location) and access to diverse bird-filled habitats were also important germs. (The excess of the above figures over the number of respondents is due to the offer of multiple answers.)

The ages at which the respondents began birdwatching and then took a part in field ornithology were also investigated. In the cases of 20 who started in the 1920s and 1930s, their first acts were made on average at only five years and their

"Nature study was in its infancy during my schooldays; it was probably the acquisition in about 1930 of my late grandfather's war-time 10 x 20 Ross binoculars and the proximity of woods surrounding my school in Glasgow that prompted my interest. I do know what then nearly put me off watching birds: it was when I discovered to my horror one evening that the Treecreeper I was watching with the glasses was working up a tree trunk, behind which two very senior pupils were engaged in amorous dalliance!" Peter Cunningham, *in litt.* (2000)

Some ageing but no sexing in this observation.

"When I moved to Sussex later in 1947, I contacted John Walpole-Bond and had many outings with him, during which he taught me much about finding nests, but failed to rekindle my early interest in egg-collecting. Indeed, Grahame des Forges, who also used to go out with him at that time, and I had an understanding with J. W-B. that he would not take eggs from nests he found with us, or even return to the site with other collectors. Those were the days when one could find five Red-backed Shrike's nests in Sussex on a single outing." James Ferguson-Lees, *in litt.* (2000)

"My earliest ornithological recollection is of being lifted up, at the age of four, to look into a Song Thrush's nest, and the sight of the lovely blue eggs against the uniform grey mud nest lining made a deep impression." Phil Hollom, *in litt.* (1999)

first contributions to the fuller discipline followed nine years later. No other age class was as ornithologically precocious. The 16 who started in the 1940s and 1950s began much later at 12 years but then took only five years to make fuller contributions, having clearly benefited from the rapid post-war development of the ornithological infrastructure and improved interpretation. The ten who started in the 1960s and 1970s again began rather earlier at eight years but then took another eight years to complete their arrival at field ornithology. With such small samples, no further discussion is warranted but two further allied observations can be made. Clearly there has never been a bar to a teenage involvement in purposeful observation, while in terms of the birth decades in the *Who's Who* sample of 200, the most prolific entries come from the 1940s and 1950s. Together, these two decades had produced 98 ornithologists (49% of the sample), another testament to the post-war surge of enthusiasm. Asked what caused their birdwatching to develop into the fuller discipline of field ornithology, 19 of the 44 respondents put first the importance of observations shared with peers, particularly in a club or local society with purpose. Thirteen also recalled the influence of the active ornithological institutions, naming particularly the BTO, the fondly remembered Junior Bird Recorders Club (of the RSPB) and the Edward Grey Institute of Field Ornithology. Also important were the real pleasures of mapping and data assembly (nine mentions), a good patch of habitat (seven mentions) and inspiring books and correspondence (three mentions). The last influence is more fully displayed in my chapter entitled *Stray Feathers*.

Asked how much public awareness there was of birdwatching when they made their start, eleven of the 13 respondents of over 80 years of age opined that there was little or none in a nation with a widely callous attitude to wildlife in the early 20th century and with the sparks of amateur natural history still confined to the literate, better-educated classes. Phil Hollom recalled public lectures as early as 1921, vividly in the case of Captain C. W. R. Knight who presented his own bird photographs as illustrations, but Eric Simms had been the only boy interested in birds in a school of 800 other pupils. Asked also how the hobby of birdwatching had become defined within general natural history, the elders produced no cohesive answer for the period before the foundation of the BTO. Keith Shackleton and Bert Axell saw its emergence as in part the sublimation of hunting or collecting, while Phil Hollom saw it as a wholesome activity, the easiest for the amateur to undertake and also a safe adventure encouraged by parents, themselves inspired by other, more-sterling pursuits such as the polar expeditions. Six agreed that the contribution of the ancient collectors of specimens and eggs had been significant in terms of establishing the basic science and much of the sporting lore, but also that it had been reduced by the emotional turmoils of the two World Wars, hardly indeed surviving the Second. The conversions of Abel Chapman and Peter Scott were given as classic examples of this change of ethos and were far more approved than the snooty attitudes of the few professional ornithologists that had been left in charge of the legacies of Rothschild and Bowdler Sharpe.

Asked about the early British places of ornithological pilgrimage, the 13 elders recalled that at least from the 1920s, the Norfolk coast and Broads were already

visited regularly (by seven), followed by Kent and Sussex and particularly Dungeness (by five) and the London reservoirs (by three, all using telescopes there for the first time). Other further-flung excursions took in Fair Isle, the Outer Hebrides and particularly the then hedgehog-free, wader-thronged Uists and Skokholm. Foreign travels were more limited but by the end of the 1940s, classic localities in the USA, Canada, Holland, France, Spain and India had all been reached by the more adventurous British birdwatchers. Most intriguingly, Phil Hollom remembered an embryonic bird observatory regime centred on the Little Eye at Salthouse, Norfolk, at least six years ahead of the construction of the first full-scale British Heligoland trap on the Isle of May in 1934. Skokholm, the first recognised bird observatory opened in 1933, had been a "Mecca too far" to Bert Axell who "spent all his spare cash on rings"!

Asked about contact with birdwatchers of other nations, 11 of the 13 elders were conscious of some interchange with similar emergent tribes of birdwatchers in Germany (four), the USA (three), Holland (three) and France (one). Phil Hollom was the first ambassador for BTO concepts to the USA and also recalled being a member of a 30-man expedition of the Oxford Ornithological Society to Holland in the summer of 1935, noting particularly the boat-handling wardens already experienced in interpretation for visitors at a time when the early RSPB watchers were trudging the Dungeness shingle and chasing people away. Bert Axell

Cranes bred commonly in the East Anglian fens up to 400 years ago but with the subsequent development of intensive agriculture, they were lost. There was real astonishment at the RSPB when in 1981 a few pairs returned to Norfolk. Although not always successful, they continue with annual breeding attempts.

"Ken Williamson – God, and later a friend. James Ferguson Lees – God, and later a constant field companion. Bert Axell – thanks to him, Dungeness provided the greatest education…" Tim Sharrock, *in litt.* (2000)

An unabashed salute to three mentors of great example.

confirmed the prominence of the Heligoland saga in early British ringing talk and the occasional early exchange with European counterparts as recoveries began. The elders were distinctly starved of a topical ornithological press, with only seven recalling early access to journals of the BOU, the RSPB and the British Empire Naturalists' Association, the *Scottish Naturalist* and the odd publication from local societies. To these few media could be added only the then regular articles and letters that dealt with nature in *The Field* and *Country Life*.

Asked to name their particular heroes and mentors, my panel identified 162 individuals (in a range of one to 16 per respondent). Of these, 80 were local figures and 82 came from national circles. Of the latter, 22 received multiple mentions, the most frequent plaudits being given to Sir Peter Scott and James Fisher (seven each, most from the 50 to 80 year olds), myself (six from the 20 to 60 year olds), Max Nicholson (five from 70 to 80 year olds) and Ken Williamson (five from the 50 to 70 year olds). After them came Richard Richardson (four from the 50 and less year olds), followed by Horace Alexander, David Lack, Reg Moreau, Bernard Tucker, Harry Witherby, Gilbert White and Norman Ticehurst (three each, mostly from the elders). I find it unnerving that of a total of 22 heroes and mentors with at least two mentions, only four are still alive and none is younger than 71. Small wonder that today there is lament over how thin on the ground birdwatching characters have become. It should not be forgotten, however, that regional or special-interest camps also produce leaders. Phil Hollom, Tony Norris, John Raines and Robert Spencer all fall into this category of inspiring people. Outside my panel of respondents, the appreciation of Richard Richardson would undoubtedly

Tony Norris took a special interest in the Corncrake. This painting of a bird on Canna, Inner Hebrides, was presented to him on his 80th birthday.

be greater, due largely to his helpfulness to beginners and the early rarity hunters of Cley (Cocker 2001, Taylor 2002).

Asked to comment on the best features of British birdwatching at the end of the 20th century, there was near unanimity that the development of population monitoring and conservation practice, supported by amateur teamwork, professional research and latterly mass subscription, was the jewel in the crown. Thirty-nine of the 44 respondents gave all or part of this answer. Two other fields drew praise. Ringing and migration studies achieved eight mentions, particularly from the 50 to 70 year olds, and the advance in field identification skills also got seven votes, across the whole age span. Asked to pick out the worst traits, my respondents were less unanimous in their views. A loss of good manners or friendliness was regretted by 13 and a lack of welcome for beginners by three. Excessive attention to rarities drew rapid fire from 17 and a clearly associated distaste for the losses of fun, pride and independence and the converse gains in zealous committees and the recently arisen gurus that dominate the topical press came from six. The losses of wilderness, even just free-range countryside, and of common birds saddened ten. In contrast to the apparent absorption of younger minds with new English names and the splitting of species, such subjects were hardly mentioned by the panel.

Asked if British birdwatching and field ornithology was well organised, the panel's opinion were noticeably divided. Of 33 answers given by still-active respondents, 17 were positive, five unsure and 11 negative. The most major criticisms stemmed from concerns over the loss of the hobby's once broad church, the lack of standardisation in local reports (with resultant waste of observer effort and neglect by central bodies and professional ornithologists) and the lack of a youthful, forward-looking national society for birdwatchers like the American Birding Association. To gauge their ratings of British bird conservation, the 33 younger respondents were offered seven phrases often used in conversation about the issues. The foremost choice of 18 indicated agreement on the need to improve without delay the fitful and fragmented dialogue with farmers, this fault being linked by 12 to the large number of different conservation bodies and by seven to insufficient local knowledge of the field facts (of birds and agriculture). The winning of any national campaign for the restoration of balanced habitats and their common species was considered virtually impossible by six but the prospects for local battles were rated as more hopeful, with 21 respondents actually noting successes in their areas. Overall, however, only 16 of the 33 still-active respondents were satisfied with the general performance of the conservation movement.

In terms of enduring loyalty to birdwatching and conservation societies, my respondents supported an average of 4.3 local clubs and broader-span societies and charities. For the 50 and 60 year olds with peak incomes, the average number of subscriptions rose to five, with the wide influences of the London Natural History Society, the Scottish Ornithologists' Club and the West Midland Bird Club showing particularly in these age classes and undoubtedly reflecting their four-figure memberships. Thirty-one of the 44 supported both the BTO and the RSPB, with an almost complete commitment to the prime engines of bird

"Birdwatching as an absorbing hobby has been turned into a leisure pursuit. I abhor this but what I find equally unacceptable is the associated trend of reducing our wilder areas and so-called "reserves" to little more than theme parks. The RSPB and The Wildfowl & Wetland Trust are too often guilty of this. Too many "managers" are obsessed with the "hide" and "marked walkway" mentality. Yes, there is often the need to lessen disturbance, but let us please, on at least some areas, have as little formality as possible!" Eddie Wiseman, *in litt.* (2000)

A plea echoed by many other post-war birdwatchers.

"When I started birding at Tophill Low 20 odd years ago, I was the youngest there. Now, all those years on – I'm still virtually the youngest. This cannot be good." Frank Moffatt, *in litt.* (2000)

"British birdwatching is fantastically well-organised. Where else could we get 2000 twitchers into a supermarket car park on a Saturday morning? And where else could we get 2,000 birders out there at dawn in May and June counting birds in km squares for a mass-census!" David Parkin, *in litt.* (2000)

Do you watch the birds in your garden?

Do you worry about their future?

If so, you probably help wild birds with food, water and nest sites.

You probably support conservation groups which save their habitats.

Perhaps you already keep a simple record of the birds in your garden?

Did you know that garden bird records can really contribute to scientific conservation research?

It's easy to help .....

The Garden Birdwatch Scheme has been remarkable in promoting purposeful birdwatching to legions of caring people. By 2001, 13,000 of them were contributing to conservation research and enjoying their own avian charges even more. In 2003, 14,000 sent in counts of the declining House Sparrow.

"Co-ordination between bodies could be improved and divisions of responsibility more clearly defined. A 'Council of Ornithological Institutions' meeting, say, annually could identify areas of overlap and inadequate coverage." Allan Vittery, *in litt.* (2000)

An idea echoed in the comments of several other correspondents.

conservation in the 20 to 50 year olds. Thirteen of the more elderly remained loyal to the BOU but only seven of the whole panel supported the re-constituted Wildfowl & Wetlands Trust, which in its original guise had been as popular as the RSPB. Club and societies with international remits attracted support from 25 panel members, suggesting a much higher incidence of allegiance than that indicated by the *Who's Who* analysis.

Asked if they thought that the habit of purposeful birdwatching had passed its peak of allegiance, 15 of the 33 non-elders thought that no such event had occurred and furthermore they listed 11 factors favouring its continued growth. Of these, the take-up of the Garden Bird Watch by 13,000 enthusiasts was the most encouraging and obvious event, particularly as one in ten go on to full membership of the BTO. Interestingly in much of the associated commentary, there was a lack of faith in trying to force-march recruits into purposeful birdwatching. The real trick would be to offer via today's multi-media and leading personalities like Bill Oddie a constant inticement to the greater fun and achievement available from a fuller commitment to ornithology. An idea of Tim Sharrock, the subsidising for young observers of stays at bird observatories, exemplified the panel's view of what was currently missing in birdwatching or birding. Perhaps in the last example of an easily begun and constantly fulfilling apprenticeship, there is a counter to the worry of the ten respondents who agreed that birdwatching was waning. Reminded of the fixed ceiling of about 10,000 observers willing to do recent survey and atlas work, 23 of the 33 non-elders were distinctly more pessimistic about the expansion (or replacement) of this precious workforce. Five were also concerned about the increasing age of people undertaking the most demanding activities of ringing and wildfowl counting. Of the 25 explanations offered for this state of affairs, 11 featured again the lack of good training, shared excitement, close interpretation and plain leadership.

It is high time for me to introduce a lighter note to this display of the making and modes of purposeful British birdwatcher. So finally here are some findings from an earlier sample of the behaviour of some senior birdwatchers and ornithologists, taken by Andy Richford at the BOU/SWLA conference held at Dartington Hall in September 1996. It was undoubtedly biased by the conference theme of 'Birds in Words and Art' but the answers on 51 questionnaires, returned by attendees who averaged 57 years of age and of whom 41 were male, are nevertheless intriguing. Their most frequent shared behaviours were joining ornithological societies (100% supporting on average 6.5 clubs or societies), travelling the world for birds (64%), list-keeping (48% for Britain or other geographical area, 44% for the world or life, and 22% for each year), and at least one twitch after a rarity (44%). In the manner of *Desert Island Discs*, the attendees were also asked what luxury items they most desired in association with their hobby. The chief wants were endless booze (20%), waterproof noting materials (18%) and a good telescope (10%). Among the long tail of highly idiosyncratic wishes, a kind woman or wife (6%) and a really comfortable bed (4%) were also listed, but separately!

Driving to Lapland. First brew near Lake Vanern, Sweden, 1st July 1957. Note on author then popular Grenfell Jacket with sewn-on fur collar; car still clean. JC.

The coveted crossing of the Arctic Circle, south of Turtola, Finland, 3rd July 1957. Left to right, Joe Cunningham, Peter Naylor, author. Note last tarmac then 200 miles to south; car filthy.

Peter Naylor filming ducks and phalaropes on Ekkeroy Island, 5th July 1957. Note wind-up Bell and Howell cinecamera. JC.

Joe Cunningham in habitat of Bean Geese and Pine Grosbeaks, near Pasvik River, Norway, 8th July 1957. DIMW.

Free accommodation in last of the houses built over 3,500 years on promontory of Pasvik River. Note permafrost damage. JC.

Free boat, Elenvand, 9th July 1957. JC.

A Cambridge Bird Club expedition to Lapland with an original Morris Minor lent by Macfisheries Ltd.

The only answer to Finnish mosquitoes. Black stetson (from Canada 1956), bee veil and gloves. JC.

The bridge to Russia blown up due to the 'Cold War', Nyrud, Pasvik River, 7th July 1957. Note fish poached from Russian water were tasteless. PEN.

Making camp at Uluun Nor Lake in the Gobi Desert on 30th May 1980. DIMW.

The first-ever bird tour of Mongolia, with Birdquest in 1980. The high camp in the Gobi Altai on 2nd June 1980. Even with hot water bottles, feet nearly froze but there were Altai Snowcocks on the ridge. DIMW.

A camp fire was nowhere more welcome than in the Gobi Desert at dusk. Third from left, Mark Beaman. DIMW.

Just occasionally you can still walk in true wilderness. The author strides out across the salt and silt pan of Uluun Nor Lake on 30th May 1980. There could be water and birds in the distant reeds... H. D. Altmann.

More Than Pathfinders

Roman threads on Swallows' legs – the first British recoveries of marked birds –
a Danish breakthrough – the 20th century explosion of ringing – the British Birds
Marking Scheme – Ringing's Girl Friday – a Brambling in the hand –
two English Meccas – two men of oases – changes in catching techniques –
the miraculous mist-net – veteran performances – international co-operation in
Kazakhstan – a Sussex benefactor – the widening science of ringing

The purposeful marking of birds goes back at least to Roman times. Fisher and Peterson (1964) repeated the remarkable tale of how in the second Punic War, about 210 BC, a parent Swallow was taken from her nestlings by members of a garrison besieged by the Ligurians and smuggled out to an officer in a relieving column. He tied a knotted thread to one of its legs, the number of knots indicating the number of days that would pass before the column's attack, and let the bird fly back. Another Roman recognition of the Swallow's homing ability was noted by Pliny in the 1st century AD. To deceive the local bookies, birds with coloured threads on their legs were used to transmit the winners of chariot races from Rome to Volterra. About 1250, a German Cistercian friar noted that a man attached the following question on a small piece of parchment to a nesting Swallow, "Oh, Swallow, where do you live in winter?" The answer came in the same manner the following spring, "In Asia, in the home of Petrus". Other forms of bird marking appeared in the 13th century but with the identification of ownership as its aim, particularly in the case of the precious birds of falconers, perhaps some of their prey (said to be awarded medals for providing great sport) and certainly domesticated fowl.

The potential of bird marking in the study of their seasonal distribution was not probed again until the mid-18th century. Then there were still many proponents of avian hibernation, particularly in the case of supposedly pond-diving, mud-encased hirundines, but another German, J. L. Frisch, devised a test of such fallacy. He tied threads dyed in water-soluble colour to the feet of his breeding Swallows and found the colours still fast in the following spring, a state incompatible with immersion in water and mud. In Britain, the earliest find of a marked bird comes apparently from Trotton in Sussex (White 1789). In the dreadful winter of 1708-9, "a duck … with a silver collar about its neck, on which were engraved the arms of the King of Denmark" was shot. This incident was apparently investigated by Richard Kearton in 1902, who considered the duck to have been "in all probability a Cormorant, used for catching fish" neck-haltered in the Far East fashion. Ninety years after the strange duck came the first recovery of a truly wild bird. A Woodcock was shot on a Northumberland estate where a year before it had been caught in a rabbit net and marked with a brass leg-ring (hence also the first record of avian site fidelity).

In the 19th century, pioneers of migration study in America, France, Germany, Finland, England and Scotland tried marking members of at least a dozen bird

Until the creation of reservoirs and gravel pits, Cormorants were scarce inland. Nowadays much commoner, they court the displeasure of anglers. Quite when the continental race (as here) entered England was missed but ringing has proved that it has come from six European countries. For more like facts, see *The Migration Atlas*.

families but the first systematic scheme was the invention of a Dane, Hans Christian Cornelius Mortensen. In his second attempt in 1898, he used for the first time the crucial combination of lightweight aluminium rings and their imprinting with both a return address and the all-important serial number. The last innovation allowed a truly individual register of each capture's identity and its associated temporal and geographical stations and biometric details. The way to the mass ringing of birds and the fully intelligent analysis of their recoveries was open. The first steps along it were trod by the Germans, who gave up shooting and began trapping and ringing first at Rossitten in 1903 and then on Heligoland in 1909. Only five years later, schemes based on the Danish model were in operation in North America, 13 European countries and Australia. Two British exercises began simultaneously with the 'conversion' of Heligoland in 1909, one centred on Aberdeen University and pioneered particularly by the then undergraduate A. Landsborough Thomson and the other encouraged by H. F. Witherby, using the platform of *British Birds* and with the recovered rings directed initially to "Witherby, High Holborn" but later to the British Museum of Natural History in London. Human loss in the First World War forced an end to the northern exercise and all early ringing activity was brought together under The *British Birds* Marking Scheme. The significance of the ongoing stewardship has been little saluted in recent decades – few today would think that the Ringing Unit of the British Trust for Ornithology was anything but a primeval institution – but Landsborough Thomson and Witherby gave respectively over 50 and 27 years of service to a classic co-operative study. In the late 1920s, Witherby acquired the assistance of the 'Girl Friday' of British ringing, Miss Elsie Pemberton Leach. The daughter of a much-decorated General, she had grown up amongst swishing trout rods and speaking hounds but inspired in her thirties by Norfolk's tern guardian and first-ever ringer Miss E. L. Turner, Elsie took up birdwatching. A meeting with Witherby resulted in her taking on first a share of and second nearly all of the administration of the marking scheme. Blessed with a mind that grasped and retained to an astonishing degree the ever-multiplying details of the registers, Elsie became indispensable and never more so than in 1937 when Witherby asked the BTO to take over responsibility for what by then had become the Ringing Scheme. Elsie agreed to serve the new Bird-Ringing Committee as Honorary Secretary, becoming the sole or part author of its reports for the next 16 years and continuing as the monitor of foreign-ringed bird recoveries until 1963. Given the touch of Agatha Christie that she showed in her search for full recovery details, Witherby coined the apt phrase "Relentless in pursuit" as her motto but she also courted the accusation of treason during the Second World War when to trace recoveries in Axis territory, she used the still-open diplomatic channels of Dublin. Three times the code-like details of her enquiries raised the eyebrows of Military Intelligence but as Chris Mead related, "She was clearly beyond reproach!". Too little known outside the ringing universe, Miss Leach would have been delighted to read the *Migration Atlas*. Its fundament belongs to her, Witherby and Landsborough Thomson and all the keen ringers that their examples have so inspired.

My first experience of bird catching and ringing came in April 1950. Led by 'Eek' Turner, my second French Master, quarter-mastered by my Dad, three eager teenagers were let loose on the Isle of May from 21st to 25th April. The catching equipment of Scotland's first and Britain's second oldest bird observatory then consisted of one seemingly huge wire funnel, known as the Bain Trap after its chief builder (and the first erected in Britain of the classic, full-scale Heligoland model), and two movable open wire cages, christened Dipt and Dupt for reasons lost from my memory. In the Low Light, by then the home of the rarity-collecting Good Ladies and less lethal migration students for over 30 years, ringing instructions and schedules and a selection of shiny aluminium bands were all ready. Off we rushed in search of migrants, only to learn the hard lesson that a spell of strong and persistent north-westerly wind was not a classic vector for traps full of birds. The desert of ringing opportunity was made even worse by the many accounts of exciting captures in the observatory's excellent logs, which provided our evening entertainment.

The Bain Trap on the Isle of May, April 1950. Seemingly just older than the first large trap on Stokholm. WSW.

At last, the wind relented and a few small birds appeared. A Wheatear caught in Dupt, which proved to be an early Greenlander, was a start but a leaf warbler eluded every driving ruse. Then at long last, the Bain Trap worked its Heligoland magic; a late sweep of its field on the 24th flushed a small dark bird with a white rump, oval not square in shape. Not another Wheatear. Mercifully, it flew into the mouth of the trap, down the funnel and plop, down into the catching box. Out of the trap's side door, we barged; arm up the removal sleeve and out in my nervous hand came a superb cock Brambling in full breeding dress. Its head all sheeny black, its shoulders and breast drenched in orange, and these strong colours set off by satin white, it was better than rare; it was absolutely stunning. Our ringing starvation was over and because the plumage contours of the species are so clearly delineated by colour, the Brambling tutorial in plumage topography was easy to take. The *Handbook* description lived. The Isle of May Bird Observatory had, in the nick of time, done its job of education. We could go home and be properly fed by mothers.

On the 24th April 1950, a cock Brambling flew straight into the Bain Trap, Isle of May, and gave its eager captors a colourful lesson in plumage contours.

The association of the early bird observatories with bird ringing was strong and inescapable. An exciting capture, such as our individually brilliant Brambling or scores of early November Blackbirds or late April Whitethroats during the then regular big falls of migrants, provided 'up close and personal' experience of full-grown birds. Hitherto, these had been the provinces of only a few afficionados skilled in the operation of strange devices like clap-nets and Potter traps. Thus while it was the overall lure of migration that impelled us coast- and isle-ward in spring and autumn, it was to the Heligoland trap that we looked for brief but compelling intimacies with birds in the hand.

The very first, small Heligoland-type trap in Britain appeared on a Welsh isle. Its erector was Ronald M. Lockley; its position was his garden on Skokholm, Dyfed, and its first captures were made in 1933. I never met Lockley but, like many others, I was inspired by his work on seabirds and particularly his paper, in a special supplement to *British Birds* (Vol. 46, 1953), "On the movements of the

"I caught a Great Spotted Woodpecker on Lerness one afternoon, by creeping up to it on all fours: a humbling experience, as the bird doubtless mistook me for a sheep! It was a good weight and could fly well when released." Ken Williamson, *in litt.* (1953)

Any ruse on the way to a ringed bird and the chance of a recovery.

Manx Shearwater at sea during the breeding season". With photographs of swarms of birds, flight-line maps and 201 recoveries from nearly 41,000 Welsh-ringed birds, Lockley illustrated the fishing saga of his subjects as no one ever before. To our amazement, he showed that those that nested on Skokholm and Skomer fed regularly in the Bay of Biscay, performing return journeys of up to 1,200 miles after their favourite food alias fat, young sardines. As the progenitor of British observatory recording and ringing practice, Lockley was the model for many later wardens and observers who also combined breeding seabird studies with migrant counting and ringing. On Skokholm, his garden trap and a full-scale sibling built in 1935 caught about 450 birds each spring and about 210 birds each autumn from 1947 to 1976. The hope of similar catches saw to the deployment of a portable Heligoland-type trap at the Little Eye near Salthouse, Norfolk, from 1933 to 1937 and the construction of another full-size funnel on the Isle of May, Firth of Forth, in 1934.

In its heyday and with Bert Axell teaching its lessons, Dungeness Bird Observatory was one of the great schools of British birdwatchers. The nearest birds are Black Redstarts (right) and Firecrest (left), yet to be driven into the Heligoland trap.

As ringing became a widespread endeavour after the Second World War, two clusters of traps became particular Meccas. These were at Dungeness, Kent, and Spurn Point, Yorkshire. Over the former, lorded the magisterial Herbert Axell, a born ringer who, not content with spending long hours on the aggressive conservation tactics needed to recover the local seabird colonies, found somehow the other hours and helpers essential to create the formidable battery of traps that harvested gorse- and sallow-haunting migrants. With his co-invention of the Dungeness pliers in 1954, Bert also made the hitherto fiddly task of crimping rings round passerine legs failsafe. With its close association with local societies and also the Ornithological Section of the London Natural History Society, Dungeness Bird Observatory became a formidable centre of ringing excellence and observer training. As the BTO's indomitable raffle-ticket seller, Mary Waller, attests, Bert could be difficult and tetchy but being one of his approved disciples became a much-desired rite of passage.

The situation at Spurn was different for although it was still the 'happy hunting ground' for migrants that John Cordeaux had so praised, it was, until the late 1970s, a long way off the beaten track of the A1 and very much the haunt of Yorkshire and other northern observers. Only once a year did it become an ornithological metropolis by hosting the annual BTO sponsored course in trapping and ringing birds. Gibraltar Point also acted in this role. For years, their chief visiting apostle was the avuncular Quaker Bob Spencer who from 1954 as the BTO's first full-time Ringing Officer doubled as Britain's most enthusiastic supporter of ringing as a scientific tool and its greatest carer for the welfare of all the birds that had their calm disturbed by capture and handling. Normally a gentle person, he played his second role to stern perfection in 1965 when he refused James Ferguson-Lees and me any form of temporary ringing permit for the second of the Jordan expeditions in that decade. Although both of us were observatory-trained, we had not applied at the appointed time for the new ringing permits introduced by the BTO to control inexpert handling. Even if we found an accidentally trapped bird, Bob's admonition was still "Let it go!" With the Azraq marsh a classic ringing site, with lines of tamarisks and watercourses all over the place, it was a blow but in 1966 we solved our problem by getting Bob and John Ash on to the expedition complement as full-time ringers. With their recent experience at the Figuig oasis in northwest Africa, they were the obvious choices. We showed them some likely net sites and went about our surveying of other oasis and desert habitats, returning late in the day to hear of each day's catch. Working in temperatures of up to 35°C, regularly clearing the panels of 15 nets, Bob and John caught nearly 200 birds a day for a fortnight. Their reward was four days' freedom to catch up with some richly deserved lifers (now ticks) of the desert avifauna. If ever I had had doubts about ringing, none survived Bob and John's consummately careful effort. Every bird flew strongly away.

The character of British ringing has changed over its near century of endeavour. The originally high incidences of nestlings and traditionally trapped birds fell as innovations like rocket and later cannon nets caught on the occasion amazing numbers of waders and wildfowl (the particular objectives respectively of the Wash Wader Ringing Group and the then Wildfowl Trust) and collapsed when the importation from Japan of fine mesh nets offered any ringer a handy device that could catch almost any cover-haunting or line-flying bird at any time or place. The well-named mist-net took its British bow at the Oxford Bird Observatories Conference in January 1956, being introduced by Anthony le Sueur of Jersey and being quickly snaffled from him by Bert Axell for a field trial at Dungeness. Mundanely, its first capture was only a House Sparrow but soon its portability and virtual invisibility was fully proved when a mere 6 x 6 feet section set against a floating barrel bagged the first Black Tern to be ringed in Britain. The Dungeness experiments continued and the mist-net soon radiated into subspecies of differing lengths and panel or shelf combinations. My own first experience of mist-netting occurred on 22nd April 1957 when the legendary Clive D. T. Minton gave several

At the Azraq Oasis, Jordan, in April and May 1966, the biggest mist-net catches were of Red-throated Pipits and Sand Martins.

Except when looked along (as here), mist nets are invisible. Erected in gaps of cover they give ringers an almost instant catching facility. DIMW.

members of the Cambridge Bird Club a convincing demonstration of its effectiveness. At the city's also legendary sewage farm, the first drive alone caught a pair of Bullfinches, two each of Sedge and Willow Warblers, and a Great Tit. No longer would the Heligoland trap lord over the other trapping devices.

The mist-net was the new catching power, versatile, dependable and ubiquitous where its poles and stays could be stuck. On 27/28th August 1959, just one net caught warblers all day on St. Agnes and tubenoses all night on Annet until the ringers slumped exhausted among the latter. As its captures of rarities mounted, the news that then still-difficult field identifications could be confirmed in the hand within minutes spread like wildfire. The numbers of ringers grew as the mist-nets availability increased, with new groups targeting not just the traditional observatories and like places but also inland localities as far apart as Romford Sewage Farm and Knaresborough's riverine jungle.

Such was the ecumenical character of the mist-net that the legendary exploits of the early trappers cum ringers began to fade. Chris Mead's favourite ringer manque was a Lake District millionaire, Dr. H. J. Moon. He much preferred the act of ringing to the preceding search and capture; so he would reportedly take champagne from his butler while awaiting the return of his chauffeur and gardener with the next bunch of chicks of moorland birds, no easy finds. Yet they made up the majority of his and his retinue's 78,000 bird haul up to 1943. After the Second World War, another ringing colossus arose. Major-General C. B. Wainwright decided that more needed to be known about the ways of wildfowl. Initially independently but soon in alliance with the new Wildfowl Trust formed by Peter Scott, the General went after the birds of the narrow arm of Abberton Reservoir with traps set along the water line. In the two decades from 1948 he and his team ringed over 28,000 Teal and 15,000 other wildfowl, in a total catch of over 100,000 birds. He not only put the catch of the Slimbridge decoy to shame; he also lived long enough to muse over 9,000 recoveries gathered at a rate of which most ringers only dream.

To the above examples of ringing lore, the new mist-net deployers were soon adding other amazing performances. Even his fellow members of the Cambridge Bird Club quailed as Clive Minton, assisted by Chris Mead and other 'heavyweight' ringers, led the wholesale capture of waders on the Wash and at Wisbech sewage farm. At Romford, Bob Spencer initiated a concentrated effort on Sand Martins. In seven years, over 250,000 of our most vulnerable hirundine were ringed and over 11,000 recoveries were made within Britain, especially in the south coast roosts, and in West Africa. Given these experiences, a more scientific penny dropped in the Ringing Unit. Ringing could be structured to tell us far more than the one-moment state of a bird and the start and end points of its following movement. It could validate the occurrence of not just species but also subspecies and even the scale of populations and their productivity. It could gather not only the elapse of time between ring placement and removal but also survival rates in both adult and young birds. As with much else in birdwatching, ringing grew up and as its objectives became more targeted towards conservation issues, the element of random hunting first fell and second was discouraged.

Among the wildfowl that haunted Abberton Reservoir, Essex, in cold winters were Smew. Depicted here are four drakes and three red-headed ducks. Regrettably there has never been a ring-recovery of this delightful bird.

To recapture flavour of the old days, some ringers migrate. One of the most constructive ringing performances in both avian and human terms that I have witnessed in recent years was the co-operation in Kazakhstan in May 2000 between the Flamborough Ornithological Group (FOG) and the multi-faceted ornithologist Lars Svensson and the Institute of Zoology in Almaty cum Chokpak Ringing Station.

Already veterans of Siberia and Mongolia, FOG had determined not to go on just catching the odd gem along a tour route but to do something real to assist the financially beleaguered but still amazing Kazakh initiative in migration research. The Chokpak map of recoveries is a mind-blowing revelation of how birds twist their way through the Tien Shan mountains, with the outwards fans of movement both north and south of the famous pass dotted all over Asia. Their catches can be incredible, up to 5000 birds a day, especially if the wind is easterly in spring and westerly in autumn. Sadly in 2000 our wind at Chokpak was not favourable but after a second welcoming meal on the first night and through a rising haze of Islay malt and Almaty vodka, 11 Common Nightjars of three Asian races were suddenly thrust under the yurt's light and our instantly alert gaze. Funnelled by the pass, three taxa of a still-radiating species had come together to show us their slightly but distinctly different colours, receive their rings and be gone by dawn. They had been driven into the biggest traps that we had ever seen, close mesh-nets strung on girders worthy of Irn-Bru and the Forth Bridge and tall enough to catch a London bus. Away from Chokpak, elsewhere in southern Kazakhstan Andrew Lassey, the irrepressible 'in the hand twitcher' Mike Pearson and the unique BTO ringing veteran Terry Dolan caught 578 birds of 66 species. Edward and Andrei Gafrilov and the other Kazakhs had never seen mist-nets deployed so adroitly as samplers of migrants and breeding birds. The contrast of techniques was fascinating – in weapon analogies, it was Kazakh howitzer versus British rapier – but the result was a model of international co-operation now being achieved across the fallen Iron Curtain. The southern Irish may have taken some part of a once-shared baton back across their sea but to the east, the Russians and Kazakhs (and who knows else) are eager for Scano-British get-up-and-go. They have a keen sense of the international history of ornithology. Let us share its future with them.

To loner-hunter-gatherers like me, ringing can seem at times a tedious bird-intrusive, over-localised business but make no mistake, its gifts to ornithology have been and still need to be immense. It bears repeating that the found paths stretch not just to destinations but far beyond into life-span, age class behaviour, moult strategies, transmitter attachment, the new molecular taxonomies and so on. Its 2,000 devotees spend their own good money on maintaining a century of excellent tradition, still on occasion good sport and thunderingly good science. Not being a ringer myself, I am shy of nominating a modern hero but by all accounts, Stephen Rumsey fits that bill. In 1985, he bought 300 acres of long-drained land at Icklesham near Rye, Sussex and promptly turned 100 of them back into marsh, growing not just reeds but also the Spotted Crakes, Sand Martins, *Acrocephalus* warblers and Bearded Reedlings that re-inhabited their restored niches.

"More recently, 22nd July, a female Red-headed Bunting arrived. It was a very elusive bird and we saw little of it, the last time being 31st July. The synoptic situation at the time of its arrival suggests very pointedly that the bird was a natural vagrant and not one of these birds which British Birds *have discovered are released from time to time in London."* Ken Williamson, *in litt.* (1953)

Proponents of extreme 'reversed migration', as the cause of vagrancy by a central Asian species, would vote with Ken Williamson but the spills of cage birds, at least in Europe, continue.

"For me, becoming a ringer seemed like a natural extension of my interest in birds. I wanted to know more about moult and migration, and becoming a ringer was the best way to do it." Jane Reid, *in litt.* (1999)

Terry Dolan, one of the almost apocryphal, BTO-approved trainers of ringers, would applaud this sentiment. He is still passing on his handling skills, for example to Kazakhs, after 50 and more years.

By restoring a reed bed at Icklesham, Sussex, Stephen Rumsey has attracted back since 1985 not just Reed and Sedge Warblers but also the elusive Spotted Crake.

By 1990, his reserve was well enough established to allow the insertion of mist-nets and by the autumn of 2001, more than 40,000 birds had been trapped. Determined to find them again in their winter quarters, Rumsey and his teams went to Senegal and Gambia three times and retrapped 300 of what can surely be called his own avian offspring. To ensure the security of his wetland creation, Rumsey got it declared a Ramsar site and has since created the Wetland Trust to preserve it for the future, so providing a model example of how private wealth can still add dramatically to the wider charitable support of birds and habitats. Not content with one remarkable piece of habitat recreation and migration study, Rumsey has recently underwritten the whole of British ringing by also ensuring that the manufacture of the rings themselves would not be interrupted.

He enjoyed what the former maker Cecil Lambourne's products did so much, that he bought the company from Cecil's son Kevan, renamed it Porzana (after his Spotted Crakes) and incorporated it into the trust. The tradition of personal generosity to British field ornithology was thus updated and the BTO and all ringers breathed a huge sigh of relief.

The efforts of British ringers continue to produce ever more fascinating and important results. Some groups work to the new tablets of the BTO Ringing Office, which has set intelligently disciplined objectives for the schemes that maintain constant ringing effort at about 120 sites and concentrate on the retrapping of adults at over 80 localities. Together these are providing crucial conservation data on long-term population trends and the now rather shaky survival rates of so-called common birds. Other teams target summer visitors such as roosting Swallows and the reviving contingent of Nightjars, while yet others continue the long-term monitoring of seabirds and raptors. An honest few at migration sites still hope for choice rarities among a Goldcrest fall although such specific targeting is now officially discouraged by the BTO. Significantly rings are now sold at prices that reflect the likely conservation value of the recoveries. Thus a ring for a fairly secure Greenfinch costs 23p, but after a higher refund from the subsidy fund of the BTO/JNCC, one for a rather imperilled Dunlin nets at only 8p.

Sometimes a tiny warbler is not another Goldcrest. The scene depicts a Pallas's (Leaf) Warbler at Flamborough Head, Yorkshire, easily caught in a low-set mist-net on the first drive and enjoyed by its ringers and other judges of avian jewels during a few minutes of captivity.

This is not the place for a review of all the discoveries of the 93 years of co-ordinated British ringing but I have three favourites. The tit specialists, of whom the chief is still Chris Perrins, have actually proved that Great Tits hunted by Sparrowhawks are fitter than those not so. Our wintering Blackcaps are not British birds that hang on in the now milder winters but native Bavarians which fly west to our maritime climate. Our once rare but now fast-spreading Bonxies (Great Skua) are not just threatening more and more of our own Kittiwakes but others as far to the north-east as Murmansk. Such findings are not just the stuff of science. They are wonders given to all those interested in birds by the three full generations in the all-time roll of over 5,000 ringers whose total catch now exceeds 30,000,000 birds. Of these, over 580,000 (1.9%) have been recovered. It bears repeating that the average ringer's individual contribution to this astonishing harvest of birds is not just years of effort but (at today's costs) a personal expenditure of well over £1,000 per recovery. Ringers are more than pathfinders, indeed; they are really generous investors in the commonwealth of man and bird.

One of the ultimate avian predators, the Bonxie or Great Skua is fearless. When one comes at you, it is you who ducks. Yet over 66,000 have been ringed by brave ringers!

Nothing in the British ringing scene prepared the Flamborough Ornithological Group for the scale of Kazakh activities.

Above: a 'howitzer' of a trap at Chokpak ringing station. Most British 'Heligolands' would get lost in the bottom third. JMP.

Left: the ringing hut at Chokpak Ringing Station and, behind, the tree belt ideal for mist-netting. JMP.

The early years in Scilly. Trans-shipment at St. Mary's, 30th September 1961. The inter-island launch 'Kittern' unloading from the Scillonian's side-doors. Facing camera, Bob Emmett (left) and Peter Colston.

A St. Agnes welcome. Brian Newport (left) and Lewis Hicks, latter the ever-helpful landlord of (and boatman to) the bird observatory.

The St. Agnes promise. The wonder of Britain's first trapped and ringed Hoopoe caught in the expression of Francis Hicks, March 1958.

The 'Undaunted' taking observers and ringers off Annet, 4th October 1964. Francis Hicks takes the strain as Lewis and Alice come alongside the landing shelf. Young Petra Wallace sleeps on. REE.

Above: the view from the St. Agnes Lighthouse showing Lower Town Farm, home to the observatory, and the Big Pool, with the Browarth Point fields to the right. DIMW.

Right: two of the birding gurus of the 1960s and 1970s. Peter Grant and (behind) Tim Inskipp, their respective specialities being gulls and the birds of the Indian Subcontinent. Janet Coram.

Above: beyond the last house of Malin More lie Rathlin O'Birne and North America. The inshore waters can throng with cetaceans, tuna and seabirds. DIMW.

Left: the gate to 'Magic Garden' opens on to the last sycamores and vegetable plot before the Atlantic. No wonder a few steps through it and a Scarlet Grosbeak or a Melodious Warbler can be yours.

Good companionship and teamwork are essential to migrant searches. By their cottage in Glen Colmcille, Dave Allen, the author and Anthony McGeehan show their bond. AMcG.

The tales from the 'Rocky Bird Observatory' have attracted other enthusiasts. Here Gerard McGeehan, Vernon Carter and Davie Steele enjoy a Yellow-browed Warbler at close range. AMcG.

In Ireland miracles still occur. Here a Robin checks out Dave Allen's hat and beard as a winter territory. AMcG.

New migration watch points can still be pioneered, witness recent events in westernmost Donegal and on Barra.

Renewed Global Pursuit

New vistas of lands and knowledge – three engines of travel - National Service – foreign holidays – expeditions – the foremost collector of world species – the Mountfort saga – the sad fate of the Azraq Oasis – the explorers of the Middle East – Richard Porter as communicator - changes in establishment roles – more individual hunts – the origin and growth of the bird tour

As in many other fields of human recreation, the British concentration on the conduct of the Second World War induced a lag phase in birdwatching and field ornithology. The loss of observers was less dreadful than in 1914-18 but it was the second in three decades, while the many restrictions on travel and habitat access and above all the destruction or disruption of much of Europe's logistic and scientific infrastructure presented a daunting prospect to those anxious to take up again the initiatives of the 1930s. Nevertheless new ornithological forces had undoubtedly simmered under the heavy lid of war. In 1945, the Witherby *Handbook* was still only four years old and not a few members of the services returned with not just de-mob suits but also with vastly increased understanding of the internationality of so-called 'British' birds. The solace afforded by birds as Man's most visible wild neighbours never beckoned more.

I was nearly 12 on VJ day, my attraction to birds already being fashioned into a hobby. The youngest of my schoolmaster mentors was Christopher Mylne, son of the totally dreaded headmaster of Dalhousie Castle Preparatory School but

Military service along the oil pipelines of the Middle East was often enlivened by falls of migrant birds. Here, near the Jordan/Iraq border, a flock of Bee-eaters considers an irrigated garden as a roost site.

excused condemnation of this relationship due to his development and praise of my nature study. It was he who first indicated to me the full dimensions of avian science and its amateur portals. Two years later I was the newest member of the Junior Bird Recorders' Club, then the entry point for children to the RSPB. Somewhere in *Bird Notes* shone my first published record and with it came a hugely increased sense of belonging and contributing. One more year and Dad delivered my own copies of the Witherby *Handbook*'s five volumes. My eye fell particularly upon one set of texts, entitled "Distribution Abroad". In them was not just mentions of the still-extant, not yet despised British Empire but clue upon clue to a half-planet full of birds. Spared the distractions of television set and computer, both still in their infancy, it was time to take some sort of alternative wing after feathered beings.

The main engine of the immediate post-war dispersal of young British birdwatchers was the 15 year continuance in the unhappy cold peace of military conscription, so called National Service. For 18-year-old males of reasonable health, this was a compulsory two-year sentence – three if you aspired to a naval commission – that conveyed essentially two extremes of experience. The poles were the boredom of a mundane posting to clerical duties or similar repetitive routine and the exhilaration of the various confusions in late Imperialistic mode that came from fighting communists, suppressing or keeping apart rebellious natives and practising for instant death if the H-bomb was dropped. All such extreme behaviours were further compounded by the turmoil of delayed adolescence characteristic of the mid-20th century. Most of my birdwatching peers went for the exhilarations and found themselves never bored in places as far apart as Belize and Hong Kong, Canada and South Africa, witness their resultant accumulation of knowledge and even some papers in the then less tense narrative of the *Ibis* of the 1950s. Whether there had ever been 'a good war' for anyone was debatable but ornithologically fulfilling National Service certainly occurred. The inheritance of the global vision of the great Victorian stocktakers may have been accidental but it was no less real, being also importantly vested increasingly in field observations.

The second engine of the post-war development of international birdwatching abroad was the resumption of foreign holidays. (No one had heard of tourism then!) These became re-advertised in the less-blitzed nations of Europe from 1948 and adventurous family after adventurous family went abroad while the wartime refuges of peripheral Britain and Ireland lost custom to countries such as the relatively unravaged Austria and nothern Italy. With travel agents like Thomas Cook still worthy of their reputation of getting you to the most out-of-the-way places given only a reasonably full purse and patience, some vestige of the pre-wars European tour was recoverable. Getting from London to Achensee in northern Austria in September 1950 took 27 hours, of which 13 were spent in the badly misnamed Arlberg Express, but the continental scenery and above all the birds were made all the more sensational by the mounting anticipation. Eight lifers in a first European day-out was just the start of my personal benefit. Similar experiences were largely confined to the upper and re-emergent professional manager classes and no cohesive record of them has survived.

Concurrent with the retrieval of continental holiday destinations was the re-emergence of the rather more risky but ornithologically greater achievement of place and study bound up in the magic word 'expedition'. I first associated such adventures with the uninhabited islands of Scotland, as so lyrically portrayed by Robert Atkinson in his *Island Going* (1949), but after only two fortuitous winter lectures, I realised that expeditions could also get one to areas as seemingly remote as Lapland and as truly so as the Nepal Himalaya. Regrettably, although there have been fitful attempts to keep one, no full register of modern British ornithological expeditions exists. To give some measure to the phenomenon, I have therefore analysed the data on expedition membership noted in *Who's Who in Ornithology* (1997). The source is undoubtedly incomplete but of a sample of 283 entries, 50 people (18%) had gone on 170 expeditions (on average 3.4 each). The most favoured regions were Europe (20% of destinations, with Arctic localities still attracting 7%), Africa (15%, with Commonwealth countries pre-dominant), Southeast Asia (5%), the Middle East (4%) and then a tail consisting of Central Asia, South America, North America, the Atlantic islands and Antarctica. Within the cell of 50 expedition members, I spotted 16 whose frequency of travel and destinations rated the title 'intrepid traveller' and I scanned all of *Who's Who* for others of similar derring-do. I found 36 more, with a remarkable average of 12 expeditions to their individual credit. Their most favoured region was again Europe (but with a much-reduced share with 10%); close second came the Middle East (9%, often making repeated visits), with Africa (6%) next and then a long tail. From the late 1960s a truly global reach became obvious. From the same time, it also became clear that expeditions or far-flung travels were being driven not just by diverse study goals but also increasingly by personal acquisitions of a world list.

In 1994, the largest personal 'world list' among the Britons entering the listings of the American Birding Association (ABA) was claimed by Alan Greensmith. On 27 October 2001, Alan confirmed to me that his world list was indeed likely to be the highest in Britain, at over 7,160 species or 76% of the current assessed diversity of the world's avifauna. With 32 years of Odyssean journeys behind him, Alan is not resting on his laurels. Still he constantly updates the world's list for discoveries of new real or newly split species and researches countries to be explored and revisited. Already he has been to over 60 nations. This span is matched by Graham Speight, another supreme birding itinerant, but neither he nor anyone else appear to come within 600 species of Alan Greensmith's global haul. Surprisingly not one of the birders cum world twitchers so lauded by Mark Cocker in *Birders: Tales of a Tribe* (2001) comes near their performances. Even more strangely, Alan is not even mentioned in the Cocker anthem to the hedonistic pursuit that on its biggest stage he virtually pioneered.

With no tinge of regret, Alan Greensmith admits to the lack of any formal publication of his myriad observations, but for many others, an expedition remained incomplete (and unsatisfying) until its report and any other necessary papers had been written and published. From one man in particular, you usually got a book as well. His name was Guy Reginald Mountfort. While other stalwarts like Peter Scott, Bill Bourne and James Ferguson-Lees took part in purposeful expeditions

One of my eight Austrian lifers was a cock Red-breasted Flycatcher. So charming is this species that the next one is as good as the last (and first).

Just occasionally a camera catches the spirit of an avian being. Here a Red-breasted Flycatcher looks demure, flicks its wings and cocks its tail. DIMW.

"I read Mountfort on Donana and was very gripped by the book but was less taken by the three members of the expedition that I met at the time. They all came over as rather self-satisfied." Colin Bibby, *in litt.* (2000)

It was difficult for Guy Mountfort's early expedition members not to show some elitism. They had been picked for just that.

"I recall a talk by Richard Porter on the Bosphorus. I was taken by the fact that he could drink as much beer as we could and was prepared to talk to us after his lecture. (I guess I am about ten years younger than Richard and ten is a great many years at the age of twenty.) I was amazed at the study in what seemed a very far off place." Colin Bibby, *in litt.* (2000)

as early as 1949, their post-war resurgence as a much-envied genre of field ornithology was one of Guy's main achievements in a private life devoted to birds and conservation. Publically, Guy was an advertising man and when I met him first in 1952, it was not due to any ornithological fate but rather to my father's appointment of his firm Ogilvy and Mather Ltd as the creative agency for Macfisheries Ltd. After a shared visit to Fair Isle in September in 1954, however, I went on to enjoy in return for three decades of various labours the benefits of knowing one of the boldest fixers in all of British birdwatching. If he wanted something, Guy did not creep about; he started almost invariably at the top, calling freely upon kings and viceroys for support and working on down only when absolutely essential. Honed in an incisive industry, his clear vision in prioritising objectives and his ability to organise supporting logistics produced a remarkable series of conservation-minded expeditions. From 1952 to 1967, Guy pioneered multi-disciplinary explorations of remaining wildernesses – World Heritage sites in modern parlance – from Spain east to Pakistan, specialising in creating media-worthy photography of their achievements and so ensuring maximal publicity for them. The Mountfort saga ended when the first of the WWF campaigns to 'Save the Tiger' required his concentrating presence and it was sadly ironic that Guy's longevity should have encompassed another even-greater crisis in the great striped cat's right to exist. Of his expeditions, the three to the Coto Donana in 1952, 1956 and 1957 were instrumental in setting a new standard for the practice of individual natural history studies within an attempt to describe an entire ecosystem. The list of expedition members stands as a roll-call of the remarkable breed of field ornithologist and bird photographer available for service in the immediate post-war period. Guy's selection of such luminaries as the Viscount Alanbrooke, Sir Julian Huxley, Max Nicholson, Roger Peterson and Eric Hosking was in no way accidental. Pulling strings as high as the naval sheets attached to Lord Mountbatten, he delivered a team effort in bird and habitat recording that was soon the talk of all alert birdwatchers. Guy may have gilded the occasional lily in his writings, but his instincts for science and conservation were totally sound. I remember feeling considerable envy for those who went to the Coto but having assisted with the early revisions of the original *Field Guide*, such mean emotion was completely dispelled by an invitation to join Guy's 1963 expedition to Jordan. The chance to join some desert-loving English and also importantly to test 20 years of field study practice in a real expeditionary context was irresistible. An exposition of mature ornithology beckoned and was eagerly consumed.

Initially it looked if the cultural and conservation promises for Jordan of the two Mountfort expeditions in 1963 and 1965, and a third conducted under the International Biological Programme in 1966, would be fulfilled. By 1970, however, a National Parks programme for all of the country and a Desert Research Station at Azraq, the latter briefly directed by Bryan Nelson in 1968-69, had sadly foundered in the wake of continuing conflicts of the Middle East. In the last three decades other human priorities have had to be served and the "perfect paradise for birds", as Meinertzhagen described the Azraq oasis, has become almost completely dessicated. Even with some recent reduction in water abstraction, the chances of

the freshwater pools and reedmarsh regenerating to anything like the magical wetland of the 1960s and before are remote. So most of the Azraq story is made up of dying memories, most enshrined in Guy's third and, to me, best book *Portrait of a Desert* (1965), a series of papers in the early issues of *Sandgrouse*, the journal launched in 1980 by the Ornithological Society of the Middle East, and the actual field registers deposited in 2000 in the OSME archive.

After the Azraq adventure, it was mostly to OSME and its predecessor the Ornithological Society of Turkey that people looked for more news of distant watches and even more intrepid expeditions. Chief among the early exploits were the re-investigations of soaring bird passage over the Bosphorus. This had been first noted by Belon in the 16th century, but was not even sampled until Ian Nisbet and Chris Smout did so in 1955 and not completely counted until 1966. Then followed notable explorations of central and eastern Turkey and of Iran through the 1970s and early 1980s, and of Yemen from 1979 to the early 1990s. Among the leading observers and scribes that visited such relatively impenetrable regions were Richard Porter, who, together with Ian Willis and Steen Christensen, also published a remarkable insight into raptor identification, Duncan Brooks, Rodney Martins, Michael Jennings, Michael Rands, the irrepressible Phillip Hollom, Lindon Cornwallis and Derek Scott. The last pair's adventures in Iran were the stuff of legend, with Lindon earning a PhD for dividing Wheatear ecologies as no one else, and Derek counting more wetland birds than a whole platoon of Wildfowl Trust members. Thus was a remarkable harvest of habitat descriptions and ornithological facts gathered in. The compilation entitled *Important Bird Areas in the Middle East* (Evans 1994) stands as a splendid monument to all who have watched birds purposefully in arid lands.

The choice of just one individual to exemplify more fully the contributions that came from those mentioned in the last paragraph is not too difficult. It has to be Richard Porter, an urgent field ornithologist and bird conservationist who has resumed after long service to the RSPB his prefered role of edge explorer and communicator. Richard first yielded to wanderlust in 1966, going to Turkey for eight solid months. It had taken him two years to save enough funds to reach

Another spring day ends at Qa El Azraq, Jordan, and its shallow, food-rich, saline water provides migration fuel for White-winged Black Terns and Ruffs (nearest), Red-necked Phalaropes (behind terns) and Black-winged Stilts (behind Ruffs).

On the eastern scarp of the Dead Sea, Jordan, Long-legged Buzzards (lowest two) breed at a time when their Asian cousins are still migrating north.

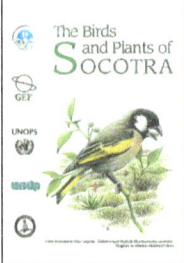

The covers of the booklets initiated by Richard Porter and other OSME members to bring an appreciation of birds to the children of Yemen and Socotra. The Environmental Protection Council of the Republic of Yemen endorsed the enlightening exercise.

Among the British teachers at Ibadan University, Nigeria, was John Elgood. It was he who described in 1958 from specimens collected in a campus garden a new weaver, the Ibadan or Elgood's Malimbe.

long-dreamed-of habitats and birds. Like the others of the 'pack up work and go' brigade that started to shrink the ornithological world anew in the late 1960s, Richard still glows when he recalls the releases of pre-pumped adrenalin that came during first meetings with superb birds such as Long-legged Buzzard and Desert Finch. He is most proud, however, not of his list length but of his part in the development of The Ornithological Society of the Middle East and particularly its two expeditions to Yemen in 1985 and 1993. The last showed how small but energetic teams could enjoy wonderful birds and make important conservation findings, especially in the prescription of Important Bird Areas (IBAs) in countries rarely penetrated and often politically atourist. Richard's fondest achievement was to initiate an educational programme for native children. With OSME's and other sponsors' help, two succinct and well-illustrated books were written in Arabic and sent to all the schools in Yemen and Socotra. They formed the backbone of a two-year learning project which was in turn supervised by Yemenis trained in conservation interpretation by BirdLife International. Richard's concept for this scheme is worth quotation, "Without a simple book in the language of the country – like our own *Observers' Book* – you have nothing on which to hook the promotion of birds, the recreation in watching them and the reasons to conserve them". In January 2002, I was delighted to see a copy of *The Birds of Yemen* pass from the hands of Andrew Grieve, then chairman of OSME, to those of a mighty adult Zanzibari in charge of a 'Sun Farm' and its 1,600 Holstein cattle in northern Oman. As that place just happens to double as the single most bird-magnetic habitat in Arabia, a friendly, better-bird-informed manager made every sense!

In distinct contrast to the buoyant amateur expeditions now increasingly assisted by the new regional bird clubs and other international bird conservation budget-holders, the endeavours of traditional institutions like the British Ornithologists' Union and the British Museum of Natural History have faltered. A flurry of work on oceanic isles at the time of the BOU centenary and some fascinating discoveries from African mountains caught some imaginations but, in general, financial restrictions have cost the few remaining columns of the old establishment the catalysts of ornithological exploration. In one respect of global ornithology, however, the BOU does deserve praise. This is in its faithful publication of a series of check-lists for islands, countries and regions that would otherwise lack organised ornithological history. This series is particularly strong in its description of west African avifaunas but its total span of over 20 titles takes in or will encompass other places as far apart as the Bahamas and the Philippines. A good example of a really informative BOU Checklist is *The Birds of Nigeria* (1994). This sketches in the early ornithology of Africa's most populous nation, which from 1872 featured only ten collectors until the arrival of the redoubtable Dr. W. G. Serle in 1930. It then demonstrates that the increasingly cohesive description of the avifauna from 1945 to 1993 depended on a hardly greater number of expatriates, mainly in district service or education. Sadly but not unusually for a former British territory, only two Nigerians (from a nation of about 96 million people) are named in the total of about 90 observers and conservation-minded people credited with the latest summary of bird records.

The individual thrusts of small groups of explorers can still promote real excitement and admiration. One of the mysteries of the recent membership of the Cambridge (now, to the distress of its elders, Cambridgeshire) Bird Club was ended in November 2000 when, at its '75th' anniversary, the missing university cadre re-appeared in their new guise as Cambridge Student Birders (CSB). They went on to dazzle their more staid birdwatching siblings with news of extinct species rediscovered in the Philippines and new ones found in Colombia. Equally as impressive as their birds was the CSB's use of the new information technology for targeted fundraising in the offices of corporate sponsors. Another important

Chiffchaffs throng in the mangroves and other green cover of Oman but some look very different from British breeding birds. Could they be Mountain Chiffchaffs? Research will continue ...

Except by rare invitations from state ornithologists, no British observers could venture into Siberia before Birdquest's trail-blazing tour in the late spring of 1980. The joy of seeing such gems as Red-flanked Bluetail (fore), White-throated Needletails (in the sky) and Swinhoe's Snipe (middle right) was unconfined.

dimension added since 1980 has been the re-penetration of Siberia and Central Asia. With vagrancies now known to stem from as far east as Lake Baikal, the opportunities to study the cockpit of the central and east Palearctic avifauna have been taken with relish by groups as disparate as some dedicated RSPB field managers on busmen's holidays and the fiercely independent ringers of Flamborough Head. The foremost of the latter, Andrew Lassey, has developed a formidable reach through Russia and the Asian countries to the south and, with a combination of observing, trapping and recording skills, has honed effectively a new kit for field taxonomy. Since 1998 as described earlier, this has been deployed alongside the expertise of no less an authority than Lars Svensson, who is still intent on improving his 30-year criteria on passerine identification by the direct investigation of breeding communities, not just of another isolated vagrant. It also bears repeating that for the remaining Soviet ornithologists, skill exchange, the offer of scarce equipment and cash employment will bolster their security and the promise of more international conservation.

The view south from the escarpment at Estartit, Costa Brava, Spain, in May 1960, long since obliterated by excessive tourism. Depicted birds include Spectacled Warbler (bottom left), Bonelli's Eagle (top left), Alpine Swifts (top right) and Crag Martin (bottom right).

Expeditions have always been the stuff of far-flung ornithology; they also make for great slices of life on or near an edge but they are not to everyone's taste. Not long after the emergence of the package holiday in the late 1950s and 110 years after the first tour of the Pyramids, a marketeer had the thought of selling the satisfaction of a special interest in addition to the other goals of mostly Mediterranean sun, sea, sangria and sex. So it was that the original *Horizon Holidays* decided to bring their 1960 season forward by offering birdwatchers a fortnight in late May at the Club el Catalan, Estartit on the Costa Brava. It cost £32 per head, all found (£428 today). The drummer-in and leader of customers for *Horizon* was Clive Minton, already well known as a leading ringer and with a wide circle of acquaintances. With my envy of those who were exploring the Coto Doñana already high, Clive had no problem with my Iberian seduction. Complete with our new wives, their eyes aglow with the thought that birdwatching might for once benefit their overall tans, we and 12 others launched ourselves into the first-ever bird tour. Among its benefits was to be a bus for nearly daily use. What bill *Horizon* eventually received, for journies as far as Barcelona to watch the incomparable Paco Camino in the bullring and more properly to the western and central Pyrenees for montane birds, we never learnt nor cared but it must have constituted a steep learning curve in bird tourist behaviour and its associated cost. Such was the wealth of the north-east Iberian avifauna, and of our tales of it, that *Horizon* got at least four more years of birdwatcher custom but alas the original company went down in the first-ever convulsion of the package holiday industry. More importantly, for Spanish ornithology, all the leading observers over the five years co-operated in a description of the birds of the Costa Brava. Bryan Sage saw it through to the pages of *Ardeola*, the Spanish journal equivalent to *British Birds*.

Cruelly, like so many other demi-paradises discovered in the 1960s, Estartit and the wonderful five miles of coastal semi-wilderness between the village and the Rio Ter has long since disappeared under the avalanche of concrete that facilitates mass human kippering and has obliterated much of the Spanish coast. Where once 'flying dragons', alias Great Spotted Cuckoos, flew, no such birds exist now for British or other European birders to enjoy. Nevertheless the increasing exploration of other bird-rich destinations has provoked the development of an amazing industry of bird travel. Its essential service of guidance conveys almost guaranteed and certainly instant step-ups in the experience of first the European and nowadays almost all of the world's avifaunas. Following on the early lead of Lawrence Holloway, who initiated the first specialist company *Ornitholidays* in 1965, offering for example the early wonders of Scilly birdwatching to more than the St. Agnes pioneers, the number of British and international companies providing bird tours at home and abroad has grown to at least 90. Required to declare likely bookings to secure ATOL insurance bonds but not yet committed to sharing sales statistics, the individual companies seem unsure of their total marketplace. From a sample of eight actual performances for 2000 and 2001 (to October), I estimate that the annual total of tour places sold from Britain is close to 10,000 and that the mailing lists of current and recent past customers contain close to 77,000 British and foreign names and addresses. Thus the habit of being guided to new species in exhilarating places may now be more common than that of joining local bird clubs and county societies, a sign of the times of birding rather than purposeful birdwatching.

There is general agreement that *Naturetrek* is the market leader in terms of 'bums on seats', specialising in aggressively priced and short-timed holidays. Overall, *Naturetrek* takes about 2,000 people on 200 tours to most of their advertised destinations. Many of the longer-timed and considerably more expensive tours are executed by *Birdquest*, *Limosa*, *Ornitholidays* and *Sunbird*, who together take about 1,950 people on about 200 of their advertised trips. It follows from these two measures that the average overseas tour run by the five top companies consists

One of the birds sought at Estartit in 1960 was Cetti's Warbler. Amazingly in the next year, the first to reach Britain appeared at Titchfield Haven, Hampshire. In a fitful colonisation, the species spread and in 2000, almost 700 males sang in 22 English and Welsh counties.

Ibisbill Tours

2004/2005

The cover of the Ibisbill Tours brochure promises customers the sight of a swarm of Sandwich Terns. The company's founder and chief guide is Algirdas Knystantus, a Lithuanian most famous for writing *The Natural History of the USSR* without the permission of the then government!

of about ten 'punters' served by one or two leaders and supported by a varying number of 'ground agents', to introduce three terms from the lingo that the industry has spawned. According to George Bennett, a veteran leader of over 150 tours, Paul Willoughby, the developer of *Bird Holidays* (from the ashes of the ailing RSPB travel service), and Algirdas Knystautus, the highly cultured Lithuanian creator of *Ibisbill Tours*, the marketplace is now "highly competitive" and "close to saturation". An increasing lack of young customers is causing some alarm. To counter this situation, most companies do offer "personal satisfaction with the whole holiday" as the premium for customer loyalty, particularly among the elders who can currently afford unusual leisure pursuits and provide most new and nearly all regular custom.

After 22 years of pioneering the toughest targets, *Birdquest* appears still to be able to sell hard labour as the necessary pre-requisite for some extraordinary birds, though some softer options are now also offered. As a member of *Birdquest*'s first tour to Siberia and Mongolia in May and June 1980, I was discomfited to be asked to acquaint Mark Beaman with the rest of the group's growing displeasure at the succession of strenuous days but I was totally astonished when real customer dissatisfaction was instantly dismissed as "the usual, two-thirds of the way through mutiny". There are many other tales, even legends of tour vicissitudes but my only real concern about the operators is to question whether they do enough for conservation. Most do willingly offer the paybacks of donations but is their stimulation of local people and bird care sufficient? Do all the records so keenly argued at each evening's log actually reach the national authorities who could use them best? Otherwise for every mishap, there are many more reports of enormous satisfaction. Witness the customer who has bought 50 tours from *Ornitholidays* in 35 years, the lady who has travelled with *Bird Holidays* 14 times in the last four years, dragging along in a rare reversal of positions a husband in her eager train, and the endless joy presented by the Desert Sparrows which inspect camel dung beside an outpost café in southern Morocco and have been more adored than any other of the world's 'speugs'!

Individual expenditure from such people and upon such encounters reaches £20,000 per annum. The number of countries on all the continents reached by the foremost of Britain's tour companies can be as high as 85 and their individual avian delights are described in glossy brochures of considerable detail and often fetching design. Prices per person ranged in 2001 from £180 for a winter weekend on a south coast estuary to a thumping £5,400 for swapping hemispheres and cruising the sub-Antarctic islands for a fortnight. If my estimate of 10,000 holidays bought in a normal year (before the terrorist attacks on America depressed air-travel from there) is close, then the total British marketplace for bird-inspired, packaged travel may be valued at between £9.5 and £11 million, a long way up from the £500 (about £7,000 today) spent by *Horizon*'s 16 guinea pigs in May 1960!

For Desert Sparrows, best find first defecating camels.

Ecotourism may yet stave off the accelerating loss of wilderness. Below: young Ugandans come to Britain every year to show the value of their wildlife to all comers. DIMW.

"Together for birds and people" is the by-line of BirdLife International. No other body has done more to catalogue the fortunes of birds around the world, witness this display at the British Birdwatching Fair. DIMW.

A wilderness gone forever? The ancient water of the Azraq oasis goes now to thirsty conurbations in the Jordan highlands. Tens of thousands of birds have lost their "perfect paradise" (Meinertzhagen).

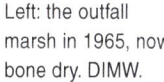

Left: the outfall marsh in 1965, now bone dry. DIMW.

Right: the Shishan springs in 1965, now much reduced.

A wilderness that is safe? No oil lies below the Wadi Sirhan on the Jordan/Saudi-Arabian border. DIMW.

In a bewildering white environment, every bird and other organism is a desert specialist.

Britain is blessed with many seabirds. They attract some of the most intrepid of all purposeful bird watchers.

Above: watching from the prow of RMV *Scillonian I*, 30th September 1961. Author on left notes observation by Peter Colston; in between Karin Wallace. Note Fair Isle sweater in its 8th year! REE.

Right: seawatch position above bridge of the *Ocean Bounty* (chair lashed to rails). DGC.

"You've just missed (a great bird)" is the sort of comment to expect when you come late to a seawatch. Andrew Lassey (top) and John Lamplough know that there are very few second chances with a seabird. DIMW.

Author 'mapping' extent of Gannetry at Roareim, 17th June 1993. DGA.

For long hours of numbed buttocks, seawatching is just the thing but when the skuas and tubenoses ccme in close, there is no pain. A brave dawn comes up at Flamborough Head. Andrew Lassey (with telescope) and the author sample the offshore passage. Philip Sayer.

For massive passages 'right under your boots', Kilcummin Point in Co. Mayo is the place. Willie McDowell was the first observer to spot the amazing seabird trap of Donegal Bay. AMcG.

Pioneered by Stephen Foster, Rocky Point, Co. Donegal, is a vicious place in horizontal squall-born snow but when phalanxes of Great Shearwaters stream by, nobody minds. AMcG.

The Booming of Scilly

The early scores of Britain's north and south-west poles – the early 20th century lagphase in Scilly – the pioneers of St. Agnes – Britain's first Northern Waterthrush – an arrival on St. Agnes – the welcome at the Lighthouse – the union of island and observatory – an Edwardian lady – an inheritance of care – wider explorations – the escalator to the full Scilly season – near-thumping controversy – retribution for closed minds – a rising tide of twitchers – incomplete science but fun for all comers – the 1999 miracles – the polar contest for new species drawn

Prior to its slaughter in North America in the late 19th century, the Eskimo Curlew (centre, landing) occasionally crossed the Atlantic as still do American Golden Plovers (all other figures). The last British curlew fell to a Dorrien-Smith on Tresco, Scilly, on 10th September 1887.

The Bluethroat is a passage migrant most associated with the east coast but one was seen in the 1956 exploration of Bryher, Scilly.

Mere narrative is not normally the stuff of modern ornithological journals and had it not been for the need to re-enthuse the subscribers to *British Birds* in the early 1970s, it is likely that my essay on "An October to Remember on St. Agnes in 1971" would never have been written. Once published, however, its mixture of self-confession and avian excitements became the anthem of what today are thought to have been 'the early days on Scilly'. The reality is that the ornithological promise of Lyonnesse had been signalled not 30 but 170 years ago.

Intrigued to discover how the north and south-west poles of Britain had contributed to the collector-ornithology of the 19th century, I checked Keith Naylor's modern catalogues of rarity records and discovered to my surprise that the Isles of Scilly had produced in that epoch 34 particularly significant observations. Before 1900, the much-longer arc of the Shetland Isles had contributed only 16 exceptional occurrences and Fair Isle not one. Why then did the early collectors and migration students go north and north-west? The answer has to be their preoccupation with the Heligoland saga, so quickly enhanced by the early Scottish and Irish returns from the coastal lights scheme. No general motivation for the explorer-collectors to go south-west with shotgun and dust shot is discernible in rare bird records.

In Scilly, only the dynasty of military lairds on Tresco, the Dorrien-Smiths, and a few other sharp-eyed islanders joined in the hunt and secured the occasional prize bird. Furthermore, their finding-rate slowed markedly after 1895. Thus there was no widely communicated consciousness of Scilly's potential to show migration and vagrancy in the first half of the 20th century. The reluctance of the more hidebound reviewing ornithologists to accept records of birds that may have enjoyed assisted passage (by ship) drew a further smokescreen over the promise of the Atlantic Approaches.

After the Second World War, a few Cornish observers, notably the redoubtable A. G. Parsons of all-weather raincoat and small Nelson-type telescope, made regular forays to Scilly, but still the records did not flow. I well remember the verdict of David Ballance and Geoff Scott of the Cambridge Bird Club who scoured Bryher in September 1956, "Beautiful place, the odd good bird like Hoopoe, Bluethroat and Lapland Buntings but nothing special; we shan't go again". Amazingly a fall of 100 Ortolans earlier in the same month failed to catch national attention. So who did change utterly the autumn geography of British migration study and rarity hunt and begin the booming of Scilly as a birding station that would attract fame across the entire Holarctic region, let alone Britain? The answer was an exemplary group of young Londoners, reared by the Ornithological Section of the London Natural History Society, more self-taught in field ornithology than most of their wider-flying peers, used to self-finding hunts in Norfolk, particularly inspired by Richard Richardson, and already devoted to serious studies such as the husbanding of the New Forest's precious Honey Buzzards. Anxious to deploy the new observatory strategems of migrant census and mist-net on a new island, John Parslow and a handful of other observers, particularly associated with Beddington Sewage Farm in Surrey, chose to pioneer the nearest isle to the Bishop Rock lighthouse, St. Agnes.

Used to bird observatories typified by gaunt Heligoland traps on treeless vistas, I can still recall the shocks provided by the group's illustrated report on their first spring visit in 1957. Every scenic slide featured flowers and tall, sheltered hedges; the cover was unbelievably thick. How could birds be found in such verdant confusion? The answer, by getting to know a small isle like the back of your hand and hunting it from dawn to dusk. When on 30th September 1958, Bob Scott and Geoff Harris found a strange plump passerine that wagged its tail but was not a pipit on the seaweed wrack of Covean and later untangled from their mist net, stuck in large shingle, the Old World's first Northern Waterthrush (from North America), the long-jammed Scillonian penny dropped. Suddenly Fair Isle, Cley, the other early observatories and the south-east English fleshpots acquired a serious competitor in the south-west.

News of even extreme rarities spread slowly until the late 1960s and so there was no instant flocking of observers to St. Agnes and Scilly. It took the combination of repeated invite and sharp arm-twist to get me and my first wife Karin there. Loyalty to a LNHS initiative and the lure of the ebullient Brian Pickess, as bird guide and island ambassador, tipped the balance. Surviving the sight of grockles being seasick in time to the waltzes played by the evil skipper of the flat-bottomed RMV 'Scillonian I' and sorting through the bewildering scrum of boats in Hugh Town harbour, we finally tumbled into the official inter-island launch Tean and reached the eastern quay of St. Agnes in the late afternoon of 25th August 1959. Aided by two scientists scarily intent on researching the fall-out of metal flakes from the upper atmosphere onto a clean oceanic isle, our rucksacks were transferred to a tractor and trailer driven by a small dark Celt of a man. "I'm Lewis Hicks", he said, "Welcome to St. Agnes. I'll take you to Lower Town. Brian will do the rest." Clinging somehow to the rear of the tractor, we bounced up, over and down the then roadless spine of St. Agnes, passing a sublime lighthouse, and were disgorged into the old farmhouse that served as bird observatory and scientist lair. It nestled in *Pittosporum* and tamarisk hedges and overlooked a cricket pitch and a sparkling, rush-edged pool. In bounded Brian and our kit thrown hastily on to our bunks, his first order of "Come on; round the island" was not to be disobeyed. Birds flew everywhere and as every corner turned or breast topped brought yet another fabulous aspect of ancient rock, heathland, flower field and Atlantic ocean, we succumbed heart, soul and head to the spell of St. Agnes. My diary awarded her

"There were good birds before the waterthrush – a Little Bustard on Gugh in March 1958 and a Rufous Bush Chat also in spring – but if you remember, that year saw the start of the Rarities Committee. I never liked that set-up and several early records for St. Agnes were not submitted." John Parslow, pers. comm. (2001)

Seventeen years later on 29 October 1975, St. Agnes delivered a second (and first official!) Little Bustard but has yet to produce another Rufous-tailed Scrub Robin, to use the bush chat's third English name in the last 60 years.

The spring magic of Scilly as captured on a *British Birds* cover for April 1977. As fog clears, two Swallows and a Hoopoe find the archipelago.

instant "gem of an island" status and for the next ten years she was never once to disappoint us.

The magic of St. Agnes was multiple and the few tens of regular observers that manned the observatory during its 12 years of life have never ceased to acknowledge their debts to the place and its people. It was as if God had willed that there would be a perfect antidote to the captivity of London and the frequently harsh weather and often few birds that the other observatories were wont to deliver. Our chief privilege was granted by the isle's leading family, the Hickses of the old Lighthouse. Not content with giving the observatory a whole farmhouse for its headquarters (at an initial and unchanged cost of 4/6d per person per night), Lewis and Alice adopted 'real birdwatchers' as extensions to their own family of three children, offering friendship and local wisdom without hesitation and asking only enthusiasm and conversation in return. So it was that the visitors to the observatory learnt about rabbiting, Scilly's endemic shrew, gig-racing, flower farming, a world-wide web of relatives, the politics of the Duchy of Cornwall, the vicissitudes of the Steamship Company and, just occasionally from Alice, some delicious titbit of Scilly gossip. Add to that the provisions of a sturdy boat rightly named 'Undaunted' for trips to Annet (by day and night), superb Guernsey cow full-cream milk, free bulbs, the *Horse of the Year* show on TV (gently undulating to the rhythm of the generator) and home-cooking at least once a visit and hopefully some sense of the very special welcome conferred upon birdwatchers will be apparent. Few of us ever saw the Hickses and the other island farmers at the back-breaking work of flower-picking, for this was at its peak in later winter and early spring, and again it seemed a God-given favour that they all had so much time for us in the late spring and long autumn. Only the supposed honour of assisting the cricket team in one of the mean inter-isle matches gave cause for discomfort.

Mention of a superior being reminds me of the feudal grip still partly exerted upon Scilly by the lairds of Tresco in the late 1950s. One of Commander Tom

In the early autumns of the St. Agnes Bird Observatory, Aquatic Warbler (right) and Sedge Warbler were seen in the same cover.

Dorrien-Smith's fishing rights was to net the Cove between St. Agnes and Gugh. On 6th September 1959, this exercise marshalled on the Bar not just the Commander's markedly upper class entourage but all able-bodied islanders and press-ganged birdwatchers. The Trescoites took a gig and strung a net across the southern cove; the St. Agnesians were split into two teams, attached to the net's end lines and commanded to heave. As the strain was taken up, a whisper came down the lines. "Don't forget to fill your wellies." The pro-gustative wisdom of this initially incomprehensible injunction became silvery-clear when the first of the incoming gill-strung fish was whisked from the net and poured down a boot. The repeated slights of hand and lower leg led to the Commander snorting audibly at the depleted spill of the net's central span. The pressed gang cared nought, especially as he had not endeared himself to Renton Righelato, the observatory's keenest ringer, by insisting that his ornithological *droit du seigneur* entitled him to the observatory's next caught Aquatic Warbler, a species not represented by a skin in the Abbey collection. No such outcome occurred, for none of the three trapped Aquatic Warblers of a remarkable five seen between 27th August and 18th September did anything but fly on, while the observatory cuisine improved for several days.

In the air rakish and impressive, on the ground lumpy and eyeless – the two guises or "jizzs" of the St. Agnes Nighthawk of October 1971.

Other conscious support for the observatory came from Don and Carmen Hicks of the Post Office and Shop, Dick Legg, whose lane and tamarisks were among the best for birds on the isle, David Peacock, who could not swim but was the island creeler and would take us to Tresco, and particularly Gordon and Ruth George, who worked for Lewis. The last couple lived at the east quay and were much liked for dispensing energising 'birdwatcher's cake' to anyone looking a mite hungry. Once they were about to feed John Parslow a mullet dinner but when a Manx Shearwater came uninvited through the nearest window and added glass fragments to the condiments, all three had to settle for bread and cheese. Due to the then prevalence of eel worm (a scourge of bulbs), two farms were always closed off from searches but with a bit of cunning and a co-ordinated watch on tractor movements, Eddie Hicks' fruit cage – a haven for Yellow-browed Warblers – could be penetrated. Otherwise, the pioneers behaved rather well, content to harvest their own birds and never disturbed by the national news brought in erratically by the weekend changes in personnel.

Although most of the islanders had an eye for unusual birds, only two displayed any real ornithological bent. The first of these was a slim, quiet but slightly mischievous Edwardian lady who lived in a cottage near the west harbour at Periglis. She was frequently visited by similar friends, usually put up (or rather down) in the isle's smallest dwelling, euphemistically called 'The Cabin' but more like a nun's cell. From her one and a bit dwellings, Hilda Quick would glide forth and observe nature; within them, she would care for some delightful possessions, distil most excellent cordials and make usually rather less good wines. Even Bob Emmett, normally not one to have his birdwatching disturbed by polite conversation but equally famed for a house full of stills in Acton, would sip the raspberry with appreciation but it was always a bad time when the gorse tip came round, even if by then the talk was yet again of Hilda's best bird finds. Chief among these was what for 11 years was taken to be Britain's first Blue-cheeked Bee-eater. This bird

The umpteenth launch of eager observers approaches a Surf Scoter north of Tresco, Scilly, in October 1975.

A Blue-cheeked Bee-eater's view of the Lighthouse on St. Agnes, Scilly.

"There were other things in that collection – like an old Saker." John Parslow, pers. comm. (2001)

Another lost 'first' for Britain?

Perhaps David Hunt's greatest find, the first Yellow-bellied Sapsucker to leave America and reach Europe was on Tresco, Scilly, from 26th September to 6th October 1975.

had made her morning fetch of milk, and Lewis Hicks' first visit to Lower Town, a sub-tropical experience on 22nd June 1951. How often did we look along the telephone wires hoping for the second … In fact, their bird was actually that, as John Parslow discovered during his examination of the Dorrien-Smith collection in June 1962. The first had reached St. Mary's 30 years earlier, on 13th July 1921, and had been reported to Major A. A. Dorrien-Smith who, since his collection lacked an adult Bee-eater, asked for it to be shot. The corpse went to a Brighton taxidermist for mounting but neither Pratt and Sons during its preparation nor the Major on its eventual return had spotted the difference. Hilda was never bothered by such twists of ornithological fate, however, and went gently on her way, doing useful things like editing the new Isles of Scilly Bird Report for its first five years or capturing on film for her own amusement the growing sizes of twitcher crowds. "For the rogues' gallery", she whispered to me, as she snapped some serried ranks enjoying a Radde's Warbler on 13th October 1973. Like H. G. Alexander, Hilda was a link to a quieter, less frenetic but no less enjoyable form of birdwatching. It seemed fitting that those who attended her funeral in the autumn of 1978 did so in the company of Britain's first ever Semi-palmated Plover, found below her house by Paul Dukes on 9th October.

The second islander drawn to birds was, and indeed still is, Francis, the younger of Lewis Hicks' two sons. For 40 years, the local continuum of the St. Agnes birding saga has come from him and the other veteran visitors who have remained loyal to the island. Inheriting his father's work ethic and fully appreciating his luck in living his life in a beautiful place, Francis remains one of the two last farmers on St. Agnes that still grow flowers successfully – against the vastly increased competition of mainland glasshouses and new bulb genetics – and maintains a steady eye and mind on the isle's birds. Past these attributes flow the succeeding hordes of birders that now come to St. Agnes driven by today's instant communications. Francis is also the main conduit for the superbly nostalgic news of the early pioneers who, though they have largely shunned Scilly in recent decades, are beginning to revisit their favourite isle of great birds, real and deliciously imagined.

St. Agnes is not all of Scilly, however, and when in 1969 Lower Town Farm became unfit for occupation even by birdwatchers, the days of the observatory quickly became numbered. A search for alternative accommodation led nowhere and as the observer community became perforce scattered into the isle's holiday cottages, much of the old corporate spirit and discipline were lost. The migration seasons were just as exciting; the great rarities continued to appear but the escalator towards the full exploration of Scilly and the amazing boom in October traffic had started to move. The generally accepted view is that the modern 'Scilly season' virtually exploded into life in 1975, when the arrival of Britain's first Yellow-bellied Sapsucker on Tresco prompted the first overnight invasions of twitchers but, in my view, its origins go back to 1964, the year of David Hunt's appointment as fitful gardener to the Island Hotel on Tresco. David, a trombonist of the last days of the jazz era and irresistible ornithological hedonist, soon re-opened an alternative axis of discoveries from the archipelago's richest habitats. For the first time, a sense of having to look regularly over one's shoulder (and not regard a visit to another island as simply a day off) began to eddy round St. Agnes. On 16th October 1966, Bernard King asked David to investigate a tiny sprite lurking in the sallow screen of the Great Pool. It proved to be Britain's first-ever Parula Warbler from North America and overnight a new urgency in rarity collection was raised. As David recounted in his racy and revealing *Confessions of a Scilly Birdman* (1985), a small scrum of St. Agnes birders led by Ron Johns hired for the first time a special boat for a three leg trip and, braving a westerly gale, secured yet another Scilly smasher. The bird was even celebrated in a reprise by Paul Holness of a Rolling Stones rif. "Have you seen the Parula, baby, lurking in the sallows?" was sung ad nauseam to those who came too late but, 35 years on, the phrase provides a charming echo of all the wonderful popular music of the time and also the growing ecumenism being shown by the changing tribe of birdwatchers. Newcomers like David Holman and Ray Turley proclaimed themselves ornithological illiterates but the former went on to be a member of the Rarities Committee and Birdline voice, while the latter became a most accomplished bird sculptor. More than birds fledged in Scilly in the late 1960s.

By 1968 and 1969, the boatmen of St. Mary's were in open rivalry for the transport demands of the growing number of late autumn rarity-intent visitors. David Hunt, by then anxious to leave Tresco and get for his beautiful wife Marianne and his two sons the better earnings available on St. Mary's, became guide to the Seabird Specials run by the independents. In 1968 two notable birds – Scilly's second Northern Waterthrush on Tresco and an amazingly tame Upland Sandpiper that settled on St. Mary's – fuelled further the charges to and between the islands. The first Scilly Bird Report, published in 1970, listed no less than 60 contributors from well over 100 visitors. In 1971, David Hunt began to lead wildlife holiday parties from St. Mary's and even more significantly the Scilly Isles Council and Nature Conservancy responded to the demands for easier land access, opening two nature trails on St. Mary's. The seeming monopoly of good birds by St. Agnes was over. Sadly our theories about that isle being favoured by the proximity of the Bishop Rock light, passing liners (in fast decline in any case), southerly exposure

"By the way, I was interested in your mention of the gait of your bird – I thought Wheatears always ran on the flat, and observation of the birds on the airport and surrounding slopes bore this out. Running, jumping and standing still all seem to be part of the repertoire! An interesting local report, from a thoroughly reliable source, is a Crested Tit last week – have haunted the area, but there are one hell of a lot of pine trees!" David Hunt, *in litt.* (1971)

Sent on a wild wheatear chase, David Hunt's comment on gait contained a typically humorous quip. The Crested Tit might have been England's third but as with another in Staffordshire in May 1999, it remained unconfirmed.

Not the original Northern Waterthrush of St. Agnes but the second for Britain on Tresco from 3rd to 7th October 1968. Oddly pale, its identification was disputed from Canada to Holland and back again.

A Least Sandpiper passes the Bishop Rock Lighthouse and the Gilstone Reef on the way to a Scilly shore. It settled on Periglis, St. Agnes on 4th October 1962.

and even back-of-the-hand knowledge had to be ditched. Rarities were increasingly shown to be ubiquitous within the entire archipelago.

Coming back to St. Agnes in 1971 after a near three-year sojourn in Nigeria, I was astonished by the rise in the number of birdwatchers. Where once a rarity might have been seen by a handful of veterans, any good bird would rapidly attract a boatload of 30 to 50 new faces and it was clear that an all-island search strategy was close to achievement. Thus while St. Agnes was still respected as something of an ornithological sanctum, any sense of its experienced observers exercising any real leadership over the archipelago had largely gone. This collapse in discipline was never more apparent than during the still-famous controversy over the identity of a smallish crake that haunted the Big Pool from 26th September to 9th October 1973. The basic cause of the dreadful affair lay not with any observer but in the imperfect illustration of Sora Rail in a then popular American field guide. Accordingly, in spite of 12 days of sightings, an increasingly heated debate had raged on and on, with a majority insisting that the bird had to be just a Spotted Crake. So it might have stayed but for the courage of Dave Smallshire and the brothers A. R. and B. R. Dean who, back in the calm of their home patches, had looked at other references and rang the Turk's Head (once the Georges' cottage but by then the pub on St. Agnes) on the evening of 7th October to urge redoubled effort. It was just my luck to be called to the phone and be stuck with leading the minority opposition of one. Girding myself with the ghostly vestiges of my former chairmanship of the observatory, I spent most of the next two days sitting in the rushes, waiting for the bird and praying for minds to re-open. The last human condition did not occur. Two senior St. Agnes observers with North American experience crept up and whispered support for the Sora, only to add "but don't say I said so!"

In the afternoon of the 8th, the arrival of Dave Flumm, Mike Rogers and Harry Robinson from St. Mary's allowed the manning of two independent vantage points and we settled to a long, patient watch. Mark by mark, we built up a firm case for Sora and when finally we were all sure that a black line did really show under the bill base, we exchanged upwards thumbs and relaxed. Later, I strolled down to the Turk's Head for the celebration. Unbelievably the furore over the bird's identity was undiminished, a new zenith of abuse being reached when the leader of the Spotted camp Dave Britton accused me of faking my sketches in favour of Sora. I threatened to throw him through the pub's door and would undoubtedly have enjoyed his flight. Fortunately, however, the stricken look on his wife's face dissolved my anger and so I sat still, listening impotently to the unfolding of the next ploy in the saga. Somehow, the Didymi (doubters) had got permission from the St. Mary's vet alias local Nature Conservancy representative, Peter MacKenzie, to trap the bird on the morrow, despite the Big Pool being a SSSI and thus supposedly hallowed rushes and mud. I trudged back to my cottage, hardly content with my birdwatching fellows.

On the 9th, God took two hands in the final game. Firstly, He delivered a cock Little Crake alongside the Sora at 08:05 hrs – an immensely cheering co-incidence – and secondly, He decreed that the evening trapping session would end in total farce. The Sora walked dutifully into the net, was there ignored by the net-minder, none other than Dave Britton himself, and wriggling free of the mesh performed one last flight to the safety of the opposite rushes. Asked what on earth he thought that he was doing, the leader of the Spotted camp could only mutter abjectly, "Sorry, I thought it was a rat". Paul Dukes announced an imminent heart attack and the Big Pool echoed with guffaws of laughter. After its unneeded brush with man, the bird left overnight for places unknown and 14 days of rather

The hard-won, not faked, field sketches of the 1973 Sora Rail on St. Agnes.

The birdwatching equivalent of a shooter's 'left and right' – the amazing coincidence of the Sora with a male Little Crake on 9th October 1973.

No longer officially a rarity, the Little Bunting (fore) is still a delight. It not only comes every autumn but may also stay to winter. If only more observers would work farmland patches as they do Scilly, they might learn that its Robin-like call sounds more often from any thicket or hedged stream that resembles its breeding niche.

"I first went to Scilly in 1938 and was immediately taken by the strange appearance of the Dunnocks there. I tried to get Meinertzhagen to go and see and shoot some but he didn't bother." Richard Fitter, *pers. comm.* (1999)

The first but not the last observer to suspect that the Hedge Accentors of Scilly may merit separation as a subspecies. For a start, the adult males never get as 'purple' as those on the mainland.

bad behaviour went into birding history. The final irony came on the 10th, when Francis Hicks and I (not the Spotted camp) had to face the joint music of a mightily disturbed Peter MacKenzie and a thoroughly wrathful A. G. Parsons, both with responsibility for bird protection in Scilly. We made peace, but I made a mental note that even the ghost of the observatory was no more. Ornitho-politics had finally reached Scilly and muddied all our feet.

The next eight years saw further rapid developments in Scilly birdwatching. Two of the best were Robert Allen's sterling solo survey of breeding seabirds in 1974 and the further improvement of the St. Mary's nature trails, the latter a thoroughly sensible way to order the now fast-increasing invasions of twitchers caused particularly by the soon-legendary 1975 invasion of American birds, of which the most watched bird was undoubtedly a Surf Scoter that fished between Tean and Tresco. When fed up with launch after launch of binocular lenses glinting at it, the duck would fly onto the Great Pool for a kip. By 1976, crowd sizes frequently exceeded 100; an upper echelon of senior twitchers had arisen, with the stars such as Ron Johns and Peter Grant given virtually guru status, and even to the outer isles the national grapevine sent nightly messages. The idyll of St. Agnes was seemingly no more; the three stressful questions of "What's about?", "Where next?" and "Can we get a boat?" were heard everywhere.

In 1977, the peak Scilly crowds were numbered in hundreds. Even Bryher produced almost simultaneously rare birds from the Arctic, Iberia and America, a mix that only a decade earlier would not have been expected from the entire archipelago. One of my log entries for that October reads tellingly, "Eventually when the departure of over 50 twitchers left a puddle clear of human bodies, a Little Bunting came out of the field centre and bathed". I did not realise at the time that this vignette of an observer crush and a confined bird had an element of epitaph in it but when that winter the Tresco agent signalled the end of the cheap cottage lets by announcing that the whole island was going to timeshare, the last reason for me not to concentrate upon the majestic cape of Flamborough Head had arrived. Back to my almost-natal North Sea, I went and in the last 23 years I have been to Scilly only twice and then only briefly. In retrospect, my main regret for the first two decades of the Scilly boom concerns the limited output of the observatory regime. We really should have published 'the book' of the St. Agnes data. John Parslow did most of the preliminary work, producing some fascinating displays of passage periods and patterns. We also located almost every pair of St. Agnes's breeding birds but the really fascinating questions like the unusually brown Dunnocks, which Richard Fitter had also spotted as early as 1938 but failed to persuade Meinertzhagen to collect, and the reality or otherwise of assisted passages still beg for answers.

Ornithological hearsay is as distorted as any other. So I approach the next 25 years' history of Scilly birders and birds with caution, drawing heavily on accounts of that period gratefully received from Francis Hicks and Doug Page, residents of St. Agnes, and Ken Shaw, regular autumn visitor from Fife. Faced with the new crowds, most veterans went eventually elsewhere but their fear that constant excesses in twitching would distort the ornithology of the archipelago proved groundless. A code of conduct for visiting birds published in 1977 held, with only

one purposefully run-over foot to show for the occasional road jams of n bodies, n x 2 human feet, and n x 3 tripod feet. The centralisation of birder stays on St. Mary's, with its wider choice of lets and its roost of launches offering respectively lower bed costs and less nerve-racking boat connections, was followed by the adoption of the Porth Cressa Centre for the nightly exchange of record claims and confirmations. There for some years Mike Rogers, resplendent in a cream linen jacket with drink in one hand and fag in the other, was ringmaster and spell-weaver. The ever-widening searches produced ever more rarities, with Chris Heard near-deified for finding Eastern Olivaceous Warblers in 1984 and again in 1985 and Dick Filby and John Brodie-Good extracting yet another new American passerine – a Philadelphia Vireo – from Tresco in 1987. Over its four day October stay, this bird was adored by 900 of the 1,500 birders who were by then autumn pilgrims to the isle and marshalled by their incumbent Birder-Saint Peter Grant.

At the end of the 1980s and through the 1990s, the expectations of continuing annual feasts were not fully met and the numbers of birders fell steadily, their visits becoming concentrated in the mid-October fortnight. Increasingly, the competitive twitchers kept their powder dry on the mainland rather than risk being marooned in Scilly with the last connecting launch and helicopter gone. Conversely 50 or so regular hunters stayed loyal to the archipelago, working St. Mary's, St. Agnes and Tresco with practised dedication. Their rewards included on 21st May 1988 the first British Caspian Plover for 98 years. In some years relatively starved of land-bird gems, the eyes of observers turned seaward. In a long-delayed complement to the early hundreds of Grey Phalaropes and Great Shearwaters seen from Horse Point on St. Agnes and the torrential passages of Gannets north of Tresco, new off-isle discoveries were made. Nowadays the Scillonian's annual dash for deep water and even hitches on shark boats provide repeated Wilson's Petrels, and at long last in 2001 an indisputable Fea's Petrel was

The most elegant of its tribe, a male Caspian Plover trips across a sand pan in southern Kazakhstan. On the edge of the steppe, three Houbaras stalk by. For the plover to reach St. Agnes, Scilly, it would have flown at least 50 degrees west.

seen. The Long-tailed Skuas and Sabine's Gulls over which we ancients drooled have become almost common fare and once again the ornithology of Scilly has become an even-broader church.

The taxonomic appearance of the birders has changed, too. Where waxed jackets squeaked, now whispers Gortex; where wellies squelched, svelte waterproof boots trip lightly; where heavy bins broke necks or at best thumped chests, lightweight roof-prisms on wide spongy neck-slings provoke no arthritis, and everywhere the all-seeing 'triffids' of coupled telescope and tripod stalk. One full kit worthy of the modern cyber-soldier costs well over £2,000 and as Doug Page tellingly notes, it is so universal that "unfortunately these days one cannot tell a birder's ability by the optics that he or she carries." Furthermore, the annually increasing issue to the troops of CB radios means that not just the peeps of pagers but also the crackles of ear-pieces disturb the once-soft Scilly air. Against all this potential for excessive behaviour, the members of the new Isles of Scilly Bird Group maintain a modicum of order, and rate the manners of the October hordes to be generally good. Indeed, most reported transgressions are committed by one or other of the long-lens photographers who push vagrants too hard in order to sell their images 'by the 100s' at the Scillonian Club, the new evening hub of buzz. All this made for human fascination but one question began to throb. Would the avian cornucopia ever tip out another delirium of rarities?

An affirmative answer came in the autumn of 1999. In a return to classic October form and with the blessed St. Agnes once again the epicentre, the highest quality selection of rarities ever to appear together in Britain drew 2,500 birders to Scilly. Most of all, they enjoyed the first-ever 'left and right' of Siberian and White's Thrushes from Asia and 'the bird of the Millennium', Britain's first Short-toed Eagle, which on its first migration had drifted north-west from Iberia and not south to Africa. First spotted on 7th October by the St. Agnes veteran Paul Dukes, seen next by Tim Cleeves and named in a still-reverberating shout by Ken Shaw, the snake-eater did not skulk like the thrushes but sailed regularly over

"At about 11:00 on Monday, 11th October (1999) with the skies clearing and the day warming, the great bird left the Eastern Isles, circled up to ... 1,000 feet over Tresco, crossed St. Mary's again, and ... (headed) off south-south-east into the far distance". Tim Cleeves writing on the departure of the first Short-toed Eagle for Britain from Scilly in *Birding World* 12:48.

eager eyes, even some on the incoming 'Scillonian III'. Ken's call on the 7th was loud enough to reach without device assistance Gugh, where the assembled masses waiting for the Siberian Thrush frowned, winced, momentarily dithered and then almost to a man rushed back across the Bar. It is said that the "sheer panic and pandemonium" will live in the memories of the witnesses to the charge for ever. At 11:00 hrs on the 11th, clearing skies and warmer, rising air allowed the eagle to circle up to 1,000 feet and set wing south-southeast to France. Meanwhile, the still-untapped potential of every last Scilly islet was once more demonstrated on Rosevear on the 15th. A classic gem of birdwatching was provided by Francis Hicks who took six observers to its covert of tree mallows in which Black Redstarts and Firecrests provided a frieze for a Radde's Warbler from Siberia, one of 12 on Scilly in the month. St. Mary's riposte was another Mediterranean refugee, a Blue Rock Thrush on the 14th and 15th attended by up to 700 again breathless voyeurs. The luxuries of the 1999 season meant that some competitive twitchers had to make three round trips in order not to fall behind in the race. With St. Agnes taking the blue riband, the crush there was total, with paths and road blocked with bodies but as ever, the islanders were very tolerant of the manifest desire and eventual ornitho-orgasms. The pub throbbed and its keepers purred. The boatmen smiled on and on.

The first Radde's Warbler for Scilly was not found until 1971. The biggest *Phylloscopus*, it always impresses but for sheer beauty, Firecrests win every time!

The latest news from Scilly is of continuing evolution in its ornithological order, with the now substantial residency of observers seeking independence from their Cornish links and the whiff of a rift on St. Mary's. Hopefully, however, politics will not spoil the unsung bequest of the magic isles to British field ornithology, the emancipation over 45 years of hundreds of keen observers who have shared the hunts and trophies with almost total joy and no small inheritance of skill and knowledge, and have also added a whole month of late tourism to the Scilly economy. I felt particular if only vicarious delight in 1999 when the latest fantastic flood of rare birds lapped around not just the newest Scilly recruits but also a few St. Agnes veterans of 40 years' service, notably Pete Colston. Last but not least, birdwatchers who manage to fit in breeding with birding are still taking their children, from babes to teenagers, to one of Britain's most magic places, even perhaps England's best.

By the close of the 20th century, the contest between the two extreme poles of British migration study had been fought to a draw. Shetland (including Fair isle) and Scilly have each added 27 species to what I must now call the combined British and Irish list of 548 officially accepted live species (*The British List*, BOU 2000; *Checklist of the Birds of Ireland*, Irish Rare Birds Committee 1998). The new total advances that of the Witherby *Handbook* by about 125. With finds still being made annually, 'splitting' taxonomists in the ascendency, old records being rejected or renamed and debates on categorisation, the list(s) look oddly imprecise compared to the more conservative past editions. At least 25 more taxa have claims to full places and before our dancing eyes, the lure of 600 British ticks is swung ever more gaily. Meanwhile, with more American species than Palearctic ones likely to appear in future, the bets must be on Scilly and the south-west to recover their ancient ascendancy and take the leading baton in the overall relay race for rarities and new species. Eventually the universal law of diminishing returns will apply but, meanwhile, enjoy.

Jolly Jack Tars

Gales of skuas − seawatching − unique perceptions − new identifications − a stormy doctor − seabird stocktaking − new vistas − the great white cape − pelagic ventures − ongoing research − the threat of industrial fishing.

British experience of dramatic seabird passages began suddenly in Tees Bay on 14th October 1879 with the memorable 'skua gale' which created "an amount of interest and speculation amongst ornithologists equalled only by the famous irruption of Pallas's Sandgrouse in 1888" (Nelson 1907). In fact, this first noticed exceptional presence of skuas in the North Sea has been detected as early as the 8th October, when T.H. Nelson himself shot three out of 50 Pomarine Skuas flying north-west past his boat. Word of his booty spread like wildfire and during the following six days, many gunners went out after the birds. On the 14th, between 5,000 and 6,000 skuas passed or settled on the Redcar spur. They were "apparently exhausted from battling against the storm" and "as they afforded little sport, the shooters abstained from firing at them after obtaining a few specimens". On the 15th, the movement continued in the morning and it resumed again in the gale's last burst on the 17th, by which time the birds had also been observed and attacked at Scarborough, Filey, Flamborough Head and Bridlington. The vast majority of the birds were indeed Pomarines but among them were also "a considerable number" of Buffon's (Long-tailed) and "a few" Richardson's (Arctic) Skuas. The force of the gale drove many inland. Large skua passages occurred again, in the teeth of savage north-west to north-east blows, in at least another eight years from 1880 to 1911. That of 1886 was considered by fishermen to be as large as that of 1879. The arch but not totally unthinking collector of British-taken specimens, E.T. Booth, opined that the movements were not unusual. It was only their onshore compression that made them seem so. Surprisingly, in view of the closely associated discoveries of South Atlantic tubenoses and particularly the single-day count of hundreds of Sooty Shearwaters off Flamborough Head by the taxidermist Matthew Bailey, there was no general grasp of the phenomenon by the senior British ornithologists. With the decline of collecting, the early witness to large seabird passage and presences ceased.

In contrast to the old chases after skuas, latter, day seawatching is a relatively passive pursuit. Getting to the watchpoint can be hairy, and eventually a squall will come in to defeat all waterproofing, but once safely ensconced, you just sit on your arse and scan the sea or ocean. Sometimes marine beings pass by, sometimes they do not. The real trick with sea-watching is topographical. Find a watchpoint where onshore wind compresses birds into a half-close stream and then the scanning and staring will present great excitements, even qualitatively scientific ones in that certain identifications can be made. The full quantative assessment of coastal movements remains virtually incalculable in a reflection of the universal difficulty of measuring seabird populations dispersed over heaving water. The commence-ment of modern seawatching is not accurately dated but certainly by the mid-

In the 1970s, there was renewed interest in the skuas of Yorkshire waters. With their identifications then incompletely covered in any book, sketches after sketches were made until their different builds and plumage patterns began to appear. Shown here are Long-tailed (four at left), Arctic (middle top three), Pomarine (three at right) and Great (bottom). The white-bodied morph of the Long-tailed, resembling the Greenland race *pallescens,* was a particular prize.

Occasionally seawatch points can exhibit other arrivals. At Rocky Point, Donegal, a patch of blown scree has recently delivered what may have been the first Dunlin from north-central Canada in addition to Arctic Ringed Plovers and Lapland Buntings.

1950s, its adventure and enjoyment began to be reported in *British Birds*. Initially the by-products of the growing number of post-World War II observers that were pioneering new coastal watch points (and found themselves starved of other migrants ashore), records of seabirds off north-west Ireland, north Norfolk and along the English Channel coast soon caught increased attention. One exciting movement and more observers were hooked, but there was a problem, the lack of a good textbook on sea-bird identification. The pioneer guide, W. B. Alexander's *Birds of the Ocean* (1928) contained little comment on flight action and distant character; R. C. Murphy's mighty, much more helpful *Oceanic Birds of South America* (1936) was virtually unknown outside a few museums and private libraries. James Fisher and Roger Tory Peterson laid plans for a new seabird guide as early as 1953 but nothing came of them. Not surprisingly, some of the early claims of shearwaters at over a mile offshore, even when made by luminaries like Ian Nisbet and Phil Redman, were regarded as incredible and some sense of hopelessness resulted. New proven perceptions were needed.

Particularly after the foundation of new observatories facing the Atlantic on St. Agnes in Scilly and Cape Clear off south-west Ireland in the late 1950s, apprentice seawatchers found themselves looking in a new direction, literally to the south-west. Year by year, little by little, a new litany of Atlantic seabird identification began to be heard and written. Rather empty attempts to see plumage marks were replaced by repeated glimpses of surprisingly different shapes and flight actions in shearwaters, suggestions of varying wingbeat depths and rates in divers and other important nuances of general character. The realisation dawned that the essence of seabird diagnosis was rather different from that for landbird identification. Given much trial and error and constant practice, flying marine beings could be named with certainty and at well over a mile. It is arguable that 40 years on, even the most expert seawatcher is not foolproof and certainly there remain persistent traumas in the record review committees, particularly affecting the small shearwaters, most petrels, the smaller skuas and the taxonomically messy gulls. Nevertheless, an overall progress in seabird identification has been secured. It has been enshrined most notably by Peter Harrison, who in 1972 set out single-handedly on a veritable odyssey of worldwide observation. This resulted in two excellent guides: *Seabirds* (1983), for which he was both author and artist, and *Seabirds of the World* (1987), for which he was author, artist and whipper-in of 741 instructive plates from 123 photographers other than himself. Peter's passion, unleashed by his first thrilling seawatch at St. Ives, had given us all just the books that we needed to make real progress.

Seawatching from any of the Channel Isles is hard work but I owe my second visit to Guernsey to a chance meeting with another great voyager after seabirds, Jim Enticott. His co-operation with David Tipling in the *Photographic Handbook of the Seabirds of the World* (1997) was to give us an even better store of seabird photographs but I did not know this when we met. In the late morning of 10th October 1988, I was working a small piece of failed crops where I had previously found two Lapland Buntings. They did not show but a slim, dark-mopped, keen-eyed birdwatcher did. Within minutes, we had dispensed with the dawn's few seabirds off Chouet and were deep in conversation about some supposedly impossible tubenoses that had been reported off Flamborough Head since 1978 and their actual appearances in their home range the South Atlantic. It was one of those magic, ocean-shrinking conversations that only manic sea-birders can instantly have. Inevitably, we moved on from desirable birds to fellow enthusiasts and were soon swapping tales of the foremost British marine ornithologist of our ken, Dr. W. R. P. (Bill) Bourne. Publically overshadowed by James Fisher until the latter's early death in 1970, Bill was at first meeting as much an enigma as the mistiest stormy petrel. Quick to disapprove departures from ornithological manners (especially of his own syllabus), as I had learnt as early as 1955, and not prone to approve adventurous identifications, Bill has nevertheless a totally kind heart and had been, we agreed, the single most influential brain, powerful voice and itinerant body in the more scientific investigations that had erupted from the increased general joy in seabirds. In conservation terms, these were far more important than fresh experience of 'skua gales'. With many of the spectacular seabird cities of Britain and Ireland as thronged as any in the world, a burst of renewed island going in the 1950s – in the wake of Ronald M. Lockley's saga and Robert Atkinson's inspiring voyages west and north of Scotland from 1935 to 1946 – evolved in the next decade into an attempt on the seeming impossible, a complete count of all our seabirds. Hence the formation in 1966 of the Seabird Group and the amazing adventure from 1969 to 1970 of 'Operation Seafarer', the latter timed to coincide with the field-work for the *Atlas of Breeding Birds in Britain and Ireland* (Sharrock 1976). Although accurate accounts of individual colonies had been made for over 100 years, the opportunity beckoned for a truly national census. It was fully taken and in due course along came *The Seabirds of Britain and Ireland* (Cramp, Bourne and Saunders, 1975). The pioneering work of John Ainslie, Dougal Andrew, Dr.

One of the supposedly impossible tubenoses claimed for Bridlington Bay, Yorkshire, was an apparent Madeiran Petrel on 6th November 1982. Oddly, the submission went astray but the species remains a prime target for seawatchers.

A favourite seabird of the late 20th century was the Black-browed Albatross that looked for a mate at two Scottish Gannetries in most years from 1967 to 1995. At Hermaness, Shetland, it even built a nest!

J. Morton Boyd, Dr. Frank Fraser Darling, Dr. Peter G. H. Evans, Dr. M. P. Harris, David Lea , Robert W. J. Smith, Dr. W. E. Waters and Kenneth Williamson – to mention only ten more of the stalwart seabird enthusiasts who had set sail over the previous four decades – had its culmination. It demonstrated that Britain and Ireland hosted a third of Western Europe's seabirds, a wonderful natural asset and a huge responsibility.

In the wake of the national census, seabird studies multiplied. Battling all weathers in sturdy converted trawlers like George Scott-Robertson's splendid 'Ocean Bounty', enthusiastic teams box-searched and counted the offshore distribution of birds. The dedication matched that depicted in Nicholas Montsarrat's *The Cruel Sea*. Alistair Simpson, peerless skipper of the 'Ocean Bounty', tells delicious tales of student crews forced to watch and count in all weathers, if necessary lashed to the superstructure of the bridge, and of their ornithological bosuns allowing no scrimshanking, even for total seasickness. Switching from a few autumn skuas off Kent, Dave Davenport and other fanatics discovered that their onshore spring compressions off the Outer Hebrides could be numbered in hundreds. At Dungeness, now equipped with a bubbling outfall 'patch' courtesy of its nuclear power station, terns and rarities as extreme as Slender-billed Gull were examined as never before.

Rather strangely as Cape Clear, St. Ives and other Cornish headlands and even Welsh capes presented more and more views of the late summer and autumn hordes of large shearwaters and rarer tubenoses, the North Sea slumbered. Small passages of seabirds were still regularly noted along the east coast but the recapture of the full potential of the 'German Ocean' did not take place until a few committed seawatchers adopted Flamborough Head as their watch point in the mid-1970s. Led by Andrew Lassey, dawn-to-noon and, if necessary, all-day watches soon showed that the six-mile-long, great white cape could trounce Filey Brigg and Spurn Point for both numbers and diversity of seabirds. Never mind the blazing, straight-on rising sun, give the Flamborough Ornithological Group a northerly blow and its small battery of keen eyes, binoculars and especially the fast developing telescope technology did the rest. From 1976 to 1991, record after record seabird passage were registered for Britain and the North Sea. Stubbornly the Pomarine Skua failed to appear at anything like the scale of its 19th century movements but

In autumn, terns swarm over the Dungeness 'patch' and much has been learnt there of their identification. Initially young Arctic Terns (two fore left hand) were mistaken for the true young Roseate Tern (fore right hand). The 'patch' was also one of the conduits to Peter Grant's elucidation of gulls.

otherwise the fortuitous combination of food-rich currents offshore and the trap-like topography of Bridlington Bay allowed the Flamborough enthusiasts to make real advances in the quantification of many species formerly considered uncommon waifs, witness their literally thousands of Little Auks and Little Gulls.

Equally as tardy as the realisation of all the seabirds attracted by the currents and foods of the western North Sea was the fuller probing of the offshore waters of the western seaboard. Although the massing of gulls and tubenoses over discarded waste and offal had long been known to seamen, and chumming (the application of fishy, oily 'porridge' to the sea's surface) had been exploited by collectors off South America as early as 1913, the bird-magnetic phenomenon was not fully exploited by British birdwatchers until the late 1970s. From then, however, the news that the combination of various marine lunacies, let alone stomachs evacuated on top of chum, could produce wondrously close encounters with such hitherto ethereal species as Wilson's Petrels led to the final seabird

No recent passage of skuas has matched those of the late 19th century but on 19th October 1991, 485 Long-tailed Skuas moved south past Flamborough Head, Yorkshire. The count remains the biggest number of birds ever seen in the North Sea.

indulgence. This was the seemingly magic 'pelagic trip' or straight diesel run (steam or sail long superseded) out to the deep water of true ocean. Nowadays the original legendary exploits in leaking hulks and under drunk skippers have become much safer with regular voyages in vessels as comfortable as the latest 'Scillonian' and huge P & O ferries, but the thrills remain as huge as ever. Holy Grails as revered as Little Shearwaters and Fea's Petrels can now be almost certainly obtained for the price of a mere ticket. The last, once-impenetrable marine bastion has been breached and with the testing of new coastal watches from places like Kinnaird Head (Grampian), Frenchman's Rocks (Islay), Rocky Point (Co. Donegal), Kilcummin (Donegal Bay) and the Bridges of Ross (Co. Clare), Strumble Head (Dyfed) and Gwennap Head (Cornwall), the circumference of marine observation is almost complete. Some objectives still tantalise – the long haul to the first certain South Polar Skua from the Antarctic region is still incomplete – and thus new-comers to seawatching can still make discoveries.

Harbours can encompass more than boats. From the 1970s, Killybegs in Donegal became the largest fisheries complex in Ireland. Nowadays mighty trawlers pull in hordes of gulls from as far away as north-west Canada, Iberia and Holland. Here a Long-tailed Duck and Black Guillemots share the lough with Iceland Gulls and thousands of others.

Providing the core of conservation intelligence, the Seabird Group has continued to attract freely given volunteer effort in Britain and Ireland, also producing with its Dutch partner, the Nederlandse Zeevogelgroep, a new journal entitled *Atlantic Seabirds*. Its ongoing achievements have included a second national census of breeding seabirds in 1985-87, published in *The Status of Seabirds in Britain and Ireland* (1991) and the Seabird Colony Register begun in 1985. It is currently intent on a third major exercise designed to update completely the register and so allow a third total measure of our seabirds between 2000 and 2003. Add to these the organisation of international conferences in Britain and Europe, dealing mainly with the threats to seabirds and the marine environment, and the granting of aid to members working on research and ringing projects as far off as the Azores, and it can be seen that the output of the three hundred or so members of the Seabird Group has been both Odyssean and Herculean. With the steady hand of Mike Harris on its tiller, the Seabird Group looks well set to continue with its crusade for seabird conservation. As with landbirds, the lines of the big battles have been drawn but not successfully engaged. The most complete breeding failures experienced in northern tern colonies have been simultaneous with the increasing 'hoovering' of sand-eels to the point where a tiny fish totally ignored in my youth is now subject to the largest single-species fishery in the North Sea. Thirty years after the 'Torrey Canyon' threatened Lyonesse, oil spills continue still. There may be more Puffins in the Firth of Forth than ever before but the Jolly Jack Tars of British birdwatchers cannot go off watch.

It will be a terrible indictment of modern fishing man if he extinguishes the blithe spirits that have kept brave company with him for thousands of years. It is their right to inhabit the marine wilderness that surrounds our lands. Our harvest of it is only a privilege and all abuses must be monitored and proclaimed. Further, to invite even the nowadays reduced wrath of Bill Bourne is not a pleasant prospect!

True luminaries like Bill Bourne are sometimes immortalised in the Linnaean nomenclature of birds. His name will be attached for ever to the strange heron on the Cape Verde Islands. Taken to be an isolated form of the Purple Heron, it is named *Ardea purpurea bournei*. Bill was the first person to spot its odd appearance and bring back a rather 'high' skin.

Mother Superior

Corporate speak questioned – the Norfolk birth of conservation – the denunciation of the plume trade – the female and male propagandists – the birth of the RSPB – the first braves of protection – other early initiatives in persuasion and protection – the neglect of the Dungeness riches – their recoverer – one of the last Kentish Plovers and other gems – reserve and observatory under one hand and mind – their transfer to Minsmere – a search for England's first Citrine Wagtail – recommended Bibles of practical conservation – the growth of the RSPB – the incidence of domestic support for conservation – other measures of wildlife support – changes in Directors and direction – a conference vignette – working with the Society – new description of the real task – the Baroness – field confidence, office nerves – the continuing challenge.

Ask a staff member of the RSPB about its current role and it is more than likely that you will get a practised answer. Stemming from a recent headquarters addiction to the fancier outer regions of marketing, phrases like 'action for birds' and 'for birds – for people – for ever' – the latter the Society's current motto – and 'managing people for birds' (or is it 'managing birds for people'?) are offered. It is all to do with the standardisation of communication values to prevent the distortion of the Society's ethos and message. What would have Margaretta Lemon, Fred Austen and Jack Tart made of such practised words or the paid thought underlying them?

While some measures of bird conservation are as old as Deuteronomy, it was not until the second half of the 19th century that specific law and local action came together in Britain to confer real protection to birds in the field. Some early cares for wildfowl and seabirds have already been described in two previous chapters. For the creation of safely tenured and managed reserves, we have to thank a few ancient clerics – the Swannery at Abbotsbury was known in 1393 – and, much later, the outstanding naturalists of Norfolk. It was they who founded the Norfolk Naturalists' Trust in 1869 and the Breydon Society in 1888, the progenitors of today's fast-growing county Wildlife Trusts. What was needed, however, for a national conscience to take active root was a cause célèbre and some impassioned advocates. The former was first publicly expressed by Alfred Newton, who in a letter to *The Times* on 28 January 1876, denounced the use of birds' plumage in female fashion. The latter corps was formed by lady propagandists who became formidable in opposing the mores of their 'feather-bedecked' sisters and formed vociferous bodies such as the Plumage League and the Fur, Fin and Feather Groups. The exact moment when the RSPB was born can be debated but clearly the society itself believes that its origin stems from the formation of a new Fur, Fin and Feather Group at Didsbury in Manchester on 18th February 1889. Certainly 1989 was celebrated as the centenary year even though the full identity of an institution called the Society for the Protection of Birds under the noble presidency of the Duchess of Portland was not completed until 1891. Its first minimum subscription was only two old pence (57 new pence today) and in its first year 5,000 eager members paid up!

To improve their propaganda, the precursors to the suffragettes recruited the pens and (after a brief feminist-like debate on whether men should be members) the full allegiance of luminary male apostles. W. H. Hudson was one. By birth an Argentinian, untutored as a boy on the Pampas and plagued by illness in his teens, Hudson expressed himself and his passion for natural history mostly in writing. The compound product was a series of evocative works from pamphlets to whole books that popularised particularly British birds and inspired many people to take interest in them and feel concern for them. Twenty-five years on, his *Birds in London* (1898) was to inspire Max Nicholson's early urban studies; 50 years on, it was one of the seminal books in my own ornithological fledging. Among others wielding the pen in the cause of early conservation at the end of the Victorian era were Eliza Elder, later Mrs George Brightwen, with *Wild Nature Won by Kindness* (1890) her most telling book, and Rennell Rodd, with his moving poem about the depredation of Skylarks obtainable at 100 for one old shilling and three old pence (4.25 new pence each today). With print (the only medium open to them) thus well used and the further recruitment to its early ranks of such ornithological greats as Professor Newton, Lord Lilford, Sir Edward Grey, later Lord Grey of Fallodon (and the British Establishment's first public espouser of birds), John Harvie-Brown and Dr. Bowdler Sharpe of the British Museum (Natural History), the fledgling society had all the credentials for initial success. Expatriate branches flourished briefly in Germany, America, India and New Zealand. More significantly by 1893 Margaretta Smith, then Mrs. Frank Lemon, had established a sound administrative system. This potent lady had waged her own sharply targeted war against the wearers of plumed hats and similar sprays. Any such lady would find her display at church of Sunday finery swiftly followed by a written rebuke which pulled no punches in describing the cruelty of shooting breeding herons and egrets on their nests. Margaretta Lemon handed over the Honorary Secretaryship to her lawyer husband when the Society became incorporated by Royal Charter in 1904, but continued to inject her exceptional drive into Society matters as the Honorary Secretary of the Publications and Watchers Committee. So it was she who administered the addition to the Society's original roles of publicity and legislation of the first practical attempts at conservation. In 1901, a guard was appointed to protect the nests of Pintails on Loch Leven. The first-ever, regular wages were paid to two men of Kent as part-time seasonal guardians or, to give them their official job title, Watchers over the then great diversity of birds breeding at Dungeness. Fred Austen from 1908 and Jack Tart from 1909 were the progenitors of today's corps of hundreds of wardens and supporting scientists. Austen had particular charge of the Kentish Plovers and Stone Curlews; Tart was primarily concerned with the terns and the south coast's only Common Gulls. Dressed in white-topped peaked caps and mariner-like uniforms, often wearing back-stays on their shoes for greater purchase on the shingle, the two watchers defended the ness's birds over three decades. Both died in harness in 1937, having not only served the Society well but also influenced along their ways particularly H. G. Alexander, a son of Tunbridge Wells and already described earlier as the classic model for all observers, and H. E. (Bert) Axell, a youth of Rye with later claims to

The Pintails of Loch Leven, Fife, were the first birds to have their nests protected in Britain.

The difficulty of preventing bird deaths at coastal lights was nowhere more extreme than on Bardsey Island, Gwynedd. Eventually the erection of a second light source below the lighthouse drew most birds down to a safer landing. Here, however, two Wheatears and a flock of Choughs enjoy a fine day.

"In May 1920, when I was nearly five years old, I accidentally damaged a Wheatear's clutch. My fault was to cause a life-long concern for the safety of birds' nests. Hence eventually the Scrape and reed-bed management at Minsmere and my other work around the world. I had to take one of the Wheatear's eggs because it was the done thing among boys to have an egg collection. Another stimulant was finding another (then the only other) birdwatcher and desperately keen ringer in Rye, my town of 4,500 near Dungeness. Ryers, seeing us always out together, probably through us a bit queer. Ringing (from 1929 to 1975) and needing to know how and where species lived were complementary reasons for my being a keen birder." Bert Axell *in litt.* (1999)

A great conservator's summary of the start to his hobby and eventual chief work.

be the prototype of best ringing practice and reserve creation. The relationships of the quartet are fondly remembered and portrayed in Alexander's *Seventy Years of Birdwatching* (1974) and Axell's *Of Birds and Men* (1992).

Away from its first reserve and with the plume trade much reduced, the Society's committee and leading members turned their minds to other issues. They agreed as early as 1902 on innovative initiatives to educate children and young people in the benefits of nature study. In 1903, a circular letter to members became effectively the first-ever bulletin of bird news. The first two words of its heading 'Bird Notes and News' remained the title of the RSPB's first magazine until 1966. In the next decade, a partial solution to the many deaths of migrants at lighthouses on foggy nights came from the combination of first setting circular perches below the light glasses and second illuminating the perches by separate floodlights. The main achievement of the Society's first 30 years was, however, the stopping up of the loopholes in the 1880 Wild Bird Protection Act by a series of amendments and the early preaching for a positive universal conveyance of protection to all birds with exceptions such as game birds clearly defined and individually named. In the latter argument the work of two barristers, Montague Sharpe, who was Chairman of Council from 1895 to 1942, and Frank Lemon who continued as Honorary Secretary until 1935, was outstanding. With much of the Society's policy and some field practice in place before the two World Wars, its development seemed assured but in reality it was not until 1947 that the Labour Government set up a committee to propose the long-desired new law.

Sadly too the outbreak of the Second World War brought an end to watching at Dungeness. Between 1940 and 1949, it became so overrun with natural and human predators that almost none of its ground-nesting birds was breeding with any success (Axell 1992) and the investment made from 1924 to 1937 by the Society and one particularly generous member, Dungeness's greatest champion Richard Burrowes, who paid £4,026 of the £8,036 cost (£345,000 today) for the 1,243 acres, was all but wasted. In 1949 and again in 1951, surveys of the remaining birds produced results that hardly encouraged a return to the cost of watching.

Perhaps the War Office's firing ranges and the shingle wilderness should be left to fate or broken up by sale. Happily the few minds that mattered were more courageous and Geoffrey Dent as Chairman of the Watchers' Committee and Philip Brown as Director of Watchers and Sanctuaries determined that the restoration of the Dungeness bird community would be attempted. With the 50 employees of Brown's department already deployed on 12 reserves and watches, who would reverse the depredation of Dungeness?

Cometh the hour, cometh the man. He was Herbert Axell, prior youth of nearby Rye, who had recovered from a period of ill-health and was keen to change his life radically after service in the Army and General Post Office. In 1952 at the age of 37, Bert was interviewed in late March, offered the full-time wardenship at Dungeness within a week and started work on 10th April, so completing an appointment process of considerable alacrity but unknown conservation potential. The crux was to prove to be predator control, not for the faint-hearted - but Bert was never that. Meanwhile, from a membership nadir of 3,500 in the middle of the Second World War, the RSPB had begun to attract fresh support. By 1952, new members were appearing from the flux of the softening post-war attitudes to wildlife. So in spite of its slim purse, Bert's *apres* interview conclusion was that "you could feel the RSPB was forging a whole new industry".

My own experience of the RSPB's earliest cradle and first test-bed began 37 days after Bert's re-arrival upon his boyhood wilderness. My log for 17th and 18th May 1952 shows that on a pass from the Lowland Brigade, my parents gave me a weekend in the Romney Marsh. By map reading, I had found most of the localities mentioned in the Witherby *Handbook* and planned to explore the coast from Pett Level to Camber Sands and the Midrips and across the marsh to the Ness and Greatstone-on-Sea. Taking lunch at the George Hotel in Lydd, I persuaded Dad to enquire of a loquacious local as to the whereabouts of my most desired species, the Kentish Plover. His reply still rings in my ear. "Lord love you", the veteran said, "they fly round here in shoals" and turned back to his pint. The truth was that Britain's last colony of the little bicycling plover was on its last few

By 1952, the famous colony of Kentish Plovers at Dungeness was almost extinguished. On 18th May, only a single female ran over the Camber Sands.

pairs of legs and that day the 'shoals' for us consisted of just one female on Camber Sands. Among its supporting cast, however, were other lifers in the forms of two Temminck's Stints, four Nightingales and a Marsh Warbler. What then seemed the ultimate English avian fleshpot had me instantly enthralled and throughout my later London years until 1965, Dungeness drew me and many other emergent birdwatchers from the south-east back time and time again. In truth, the magnet was by then less the reserve and more the new Dungeness Bird Observatory but there was no lack of Bert Axell. Following the energetic nod of the Nature Conservancy and the support of the Hastings Natural History Society, the Kent Ornithological Society and the London Natural History Society, he collected the Wardenship of the observatory as well as that of the reserve. Thus two engines of protective bird study came under the hand of one inspiring driver and volunteers and practical help flowed on to the tip of the shingle peninsula in order to build an impressive array of traps. Dungeness was soon professing its two subjects of conservation and migration study so well that its influence on birdwatchers and ringers became national. The Axell team featured some stalwart observers like Harry Cawkell, John Parrinder and Phil Redman, but not being a compulsive ringer, I was shy of them and Bert. He will not have remembered my part in the driving of the new Heligoland traps on 25th April 1954, but I was in awe of him from that day on.

After seven years of service at Dungeness, Bert Axell's quiver of ornithological arrows was needed for another target. At the end of January 1959, the new director of the RSPB, Peter Conder, showed him the marsh at Minsmere, pointed out the increasing threat of its invasion by carr and woodland and, as the new family home, an "awful" bungalow. Joan Axell bit her lip and Bert's second journey in the saga of modern conservation began. With a much more diverse range of habitats and at last some slightly larger funds and machinery, and the usual twisted elbows, a really adventurous management plan was possible. Excavated and sluiced by the end of the 1960-61 winter, the new Scrape (shallow mud and shingle lagoon) soon became the subject of birding legends, not the least because of its prompt occupation by terns in the following spring, Its adventure was followed by the manipulation of other unproductive land and the limitation of the reed growth, both to the benefit of the Avocets and other birds and the increasing flow of human visitors. No longer just the *oi oligoi* of the Society establishment but a mix of adventurous members and their families and more and more birdwatchers came to enjoy the show. Like the converted blockhouses of Slimbridge, the Minsmere hides were windows on to a revolution in habitat management. The penny dropped in various chutes; a huge enjoyment of birds in constructed habitats beckoned as the RSPB's best consumer promise and product.

Norfolk-born, regarding Suffolk as a sissy's county, I left its wonderful heaths and marshes untrod until 1954 and was long satisfied with Walberswick as the ultimate reed marsh. What finally took me inside the miracle at Minsmere? In 1963, Bob Emmett and I had enjoyed a grand slam in exciting autumn migrants on the Suffolk coast and on 1 November 1964, we set out from London to repeat it. It was a day of remarkably warm 'anticyclonic gloom' but with light rain constant at Aldeburgh and obscuring lens and bird alike, we soon settled for a walk past the

The flagship species and indeed brand of the RSPB, the Avocet returned to Suffolk in the early 1940s. From Havergate Island and Minsmere, it has spread to 37 other localities. There are now 1,000 pairs in England. Here is a scene typical of the Scrape at Minsmere in summer.

The Portsmouth Group watching over Langstone Harbour from Farlington Marsh, Hampshire: late 1950s or early 1960s. From left, unknown, Euart Jones, Graham Rees, Peter Lebrocke, George Clay, Dave Billet and two unknowns. REE.

sea end of Minsmere and the cover of the old public hide. In the bushes near the south sluice, we met Bert packing up his mist-nets. To our surprise, he hailed us as Dungeness chums; to our astonishment, he invited us into the newly developed scrapes to see England's first (and Britain's seventh) Citrine Wagtail. My log continues the story,

> Hardly believing our luck in bird and access, we played the public relations card like mad and soon had permission to trample all over the east end of the reserve, an unprecedented *laissez-passer*. However, it soon became apparent that it was not going to be rarity first,

"Senior members of the Portsmouth Group showed me the relevance and the value of the study of birdlife of a defined area throughout the birding calendar. They taught me how to identify birds at long range by their jizz and by their calls. They were particularly interested in the history of Farlington Marshes and Langstone Harbour and I learned that birds' fortunes were often closely linked to changes in habitat. The Group were also the first real conservationists I had encountered and I can remember being impressed by their passion for the marsh and harbour and their beliefs that it should be conserved."
Eddie Wiseman, *in litt.* (2000)

A testimony to a mentor group acting independently from any national body from someone who used his learning well.

The avian miracles of Minsmere, the flagship reserve of the RSPB in Suffolk, were never more visible than on 1st November 1964. Its natives (behind) included Water Rails, Marsh Harrier and Bearded Reedlings, its migrant visitors featured two Water Pipits (nearest) and England's first Citrine Wagtail (middle right).

ordinary birds later. Although we plowtered in all directions, no odd wagtail showed. After an overdue breakfast, we returned to be greeted by "It's been in front of the East Hide all morning". Beating the 50 kilometre technique of any Olympic walker, we thrashed round and clambered up. But no bird was visible and soon we were at the trampling again. At last, sheer effort produced the desired effect, a dark plumpish wagtail though unfortunately flying away. More careful stalking and time, however, brought us increasingly lengthy and better views. By 14:30 hrs, after over six hours of hunt, we had it at a few yards under warm sunshine (60°F) in the marsh pool by Walton's Hide. Finally satiated, we retired to cement further the unexpected degree of hospitality before the run home.

Proof that Minsmere had indeed become another magical fleshpot was provided by a late Marsh Harrier and the wagtail's closer companions, Twites, Snow and Lapland Buntings and at least two Water Pipits. The lesson in the comparative benefit of 'hiding' versus 'trampling' was just as forceful. The RSPB has enjoyed the duty and service of many other heroes than Bert Axell but I believe his tales to be the most instructive of the Society's true engines of proactive progress. These can be found most readily in *Minsmere: Portrait of a Bird Reserve* (1977) and *Of Birds and Men* (1992). Some of the later history of Minsmere can be found in Simon Barnes's lyrical celebration of his year there in 1990, entitled paradoxically *Flying in the Face of Nature* (1992).

It is high time, however, to return to the fortunes of the RSPB as determined by its attraction of the mass support required to tackle the appallingly increased range of avian misfortune. Due to some shifts in member classification, it is not possible to track precisely the growth of Society support but it clearly regenerated in the late war and immediate post-war years. The Junior Bird Recorders Club was bravely launched in 1942, recruiting almost 2,000 keen members by 1952, and the adult membership reached 7,000 in 1953. It was in the following year, with my National Service consigned to history, that I rejoined the Society as an adult. Philip Brown himself signed the receipt for my £1 (£15.80 today or 59% of the current single membership rate). The most important benefit was the regular arrival of *Bird Notes*, its news of birds and its confirmation of the existence of other birdwatchers. I cannot remember attending any of the Society's well advertised illustrated talks and later film-shows, except when they were combined with a local society event, but I soon became more aware of the needs of birds and the right of wildlife to co-exist with man. The early media included radio delights like the monthly *Countryside* review, so ably illustrated in sound by Eric Simms, and the early television programmes, most notably the BBC's *Look* series from 1954 and Anglia TV's excellent *Survival* programmes from 1962. Even in black and white, the full reality of animals in their proper niches began to banish fully the zoo images of the 19th and early 20th centuries. The voices of David Attenborough and Colin Willock won our trust and more and more people joined in the defence of birds. As James Fisher pointed out, RSPB membership doubled to over 14,000 by 1961 and doubled again to over 28,000 by 1966. By the latter year, the Society was already the largest conservation body in Britain.

The growth wave not only persisted but became for a time even steeper. By the end of the 1960s, members numbered nearly 50,000; during the 1970s, they multiplied by five; in the 1980s, they more than doubled and in the 1990s, they almost doubled again. Entering the 21st century with over a million members of all ages, almost equally divided between the sexes and living in 600,000 homes, the RSPB can beat a big drum but the relative incidence of its domestic support is actually still low. Although a thoroughly laudable achievement, the Society's penetration of all British households is only 2.8%, less than half the National Trust's equivalent measure of 6.8% and nowhere near the amazing universe of anglers, which became over five times as large as the RSPB membership in the 1970s and is still four times so. No one should confer on the 'magic million' any state other than a useful attention device. In Holland, long ahead of Britain in the creation of a reserve system and the political adoption of conservation, the incidence of domestic support is 4.7%. It is also worrying that although the Society did not suffer from the flight from green behaviour that has followed recent economic debacles, its current rate of recruitment is hardly sufficient to cover natural loss by death and rare disaffection. A new formula for greater domestic support is urgently required. Without it, the Society will risk losing the race for new members to the many other wildlife societies set upon competing for greater shares of charitable giving. Since 1999, the Wildlife Trusts have enjoyed a surge of support and are forecasting a collective 400,000 adherents. Sustained locally by a strong historical county structure rather than the RSPB's initially hesitant regional devolution and variably led members' groups, they are real competitors. Conversely it will be a long haul before the Mammal Society with only 2,000 members catches up but recently the Bat Conservation Trust (BCT) has shown again how a challenge and a technically improved approach to finding nocturnal beings can attract people. They have 7,000 members in 4,500 households; as in the RSPB, half of them are men and half women but this ratio holds also for the 3,000 that take part in actual bat monitoring. Try finding sexual parity in any birding activity. The BCT's enthusiastic performance is a reminder that as in other issues, big can be impressive but small is more riveting.

Nevertheless, the RSPB is, in most years, the 20th largest charity in Britain, 18 places behind the National Trust but 11 ahead of WWF(UK) in total voluntary income (the measure used by the Charities Aid Foundation to order charitable support). Given that there are 500 other charities with millionaire incomes and nearly 150,000 such bodies in all, the allegiance to the RSPB is remarkable but what does it really occasion?

With the long-cherished logic reversal in the law protecting birds finally achieved in 1954 and egg-collecting also banned, the Society moved rapidly into a more proactive role. Initially this featured the chance fosterings of rare breeding species, most notably the Avocets which, after some uncertain attempts in wartime Norfolk, re-established themselves on Havergate Island, Suffolk in 1947 and the Ospreys which reoccupied Speyside from 1953 after an absence of over three decades. The publicity surrounding these birds and their successful return, and

Extinguished in Scotland in the 1910s, migrant Ospreys needed four decades to form again a successful breeding pair near Loch Garten, Strathspey, in 1954. Thanks to the RSPB and many volunteers, over 150 pairs are now doing well.

"You may be interested to hear that an Osprey was very recently shot at Portmore Loch by a misguided keeper. There has been a significant increase in the number of migrant Ospreys during the past few years, and it seems quite on the cards that they may try to breed again in the near future." Dougal Andrew, *in litt.* (1951)

Three years later came the famous pair to Loch Garten. Fifty years on, Ospreys have been returned to England.

"I would claim a part in growing the importance of research to RSPB which actually got a very long way without it in the first eighty years of its history." Colin Bibby, *in litt.* (2000)

even the delayed furore over the 1958 theft of the Osprey's eggs, was undoubtedly beneficial. The Avocet was adopted as the Society's logo, its spread to Minsmere having further enhanced that reserve's rich avian larder (and high visitor appeal). Its unique form, and Robert Gillmor's pen, made it too good an image to ignore. The Ospreys' most regularly used nest became the object of special watch facilities since used by nearly 2 million people. Each annual contingent contributes £1.5 million to the Speyside economy. For the first time ever, the names of special birds and their places appeared on road signs. It was an optimistic time but the Society soon had to face the daunting realisation that its full responsibility encompassed far more than a few fortuitously positioned, rare breeding birds. Factors such as the increasing maw of industrial development, the near disaster of toxic agrichemicals and the low level of ecological intelligence in national and local government all raised near universal spectres. In my anecdotal memory, it was then that the Society wrote itself a habitat protection plan and so first incurred some displeasure by seemingly denying well-travelled birdwatchers one after another favourite wildernesses. There were exceptions, such as Bert's *laissez-passer* of 3rd November 1964, but the general rule appeared to some of us overprotective. Happily the growing ranks of RSPB members felt no such irritations and their support allowed the pre-planned fabric of reserves to grow rapidly, with 25 pieces of precious habitats made safe by 1968, over 50 by 1973 and 75 by 1978. Simultaneously, the Society joined in the campaign that contained the flow of toxic chemicals, founded an effective species protection programme to assist, if not direct other law enforcers in targeting particular threats (and culprits), served (and answered to) its increasing membership and, perhaps most importantly, developed an acute approach to research into both species requirements and matching habitat management. Under Peter Conder's direction from 1963 to 1976, the RSPB began its necessary metamorphosis from an originally amateur and passionate body to an almost wholly professional, more disciplined and thus more convincing institution. I heard some of the arguments and saw many of the changes as a Council member in the mid- to late 1960s. Yet what I most remember from then

A chilling example of how a fine bird may lose its niche comes from the Capercaillie. Re-introduced into Scots Pine forests from 1837, the biggest grouse in the world did well but since the 1970s it has faced increasingly its second extinction. Among its chief threats are wetter summers and yet more miles of deer fences.

was the frustration of feeling utterly spare as an advisor to more scientifically educated field staff and equally without real influence as an informed critic of risky sales operations that were locking up a lot of funds. So in spite of the accidental delight of sitting beside my favourite jazzman Humphrey Lyttleton, I expressed this feeling at my last Council meeting and left during the consequent embarrassed silence. Piece said, best go … but the fates of my career would see me return.

The Society's next leader was the fondly remembered Ian Prestt. With an established professional reputation gained from service at the Nature Conservancy Council, Ian saw not only to the RSPB's permanent station at the top tables of government and general conservation caucuses but also its gradual adoption of a broader brief for the other wildlife that occurred alongside birds. This was to underpin the whole variety of life (bio-diversity) in the reserves and to encourage the first much-needed changes of agricultural policy. Ian's glee at the prospects of the then new set-aside scheme made for a much better memory than my own frustrations of 25 years earlier but two others are also worth recall. About 1983, I was asked by Ian to address the RSPB Members' Conference at York. His brief was, "You'll be on after the Duke of Edinburgh. We want you to tell them about real birdwatching. Do enthuse them!" From my then patchwork study of East Riding cereal farmland, I distilled a lecture entitled 'Rare birdwatching close to home'. Illustrated with habitat photographs and line-count figures, it would surely not only entertain the audience but also convert many of its members into instant BTO surveyors and county report contributors. On the day, His Royal Highness was hardly electric and I died a total death. Looking up at the serried ranks of largely female tweed, I became almost mesmerised by their incomprehension of purposeful birdwatching. Thank God for Jeffrey Boswall, who saved the day by being cheeky about David Attenborough, and arguing that in natural history films the occasional glimpse of copulation was slightly essential to understanding how animals survive. Heated debate broke out and under its cover, I slunk away. Piece not heard, best depart at speed.

I did better with Ian's next request. This was to go to the new Blacktoft Sands reserve and help Andrew Grieve to plan the location of its scrapes. The process was not as heavy as following the Duke; the lagoons could only go where there were no reeds and Bearded Reedlings, at the east end. Mind you, I was to regret their construction for they soon seduced my favourite waders into changing their roost position from my (north) side of the Humber Estuary to Andrew and the RSPB's (south) side. In particular, my failure to establish several early claims of Red-necked Stints – an adult was at Faxfleet directly opposite Blacktoft in September 1976 – was made even more stinging by the reserve's later presentation of the accepted first for Britain in July 1986. The bird's and its companions' greater safety at Blacktoft was, however, indisputable and its lagoons set against a wonderfully wide horizon have been hugely enjoyed by many birdwatchers for nearly 20 years.

1985 saw me enter a new relationship with the Society, first as a salesman of Christmas cards and second as chief organiser of my last employer's trading agency to RSPB Sales Ltd, which continues still. Preoccupation with shared commerce occluded partly my view of other Society issues but by then led by Barbara Young,

The Scottish subspecies of the Ptarmigan survives on the still snowbound high tops but for how much longer in the face of global warming and the wilt of Alpine plants? Here a young Golden Eagle flushes one of its prey species from a Cairngorm slope.

One of the RSPB's most unpopular Christmas cards – my attempt at north Norfolk's Snow Buntings and Shorelarks – and one of the sure annual winners, Goldcrests by Rodney Ingram.

"The really big failure has been our inability to influence the social and political agendas that generate perverse economic instruments." Colin Bibby, *in litt.* (2000)

"Ministers for the Environment are seldom at all well informed (Michael Meacher and John Gummer apart). And it's the Government that really matters." Janet Kear, *in litt.* (2000)

"More co-operation between all the birdwatching and conservation organisations in the country – to make the politicians realise the importance of the countryside to the health of the nation (and Europe and the World)". Moss Taylor, *in litt.* (2000)

The burning issue of a former doctor.

now Baroness Young of Old Scone, the RSPB continued to prosper. In 1998, the landmarks of 1 million members and 150 reserves were both passed. In 109 years, the passion of the few ladies of conscience and its diverse aftermath had created and secured the most powerful force in birds and conservation in general in the Old World. Some of the ghosts may have looked askance at the appointment of a career administrator to a role previously thought to require relevant field experience. Unabashed, with an MA in Classics but otherwise a standing start in birds, Barbara quickly absorbed the essentials of conservation and the Society's *id*. She even bothered to read the topical birdwatching press and, with her keen political instinct, produced a formidable piece of direction. Suddenly in posts where for too long perhaps there had been rather dusty male birdwatchers or refugees from commerce, there were young, bright, concerned, mostly lady managers. To ageing male supporters and suppliers of services like Chris Whittles and myself, the RSPB appeared to go through another metamorphosis. Effectively, it proclaimed itself in no mean sense a lobby, seeing this as a necessary position for decelerating the engines of land and sea misuse. These no longer threatened the restored rarities like the Avocet but instead imperilled in every season an ever-increasing number of ordinary, supposedly common birds. The full battle between the bird protection and conservation movement and a nation devoted to increasing consumption was joined, 25 years late, but with the dialogue with government subject to inarguable science and backed by a constituency of a size and generosity large enough to impress the politicians of all parties.

In 1997, the arrival of New Labour presented a fresh opportunity for agreed national conservation goals. Under the fortunate ministry of Michael Meacher, happily rated as sensitive to the basic issues as his Tory predecessor John Gummer, the Department of the Environment, Transport and Regions produced by 1999 a White Paper on *A Better Quality of Life* and a *Bio-diversity Action Plan*, allied to a new list of *Birds of Conservation Concern* (1996). Imaginatively, the first document contained 150 indicators of lifestyle sustainability; one was to be based on the annual index of breeding bird populations fashioned with total dedication since 1962 by the BTO and purposeful birdwatchers and now justifying every last slog of foot, eye and ear. However unified the new purpose seemed, there followed a smart reminder that people are not so, their individual courses sometimes prone to raise hackles.

When Barbara Young was offered, and after consultation, accepted a peerage from the Labour Government, the consensus along the Society's interface with senior council members fragmented. It was particularly sad to see the departure of

Lord Barber, a hero of the brave campaign to save the Ribble Estuary and adjacent wetland in 1978 (when the Society bid money that it had yet to raise), but most ranks closed behind the executive. A serious wobble was avoided. Ironically the whole episode was then made redundant by Barbara's onward ascent to first the chair of English Nature and second the helm of the Environment Agency. The Society lost its Scottish Amazon but she has stayed its true friend in higher places. Back at The Lodge, by now grown to a veritable ranch and stables of experts and administrators, Graham Wynne took over. The hand of a practised conservator was once again on the tiller.

For the moment, the RSPB remains overall in good working order but, with the length and cost of the full battle to share Britain's finite natural resources fairly between wildlife and man, it needs to grow. The recent steep surf of the recruitment waves has collapsed to a mere lapping of the shore. Under pressure of repeated questions, the numbers of teenage supporters are admitted to be impure. How can such imprecisions exist after years of expensive system development? The RSPB's goals remain uncomplicated: maximal bird benefit in the field, maximal member satisfaction from the cause, maximal tolerance from the rest of Britons, all to be addressed with cost efficiency. I have no doubt about the achievement of the first aim. There really are excellent minds (and hands) in the field-active researchers and wardens of the Society and their co-workers in the BTO and elsewhere. They really do know now to 'garden' for Nightjars; they are absolutely certain that if Corn Buntings do not raise 2.2 young per breeding pair, they will simply drain away. These and many other issues are clear and are being addressed directly not only on the Society's Home Farm but increasingly by many other landowners with respect for a balanced environment.

Regrettably, however, it is difficult not to sense disquiet within the Society's own ranks and birdwatchers at large about its internal systems. Somehow righteous pride of achievement goes hand in hand with continuing outbreaks of nervousness. Managing growth is difficult but it is odd that splendid advance in recent decades does not seem to have raised the Society's confidence in how to sustain the performance. Feeling boundless concern, researching members' attitudes (to the point almost of fearing the holders), sitting in endless meetings (with no phone calls allowed in from the more active world), attending bonding courses (one attracting the sobriquet of 'sheep-dip'), regarding leadership as secondary to co-ordination … Is this the stuff that will attract the second million members, soothe the brows of the not overpaid wardens who still have to find their own solutions (like Bert's shotgun) to ground predators, and construct new attention-grabbing attractions to local punters? I was not in the least surprised to see the RSPB's first attempt at a general web-site pilloried in a recent review. I just flinched to think of all the effort that would undoubtedly have laboured upon it and of the attempts to improve it which featured job values nearly twice those of reserve wardens. The outcome was also a sad contrast to the brilliant (but sadly brief) BBC exercise in direct on-line presentation of birds from the Belfast Lough Reserve designed by John Scovell and directed and scripted by Anthony McGeehan, working after hours. As Norman Dunbar, the Society's one-time Aussie Marketing Manager,

Researching how to restore the habitat of Nightjars at Minsmere, the RSPB discovered that 'mowing' round young trees gave the birds a niche. What seems to have helped more generally is the new felling pattern in conifer forests. Nearly 3,500 Goatsuckers still come to British heaths and plantations.

"People are becoming more and more removed from nature and appreciation of the natural world. Politicians and the media have encouraged this partition." Keith Vinicombe, *in litt.* (2000)

"The saddest loss to me is that in the populated half of the country at least, there is now virtually nothing that is both wild and accessible left. It has either gone or it has been tidied up and organised in nature reserves offering a distant and regulated glimpse from a comfortable hide. Or in more human terms that there are now so few people who have any direct connection (physical or spiritual) with the land." Colin Bibby, *in litt.* (2000)

"We have become a nation of consumers and many birders behave as consumers also." Rupert Higgins, *in litt.* (2000)

said, "You were quite right, Ian, about the endless fuss. There are thousands of ions here. If only they were all positively charged."

I am haunted by one other remark. It came from Rob Hume when I asked him what the members wanted in *Birds* "Birds, birds, birds", he retorted and turned back to the complex of messages that the Society's magazine was wont to deliver to yet another plan of communication content. Happily, *Birds* has recently lost its former tenseness. Much of it is now a good read and its central section of conservation news punches above the Society's weight.

Be in no doubt. If you do not pay a sub to the RSPB and cough up to the odd appeal, and yet take pleasure from birds, you are an utter rotter. It is by far the best defence to them and your solace in them but it has to do more and quickly. So while Fred Austen and Jack Tart will have their caps off and be cheering their successors in the field, Margaretta Lemon could well be demanding a more up-tempo performance in the rest of the Society's management structure.

With 18 out of the 25 Biodiversity Action Plan species missing their recovery targets and these including such treasured birds as Skylark, Song Thrush and Linnet, a stance that exhibits rather less Mother Superior and rather more Turbulent Priest is in order. Without some widespread improvement in the survival of birds in the wider countryside, the 170 reserves of the RSPB and their 111,500 hectares could be as inadequate as the Maginot Line in the defence of all British birds. The Society's 1,450 paid workers and 8,500 volunteers have a lot left to do.

Anthony McGeehan

Anthony McGeehan in an Ulster marsh 2001.

Born in 1956 Anthony McGeehan grew up in East Antrim and now lives in Bangor, though remaining proud of his roots in Donegal, his spiritual home. Learning his Rs at St. Marys Christian Brothers Grammar School, he inserted birdwatching into his life curriculum at the age of six. He made his first marks on the natural world in what will become Mourne Mountains National Park, building access paths up the ridges. Since 1998 he has been recreating refuges for wildlife on the Belfast Harbour Estate. Providing also the crux of wonderful interpretation, Antony has become the "City's Champion" for birds, serving as a Warden in the best tradition of the RSPB.

Before such redemption, as his family can attest, Antony became famous as an "All Ireland Twitcher" and a dangerously funny columnist for *Birdwatch* and *Dutch Birding*. With no need of kissed Blarney stone his series "Birding from the Hip" and "Total Birding" mixed sharp truth with delicious humour, both conveyed in a style reminiscent of his favourite wordsmith Garrison Keillor. Expert on Nearctic passerines and Atlantic seabirds, Anthony exemplifies how passionate amateur can still moult into field ornithologist and conservationist.

In this shot, Anthony crouches to use the most powerful weapon in today's birder's armoury, a high resolution telescope. Using modern optics, a trained eye can see on undisturbed birds the characters that Eagle Clarke first discerned from skins. What is missed (but now well seen) can be history too.

Island-going in the 1990s. The long and the shorter of Hebridean cruising vessels. The 'Alla Tarasova' complete with Tony Soper as guide and interpreter, leaving Loch Maddy, N. Uist; George Scott-Robertson's converted trawler the 'Ocean Bounty' nestling by the pier, 17th June 1994. DGA.

The 'Ocean Bounty' anchored within the Shiants. DGA.

Communing with White-beaked Dolphins.

Dougal Andrew (legendary island goer and leader of the 'OB' cruises) and author, triumphant after magic night on North Rona, 19th June 1993. DGC.

Ann Wilson rounding Stac Lee, St. Kilda, 17th June 1994. DGA.

David Wilson admiring window of chapel, Inishdooey, Donegal, 16th June 1995. DGA.

Above: Alistair Simpson, peerless skipper of 'OB', and author, leaving Tory Island, 14th June 1995. DGC.

Left: waiting for a boat: a common lot of birdwatchers wanting to get on to the next bird. From top, Peter Naylor, author, David Wilson. DGC.

Below: "ransomed, healed, restored, forgiven", 12 companions take leave of 'Ocean Bounty' at the Corpach mouth of the Caledonian Canal on 20th June 1995. From left, David Clugston, Estlin Waters, Robin Andrew, Dougal Andrew, Kate Sutton, Ann Wilson, the author, Alistair Simpson, Frank Spragge, Peter Naylor, Joe Cunningham, David Wilson. Kevin Shepherd.

The challenge of a count frequently produces only mirth from fellow observers. DGC.

The Birding Klondike

The start-up costs of a hobby – a first estimate of the birdwatching industry – expenditure on bird feeders and food – subscriptions and donations – sales of optical equipment – costs of bird travel – the changing market in bird books – recent best sellers – multi-media - sales and commissions of art – the inflation of topical news – a Dusky Warbler courtesy of the AA – other small expenditures – a total marketplace in Britain and America – an annual budget of purposeful observation - £12,000 for no win

When I wrote the only full review ever of a decade's birdwatching in *Birdwatching in the Seventies* (1981), I included a measure of the start-up cost of the hobby cum amateur science. In 1960, a pair of good binoculars, a field guide, a notebook, and subscriptions to *British Birds*, the RSPB and a county ornithological society would have needed the expenditure of no more than £50. In 1980, the same goods and contacts cost at least £115 and the by then increasingly obligatory modern telescope and tripod twice as much. By 2000, the same basic kit cost over £510 and the choice of telescopes, tripods, other ancillary equipment and information gadgets had gone so berserk that a further £1,000 could easily have flown away. Then I noticed in one of Stuart Winter's columns in *Birdwatching* an overall estimate of the turnover in 2000 for the entire birdwatching and caring industry. This was a staggering £200 million. Clearly here was the stuff of a chapter.

So I phoned Stuart and discovered that the industry estimate had in fact come from Chris Harbard. As the then Press Officer of the RSPB, the latter had published in December 1998 some measures of the British interest in birds. Entitled 'Bird barmy Britain', his article did not aspire to being statistically proven market research but it presented enough factual information for me to complete more confidently my own investigations. What follows is still not as accurate a set of estimates as the subject merits but as Chris said, "It is high time to try and scale all the activity and expenditure".

By far the biggest spend on birds by Britons is the direct provision of food to them. Forty years ago, this was spurred mainly by hard weather in winter and took the form mostly of scraps and water. Today, winter feeding occurs at up to 14 million homes and as much as 60,000 tons of bird seed and peanuts is purposefully added to the traditional scraps. With the annual growth rate of individual bird food supply companies still as high as 35%, we have come a long way from the first red bag of peanuts and no market ceiling is yet visible. Given a cold winter – bird food sells in a reverse weather equation to ice cream – the current market values of £25 million for bird feeding equipment and nest boxes and £95–125 million for bought-in food will be exceeded. Chris Whittles whose company CJ Wildbird Foods has done most to set standards for the industry continues to see a bright future for it, if only because new 'all-year dishes' can now be safely served in the breeding season.

The next largest investment in birds is made through the flow of subscriptions

"*I'm proud of the basic British appreciation of the bird as an individual being, coming through in protection and conservation, but also the passionate involvement in watching 'a great bird' – people really do enthuse over birds and appreciate them for what they are.*" Rob Hume, *in litt.* (2000)

An observation from the editor of *Birds* and thus in touch with a million members of the RSPB.

Bird feeding has come a long way since the original bird-magnetic red bag of peanuts, to which Siskins and Blackcaps (top) were early visitors. In 2002, an all-year menu of feeds and a full range of purpose-built feeders were available from companies like CJ Wildbird Foods and Jacobi Jayne. Particularly for tits, feeders have become crucial stations in winter survival strategies.

and donations to the conservation charities. Isolating the direct appeal of birds within those organisations with a broad remit for all of wildlife is not possible, but from latest published accounts, it is clear that £40 million goes to the RSPB and over £2 million to the Wildfowl and Wetland Trust and the BTO. Assuming that only 25% of the Wildlife Trust memberships are motivated by a pre-eminent interest in birds, then a further £2 million may be attributed to bird conservation funds. The value of subscriptions and occasional gifts to the 200 local societies, bird clubs and smaller ornithological groups in the truly amateur network is by comparison tiny. Even lumping in the revenues of the BOU, BOC and the modern international societies, it is unlikely to reach £1 million. This is a really small price to pay for the annual product of bird knowledge enshrined in the county and other area reports.

After a total of £165–195 million spent on direct and indirect caring for birds and the amateur association of their keenest observers, the next major expenditure is on optical equipment. Surprisingly, in view of the fact that leading manufacturers have long since targeted Britain as their biggest European marketplace, only the friendly people of Opticron could remember or would offer any kind of sales estimate for binoculars, telescopes and tripods. This was £45 million in the late 1990s. Every other company of the five that I approached agreed that the market was indeed "huge", with "about three pairs in four sold to birdwatchers" and "all now imported". To me, the last comment was particularly chilling. I still grieve not just for my own favourite 9 x 35 Ross Stepruvas, which were by far the handiest of my own four pairs used over 58 years, but also for the other trusted British brands such as Barr and Stroud, Kershaw and Wray, all, like British motorbikes, defeated by superior European and Japanese upstarts. Regrettably no one offered me any comparable figure for cameras and camcorders and the films that they consume but, based on a sample of camera and like use by 25 current or past observers with whom I have shared field work, I came up with a minimal estimate of £6 million for long lens photography of birds. My total estimate of £51 million for all optical or photographic aids is twice the £25 million of Chris Harbard in 1998 but in no other service or product field is the consumer demand of bird-watchers more constantly driven. As a reputedly mean Scot, keener on field craft than capital investment, I do sometimes have to blink as my friends and acquaintances, noticeably those without the cost of a family, rotate constantly model after model of binoculars and telescope. One of them can choose his daily eyesight assistance from an array of eight instruments worth £7,150. Indeed if you are to compete as a photographer of rarities, your lens alone can require a five figure investment. The dominance of optical equipment in the needs of bird-watchers is never more obvious than when they meet at fairs or when they read their more popular literature. Whatever its precise size, the market for optics is clearly buoyant and the flow of new products does not cease. The brand positions of the manufacturers are not researched but in 2001, the excellent optics of Swarowski of Austria had apparently gone ahead of those of the previous co-leaders Leica and Zeiss of Germany.

Also buoyant until the terrorist attacks upon America on 11 September 2001 was the bird travel market. Expenditure on fuel or fares within the UK was estimated by Chris Harbard to be over £25 million. This equates in 2001 to little more than 0.5 million filled petrol or diesel tanks or 10 fills each for say 50,000 mobile observers. Such a consumption level could not satisfy, for example, the average reader of the *Birdwatcher's Yearbook* who goes out after birds on 120 days of the year. Once again the estimate may be low and does not include any allowance for food and accommodation. The latter two costs has been assessed at £12 million in the case of the one million visitors to RSPB reserves. A less approximate figure can be attached to the purchases of bird tours and other specialised birding holidays. As demonstrated elsewhere, these amount to between £9.5 million and £11 million, these points bracketing the 1998 estimate of £10 million. Thus total British travel costs after birds could exceed £50 million, although a good part of bird tour expenditure is (or was until 2001) made by foreigners.

We have now reached a total bird-occasioned marketplace of £266–296 million but are far from the end of expenditure. As I enquired after the smaller spends, the lack of precise answers continued to dog the exercise but one field deserves extended discussion, that of bird books. In contrast to the booming market for optics, their sales are subject to mixed reports. According to Tony Swann of Wheldon Wesley, the axiom that 'bird books always sell' no longer applies. Nowadays the publishers of new titles and particularly the antiquarian booksellers have less faith in the ornithological sector of the natural history market. The danger of new sales collapsing almost into the industry's last trench, that of the identi-fication guides, is a real prospect to some. However much it would fit the recent trends in birding behaviour and ever-wider travel, the break with the narrative tradition would be a frightening step towards the dumming down of a great subject. Interestingly, although the last recession of the early 1990s provoked a flood of 20th century books into the second-, or more, hand outlets, material from the 19th century is now decidedly scarce and keen-eyed dealers are increasingly deflecting classic books to auctions.

A complete collection of some duplicates of John Gould's volumes recently made £500,000 and well-kept sets of Victorian classics like the Rev. F. A. O. Morris's *History of British Birds* and particularly Lord Lilford's *The Coloured Figures of the Birds of the British Islands* (1885-97), featuring largely Archibald Thorburn's early plates, still attract increasing bids at the houses of Phillips, Sotheby's and Christies. By contrast, other seemingly great but more modern books have fallen in value and, with demand at standstill, may yet deflate further. In October 2001, Christopher Johnson of St. Anne's Books shattered my investment strategy for Richard Meinertzhagen's *Birds of Arabia* (1954) by telling me that as he had four unsold copies in stock, he was not interested in a fifth. So much for my supposed £500 jewel of a book. Opinions on the scale of user rather than collector purchases of the second-hand book market vary widely from £100,000 to £250,000 or more, but significantly no one talks of growth.

As already noted, the marketplace for the still-increasing tribe of field guides (now covering almost all of the world's regions if not countries) and the closely related family monographs remains buoyant. Such books now form the dominant ornithological literature, informing new readers efficiently about identification and distribution but treating any other subject theme only sketchily. Among them, the European scene has recently witnessed two remarkable successes. The more artistic and inspired is undoubtedly Lars Jonsson's *Birds of Europe with North Africa and the Middle East* (1992); the more complete and disciplined is the Swedish, English and Irish co-operative effort called the *Collins Bird Guide* (1999). Both are discussed more fully elsewhere. Here I note that the latter is the runaway in current sales, already by the end of 2001 into its third reprint in English and thus close to 100,000 copies sold in that language alone. Harper Collins freely admit to giving it marketing priority but fascinatingly nearly 50 years of loyalty to 'Peterson', the Daddy of the genre, secures still for its fifth edition (1994) annual sales of close to 2,000. "People are so fond of Peterson that worn out copies are replaced immediately", smiled Myles Archibald of Harper Collins. In recent years, the only other bird book to match the fastest selling guides has been the companion volume of the BBC's *Life of Birds* television series, with Sir David Attenborough representing his screen narrative in print. Over 110,000 copies have been sold. By way of contrast, the most popular of the supposedly collectable volumes in the Collins *New Naturalist*, the Academic Press *Natural World* and the Poyser *Natural History* series did or do well to reach sales of 4,000-5,000. Given the recurrence of recent successes, it is likely that the annual turnover in new bird books will be at least £6–7.5 million. This bracket equates to about 40% of the market for all natural history titles. Chris Harbard calculated that the offer of at least 5,000 titles was more than fivefold the number in 1985. Increasing demand from older, retired entrants to birdwatching is another important growth factor.

Duncan Macdonald and Sue Fleming of Wildsounds feel that the above bird book market estimates are pessimistic, allowing too little for the end of the net book agreement, the seemingly complete inventory offered on the internet, the addition of CD-Roms, and the habit of both new and old booksellers of discounting prices. This last practice has affected even the value of *BWP*, with loyal customers and retailers amazed to see Oxford University Press react to excessive stocks by dropping its price and that of the Concise Edition to a mere 30% of the original cost. Not content with this, OUP then introduced the CD-Rom of the full series in a further disruption of normal sales. Then the discounters advertised even lower prices, although in the case of *BWP* and several other books no lower-costed stocks existed to substantiate the offers. Such fiction contributes to the marketing syndrome of the 'mirage on the net' and has hastened the retirement of trusty booksellers like Peter Rathbone, the founder of *Subbuteo Books*. One thing seems clear and that is the high likelihood of the book surviving as one of the most useful and eye-pleasing tools to the birdwatcher. CD-Rom versions of books sell at a rate equivalent to only 20% of the printed original and assist neither aesthetic enjoyment nor bedtime reading. Furthermore, some computer systems are unable to cope with the full research and overlay tasks set by users. OUP have had to add to their web-site special guidance on the use of the *BWP* CD-Rom. Wildsounds were also helpful in curing my almost incomplete ignorance of the new multi-media market in bird images and sounds. After 15 years of developing their business, they see buyers as remarkably sensitive to the evocations of the latest TV series. Increases in sales are noted after Bill Oddie's British adventures as well as David Attenborough's globe trots, while other consumers ring up for soundscapes of Norfolk to mollify funerals and of Wild Africa to spice up barbecues. Both Wildsounds and John Wyatt, publisher of the helpful Teach Yourself Bird Sounds series of tapes, recognise periodic pulses of re-purchase from their audiences but these handy in-car products and the videos of Dave Gosney and Paul Doherty, to mention only two working in that medium, are all worryingly subject to nefarious copying. No-one ventured an estimate of the total marketplace but I will venture £400,000. I cannot guess the unadulterated value of the copied tapes used to lure birds out of dense cover or into mist-nets, a guileful practice in need of some self-regulation according to Mark Constantine, Magnus Robb and Arnoud van den Berg, who are at work on a much-needed full sound guide to match the 'Peterson', 'Jonsson' and 'Svensson' texts.

Related to the trade in bird books and particularly relevant to the needs of younger birdwatchers for topical information are the journals and magazines. Excluding those that come from the price of a membership like the RSPB's *Birds* and *BTO News*, the subscriptions to and purchases of the half-dozen most read were estimated to be £2 million in 1998. With one reporting a recent major decline and at least three others showing no more than steady sales, I find it difficult to prove current sales of more than £1.7 million but even that sum signifies a remarkable advance on the small sum spent on the largely ephemeral popular media of the early 1960s. Allied to this trade are those of illustrations, paintings and increasingly the colour photographs that now dominate avian representation in the topical literature. Very little has been published on their turnovers but the annual exhibition of the Society of Wildlife Artists took £94,000 in 2001, putting £22,000 on to its previous record sale. With at least two other national shows with five-figure sales, at least five annual gallery revenues well into six figures and many artists reporting as many private as public sales, the market for newly created bird art for personal possession and display must exceed £1 million. The demand for the more classic examples of 20th century artwork is also strong. For example, Sotheby's catalogue of bird paintings for auction in November 1995 contained 314 lots, with suggested first bids totalling about £425,000. Archibald Thorburn's pieces may sell for as much as £20,000 each. A total revenue from such works of at least £1.2 million in Britain is likely but how much of that goes overseas is not known. With an early following in America, Sir Peter Scott was particularly adroit at exploiting his export potential. As he once said to my mother while showing her his studio, "This is a potboiler for America. Perhaps a few more snowflakes, don't you think?" At nineteen, I was shocked but only until I had to make a living myself.

The current market for works essentially of illustrative purpose is shadowy, with publishers paying distinctly variable rates, but given that at least 25 British bird artists earn most of their living from book work, another market of around £0.75 million is indicated. Even without any complete figure for other sectors from home decorations and Christmas cards to the emergent, hugely seductive

"I went to Slimbridge and met Peter Scott who shambled into a hide, unrecognised behind us, and in about two minutes, with a ropy looking pair of bins, found a Lesser Whitefront in a flock of geese we had been watching for about an hour. He talked with us for about an hour and took us to see his new Sun Bitterns." Colin Bibby, *in litt.* (2000)

digital photographs, it is conceivable that sales of bird art in all its forms passes £3 million. Altogether, sales of books, journals, artworks and reproductions featuring birds reach £11–13 million.

The next largest market segment is in clothing. Harbard estimated this to be worth at least £5 million and certainly the demand for specialised wear created by the constant advertising of branded products like Barbour waxed jackets and Hunter rubber boots is strong. Given that such injunction reaches far more people than dedicated observers, the above estimate may well be minimal but I could find no one who would hazard a guess.

Next, there is the cost of the modern, highly developed grapevines of bird news. These have spawned, after the first national *Birdline* created by *Bird Information Service* in 1986, a second national service in rarity news, a dozen *Regional Birdlines* offering usually a broader, more interesting span of bird subjects, and since 1991 two pager systems, one with internet and cellular phone services. Charged at premium rates of up to 51p per minute, even my minimal assimilation of the *Birdline* information offered on an autumn day full of rarities soon costs pounds but seemingly none of the service providers are able to distinguish the number of individual callers, making the total traffic in and purchase of rarity and other bird news hard to measure. After conversations with three *Birdline* operators, *Rare Bird Alert* and *Birdnet Optics*, I had few facts, but setting them against the observed use of the services in my circle of friends, I hazard the following on the scale of the new information systems. The revenue of the original *Birdline* may have achieved £1 million within its launch year and the whole system perhaps £1.25–1.5 million in the first flush of regionalisation which began in 1989. With the launch of the pager and allied systems, however, it is likely that at least £500,000 of information purchase was converted to them. Certainly during the long periods of rarity-thin tapes that followed the 1999 bonanza of twitchable rarities (those who stayed for later-coming observers), the traffic and income from both national and regional birdlines has decreased. Any repeated allegation of the supposedly huge proceeds taken by the *Birdline* initiators is now sharply denied. I worry that I have underscored the revenue of the new information systems and certainly have no way of adding any scale of traffic or proceeds arising from the 137 websites (Hamlett 2001) now available to keep yet more consumers sitting inside by their screens rather than looking outside for birds. My final judgement is that the information flows in (for free) from up to 4,000 observers and streams out (at surprisingly variable charges, so *caveat emptor*) to up to 9,500 twitchers and other bird-news-hungry people. Within the former most eager camp, the number of pagers on wrists and in pockets is still growing and will soon pass 2,500. Other information services like phone texting and daily email reports are also available but as yet have made no large impact.

I am short of tales to illustrate the birding Klondike but in contrast to today's instant purchasable intelligence, I do recall with great nostalgia the old days of the more restricted and privileged grapevines. Here is a classic example:

> November 11th, 1967. On advice from Bob Emmett that a Dusky Warbler – no less – had been put to bed at Dungeness, sleep proved

The Dungeness Dusky Warbler of 11th November 1967. The drawing of the head was made with the bird held gently in my left hand. Such intimate contacts are among the best stuff of observatory training.

difficult. So up at 0130, on road by 0145 with horribly unfull petrol tank, performed 'Mobil economy run' across Cotswolds to M4 'services' and reached Acton at 04:15. We made Dungeness by 06:30 and the trapping area by 07:00; a '*chak*' issued from a bush at 07:15 but no bird became evident. Happily, Peter Grant dug it out at 07:45 and we smartly embraced it as it worked along a line of sallows. There was an excellent supporting cast but the real surprise was the appearance of Ron Johns about 11:00. Gleefully, we had assumed him to be safely stuck in north Norfolk but a small miracle of 'search and rescue' phone calls performed by Ann (his first wife) and the AA had ended in his purchase of a cup of tea in a transport café near Newmarket being interrupted by an injunction to phone home immediately. Hence smart return south, brief kip and … lucky bugger, as ever. Together, we all drooled over the rare *Phylloscopus*, by then fully confirmed by Bob Scott's capture and in hand examination.

This was the sort of episode that was the stuff of early twitching legends, oft told and enhanced at Cley in Nancy's café or The George Hotel (Cocker 2001) and even as early as 1967 picked up by the BBC's Natural History editors. Hence on 31st December of that year, the first-ever radio discussion of twitching with John Sparks interviewing Bob Emmett and Ron in Slough and Kenneth Allsop talking to me in the studio. The contrast between our forecast of the maximum potential British year list and the actual achievement of the competitive twitchers a third of a century later is remarkable. Ron's first claim of 291 from 30,000 miles in 1967 has long since yielded to several others that have come ever closer to 400, products of even more manic flying and driving in the wake of the now ever-flowing tide of news. The changed times were never more obvious than on one of my now rare visits to Norfolk in July 1999. Puzzled by the lack of birdwatchers exchanging finds over a midday pint in The George, I asked the landlord where the traditional gathering was. "Killed off by first the birdlines and now the pagers" was his sad-toned reply and curtly he got on with serving the less fickle grockles of the day.

Returning from the above digressions, I come to the last field of birdwatcher expenditure, that on permanent personal recording. For many, this is as simple as a plain but always pregnant notebook and a pencil or pen; for others, notebooks are twinned with sketchbooks and accompanied by diaries; for some, it is all of the above and also files full of descriptions, annual record summaries and, given a photographic bent, all the paraphernalia of processed film storage and projection. I estimate that at least 30,000 observers from garden watchers upwards keep records and that their expenditure on their safe keeping varies from £3.50 to £40.00 per year. A total purchase of over £300,000 is likely.

Although my exercise in sizing the British birdwatching and birding industry is patently incomplete, I think it is good enough to support an estimate of £300 million a year at the end of the 20th century. With claims of as many as 5.5 million people with more than a passing or feeding interest in birds (and the cumulative sales of all field guides probably not far short of that figure), the estimate may still be low. In the world's most consumptive economy, the United States of

"Within certain age groups (from teens to my age, early 30's) I think that things may well have peaked. However, the already mentioned 'grey brigade' continues to expand from my experiences around north Norfolk. The increased wealth, the earlier retirement ages and improved optics and super field guides mean that birdwatching is still an increasingly popular outdoor pursuit for older members of our active society". Mark Golley, *in litt.* (2000)

The observations of one of north Norfolk's wardens – and latest scribe of Cley's birds – on the contra-flow in the appeal of birdwatching.

This painting of a dead Snipe chick from St. Kilda in July 1956 turned out to have featured the first plumage of the Faeroese subspecies, the breeding form of the Northern Isles and Hebridean Outliers. Keeping even the dustiest old records safe can be the crux of retrospective discovery.

America, \$5.2 billion (£3.25 billion) is spent by birders on goods and services related to bird feeding and watching (1991 Fish and Wildlife Study by U.S. Department of Commerce; 1997 American Birding Association Member Survey). There judged by total population size, over four times the number of people spent about ten times the amount of money on their hobby and recreation than in Britain. Clearly there is room for growth in British birdwatching but the opportunities for better marketing and hence better-informed demand will not be taken unless one of the British institutions takes on the task of its full definition. The RSPB would do well to match the role of the ABA. Its marketing department fusses endlessly over its members and their needs and attitudes. It could be a lot better informed and more confident if it took on the responsibility of informing all birdwatchers, birders and bird carers about their total universe and its evolving status. Fully practised audit procedures exist in the excellent British market research industry. At a cost of just 10 top model telescopes, only part of an omnibus research survey could remove most of my approximations in one month.

Finally, if all the above has left you rather lost for what a sensible budget of a year's birdwatching for yourself might be, here is what 240 days containing at least one purposeful observation cost to me in 1998: primary patchwork study (200 days) £800, mainly on petrol; secondary studies (three weeks in Kazakhstan, ten days in Donegal, five days on BTO surveys) £1,200, of which £650 air fares; sheer recreation (three days camping on east coast, one visit to Slimbridge) £175; attendance at two conferences £300; topical news (four journals, occasional birdline calls) £190; total expenditure £2,665. It should be noted that in this figure there is no allowance for any time spent on twitching. As a pastime, it is hellishly expensive and yet the result in single collected rarity or list length is not guaranteed, witness the hapless Adrian Webb, who in 2000 spent £12,000 in an attempt to take over the pole position of Britain's top year lister only to have his efforts thoroughly rubbished by the incumbent champion Lee Evans.

During the International Ornithological Conference centred at Helsinki, Finland in 1958, the late Horace Alexander and James Ferguson-Lees stood together at a field outing. Each in his turn gave exemplary leadership to purposeful birdwatchers. JC.

At the Society of Wildlife Artists exhibition of 1989, the author encourages two supporters towards a purchase.

At the Society of Wildlife Artists exhibition of 1989, Robert Gillmor, then president of the society, praises the work of Darren Rees. The latter's Fairy Tern has just won the inaugural RSPB Wildlife Art Award.

Birdwatchers maintain amazingly long and close links with one another. Dougal and Gratian Andrew have been attracting visitors to the Gullane home for four decades. Here they pay the author and Flamborough Head a return visit on 15th October 1993. Note Dougal's red socks, beloved also by Peter Scott.

International Connections

A first meeting with a European ornithologist – ancient falconry and its international esteems – early international discourse – the loss of ornithological isolation – a second tryst with a foreigner – shared enjoyments – common practice – the British view of foreign expertise and vice versa – Lars Svensson's best bird book – the drift of youth – beckoning globalisation – some numbers of foreign birdwatchers – the legacy of Yankee gentlemen

In April 1955, I had come of (then) voting age but was still nervous of approaching even other British birdwatchers unannounced. But for Dad, my first experience of the help available from European colleagues might have been delayed indefinitely. For the first time since his escape from Ijmuiden with part of the Dutch fishing fleet in 1939, Jackson Wallace was due to return to Holland. He took me along to enjoy what birds would appear and other cultural joys from the candle-lit Franz Hals paintings in Haarlem to scrumptious *ris tafel*. In his briefcase was the telephone number of Dr. Jan Kist, a Dutch ornithologist known to Guy Mountfort as editor of the Dutch version of the 'Peterson' guide. Finished with other duties, we reached The Hague on the 13th and after the umpteenth filial plea, Dad rang Dr. Kist. A busy professional, he was at first circumspect but then made the tactical error of asking how keen I was. "He's a fanatic", answered Dad. After a pregnant pause the Doctor agreed to be my guide, adding "One should always find time for a fanatic". Thus was the morning of the 14th a time for birdwatching bliss, my first experience of being shown another country's teeming birds by a caring expert. So also came the confirmation that my hobby and its associated science was not wholly nor even largely British. There was more to foreigners than mentions in the Witherby *Handbook*.

Informed international discourse on birds began about 2000 BC when the skilled practice of falconry spread out from one likely source in Scythia. It took, however, another 2,600 years for the world's oldest sport (or recorded discussion of it) to reach Britain. At the end of the 6th century, Ethelbert, King of Kent, accepted Christianity from St. Augustine but still showed his rapacious mien by sending to Germany for a brace of falcons that would fly at 'cranes'. Eventually he received from Archbishop Mons the two falcons and a hawk, already trained to the wrist. The popularity of the mind-focusing craft cum sport grew steadily. When Richard I went to the Crusades in 1189, his baggage train included 300 hunting birds and he brought back equipment and training ideas learnt from his foes. It is debatable if England's most intrepid monarch did ever meet Salhadin but it seems to be accepted that when Richard's galley was coming in to Acre, one of his birds escaped and ended up in the mews of his chief opponent. Salhadin refused to return it to Richard, a great slight in the days when the esteem of falcons and hawks was such that they held exceptional ransom value. From the era of the Crusades came also a remarkable written work. Drawing again on the sophisticated handling of birds in the Arab nations, the Holy Roman Emperor, Frederick II,

"Wherever I travel to, I always think it's crucially important to talk to the local people. Even a mediocre birder can be brilliant on his home turf. The Swedes with whom I worked at Ottenby Bird Observatory were outstanding." Jane Reid, *in litt.* (1999)

A pair of Gyr Falcons shrug off an Icelandic snow flurry. Such dark-phase birds did feature in old British records but today the confusion of escaped large hybrid falcons has made their claims unpopular. Nearly all modern acceptances have been of the white-phase birds of Greenland.

published the first advice indicating that patience and care could aid the relationship of man and hunting animal. In the reign of Henry III, the lure of the best feathered hunters led to a fascinating royal trade. To fulfil the English King's desire for Gyr Falcons, King Haakon of Norway sent his falconers on a two-year expedition to Iceland, their efforts eventually allowing him to exchange six Gyrs and six other raptors for reams of fine scarlet cloth. Henry sent special messengers to a Norfolk port to collect the birds with due ceremony.

The importance of birds of prey to those who ruled eventually translated into social customs which set a hierarchical ranking of species. Thus in 1400, Abbess Juliana Berne awarded in the *Boke of St. Albans* an eagle to an Emperor, a Gyr Falcon to a King and then lesser pairings of diurnal raptor and title all the way down to a Kestrel for a knave. In the early 1600s in another remarkable episode, Cormorants were trained to fish for James I and some were exported by him to nobles in France and Venice. The first *lingua franca* of man and bird began to recede in the face of the new and far more time-efficient sporting guns and hunting with birds was explicitly disapproved by the Puritans of the Commonwealth interregnum. In Britain by the late 18th century, hawking was rarely practised and today it survives in a very changed and prescribed guise, with the activities of the 1,200 members of the British Falconers Club less known to most birdwatchers than the motives of more conservation-minded associations such as the Hawk and Owl Trust.

For birds in general, the certain record of international discourse is insubstantial before the 16th century. Peter Tate in *Birds, Men and Books* (1986) regarded the work of the primitive authors, even the clerics who exchanged their references, as inconsequential. It is certain, however, that two later authors awarded points by Peter Tate for factual observation or the review of prior fiction had contact with Europeans of like minds and interests. They were William Turner, born in Morpeth around 1500, and John Caius or Kay, born in Norwich in 1510. The two were

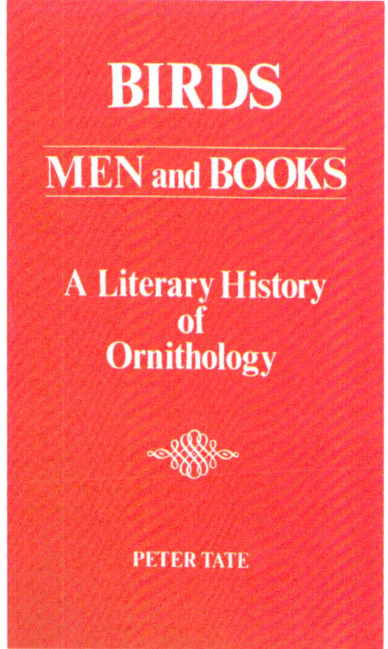

BIRDS

MEN and BOOKS

A Literary History of Ornithology

PETER TATE

Peter Tate's briskly informative *Birds, Men and Books* makes an excellent companion read to James Fisher's *The Shell Bird Book*.

seriously divided by religion. Turner was an uncompromising Protestant, forced twice to leave England to avoid persecution; Kay was a devout Catholic. What might have been a productive partnership, based on a shared Cambridge education in medicine, did not blossom but their travels in search of further qualifications and ornithological discourse took them across Western Europe to Germany, Switzerland and Italy, with both of them spending two years in the last country.

In Turner's case, there is clear evidence that in 1543 in Zurich he met Conrad Gesner who was one of the greatest natural scientists of his time. Gesner and Turner did bond, becoming frequent correspondents; Gesner also exchanged letters with Kay. Turner made another friend in the early ranks of European naturalists, Gybertus Longolius of Utrecht. In 1544 Turner dedicated his *Avium Praecipuarum Historia* to King Edward VI but the book was written in the then overlord of languages, Latin, and it was published in Cologne. Turner also undertook to edit the *Dialogue de Avibus* left incomplete due to Longolius's early death. While evidence for international co-operation in ornithology as early as the mid-16th century is thus provided, the span of certain wild species discussed was short. In the first work noted above, only 101 (less than a quarter of the actual total of British species) are identifiable. In 1555, Gesner matched Turner's output with his *Medici Tigurine, Historiae Animalium 3: Qui est de Avium natura, etc*, but after that, the other early natural histories written in Europe largely regressed into repetitive compilation with little evidence of other than solitary advances in real observation. It took a Norfolk man who also got his doctorate in medicine abroad, in Leiden in 1633, to exhibit alert and unbiased observation free of medieval nonsense. He was Thomas Browne of Norwich; examples of his genuine field observations were given in another chapter. Sadly, although he wrote sagely about life in general, the material for what could have been Britain and Europe's first model bird book remained scattered piecemeal in letters to John Ray and Dr. Christopher Merrett.

Wider travels within Europe in the later years of the 17th century allowed both Francis Willughby and John Ray to observe and understand birds in a continental context. Altogether, their independent and joint journeys took them not just to the Low Countries and France but also to Germany, Italy and even Sicily and Malta. Undoubtedly Ray's approach to classification was later to influence the Swede Carolus Linnaeus but in the early 18th century, a temporary collapse of tested science and the rash of nationally jumbled nomenclatures made the passing of the international ornithological baton hard to follow. The mists did not clear in Britain until the 1740s, when the work of George Edwards included a partial overview of European knowledge and, more significantly, the first transatlantic edition of a book featuring American birds, Mark Catesby's *The Natural History of Carolina, Florida and the Bahamas* (1754), appeared. Edwards' main treatise was *A Natural History of Birds* (1743-51) which drew on his own experience of Norway, Holland and France and lengthy correspondence with Linnaeus. It and its later revision, written up to 1764, were translated into French, Dutch and German. Thus for the first time a British pen wrote about some of the birds of the Holarctic region and its wielder exported his work. A similar renaissance in ornithological thought occurred in France, where the classifications of birds by the Comte de Buffon from 1749 and M. J. Brisson in 1756 showed for the first time glimmers of modern systematics and were partially adopted in later British books. With the advent in 1758 of the Linnaean classification and nomenclature of birds, European observations on birds acquired at last the fundament of a respectable central discipline.

In the late years of the 18th century, ornithological knowledge was readily swapped across the Atlantic, the English Channel and the North Sea. Even though neither ever left Britain, both Gilbert White and Thomas Pennant solicited and respected international opinion. The latter was a frequent correspondent of Linnaeus who recommended him for membership of the Royal Society of Uppsala, the first foreign honour conferred on a British zoologist. Conversely, the respect felt in Britain for Linnaeus, who was primarily a botanist but blessed by all natural scientists for his introduction of the binomial system of nomenclature, was reflected in the formation in London of the Linnaean Society in 1778, the year of his death. Beginning with less than ten members, the society now attracts 2,500.

By the beginning of the 19th century, all national isolation in ornithology was past. John Jacques Audubon left America five times to draw on French, English and Scottish sources of knowledge, skill and finance. John Gould's compulsions took him to three continents and as the vogue of stocktaking collection took full hold, the world became its oyster. Significantly for America, another refugee from Scotland, Alexander Wilson, produced between 1802 and 1813 the first comprehensive description of Nearctic birds. Of the then 343 known species, 264 were truthfully featured. His work is an excellent example of how the vagaries of the earlier ornithologies were at last despatched and an international network of authorities was secured. Progress to a more certain science was fast and there was a spate of faunistic accounts, particularly from Germany and Scandinavia. James Fisher considered it safe to say that by 1870, nearly every part of the civilised world had a basic fauna with due attention paid to birds. The additions to ornithological knowledge became, however, increasingly scattered in the new journals of museums and exploration societies. The days of ornithological encyclopaedias issuing from a single mind and pen were past. Finally, with the adoption of a more rounded, fully zoological and international approach to avian science at the end of the 19th century, European ornithology became less dominated by classification and stocktaking and more concerned with increasingly diverse accounts of current work on live birds.

"Haben sie Gross Trappe gesehen?", asked a very large voice of me on the Tadten plain of eastern Austria on 19th May 1964. I had and soon the voice's owner was enjoying four Great Bustards. In my then 21 years of recorded birdwatching, the towering Scandinavian was only the third non-British observer that I had ever met and the first that I had actually helped. Musing on our meeting, what struck me most was the efficacy of our separate national intelligences on relict species which had brought us simultaneously to the same edge of steppe just short of the then Iron Curtain. Memories stirred too of other Scandinavian contributions to my learning – the *Fair Isle Bird Observatory Bulletin* had been full

The open fields of eastern Austria provide a steppe-like niche for Great Bustards. In the decades of the Iron Curtain, their Western observers were closely watched by Eastern border guards.

of them – and once again I understood fully the internationality of my hobby and that the British joy in it was not singular. Other peoples were involved; other perceptions were being made.

Thirty-eight years on from the Tadten tryst, it is impossible to be chauvinistic about the practice of birdwatching and field ornithology. Almost every wealthy nation's observers meet constantly to share bird spectacles in every ecosystem of the planet. Quite when such intercourse began is not easy to ascertain. My oldest correspondents have offered no precise evidence for it before 1935, in which year the Oxford Ornithological Society took 30 members to Texel and the Naardermeer in Holland. More solitary travels to the wetlands of the Camargue, inside the Coto Donana and on to the goose-grounds of the east European steppe are also recalled for the 1930s but the first lengthy engagements with north American bird haunts appear to have been the accident of postings in the Second World War. The causes of the post-war growth in international birdwatching were touched on in an earlier chapter but for a broader appreciation of the associated science, it was the series of the International Ornithological Conferences that became instrumental to regular exchanges. As their reviews in *British Birds* make clear, the IOCs have performed well as mobile Meccas and particularly since the Basle gathering in 1954, strong British contingents have exchanged everything from erudite papers to birding chatter with colleagues from other nations.

With the publication of more national journals containing papers of international significance, the 1960s saw the beginnings of a huge growth in original and, importantly, peer-reviewed observations. The relatively small churches of the Witherby *Handbook* and like European works were quickly proved to be insufficient centres of references. It became virtually impossible for anyone but academics to stay fully in touch with the widespread chains of research into fields such as bird behaviour and taxonomy. The amateur response to this situation remains fluid but like his professional partner, the unqualified, yet serious birdwatcher is now an increasingly international being. He or she shares most the universal need for reliable identification criteria that generates accurate status and distribution data and so provides the bases of proactive conservation strategies. Throughout

Peter Scott had an eye not just for wild fowl but also for the activities that they provoke from their hunters. Here fowlers use a flare to net ducks at night on a Caspian lagoon (from *Wild Chorus* 1938).

North America, Europe and the Antipodes, the framework of his or her hobby has an increasingly standard pattern of organisation and output, still rooted in past ornithology but now harvesting present and future objectives with astonishing loquacity and instancy through the modern media of communication. The growth of co-operative ventures in ringing, distribution mapping, population measurement, taxonomy and rarity vetting has flattened the national borders of ornithology almost completely. The initiatives are remarkable testaments to the maturity and vision of both the individual observers and the national institutions who, under the still widening survey and remit of Birdlife International, accept that birds are a common resource everywhere in need of certain definition and care.

To discover how British birdwatchers felt currently about their foreign colleagues, my questionnaire respondents were asked for their ratings on past inspirations and current contributions to birdwatching and field ornithology. The greatest praise was given to the group of four Fenno-Scandian countries. Of the 44 respondents, 26 recalled their past influences with pleasure and 29 awaited their current findings with eagerness and even envy for their detailed perceptiveness. Looking within the quartet, and no doubt aided by the glossiest of new journals *Alula*, the Finns are gaining ground on the Swedes. After the Nordic nations, and in a strong position given their relatively small number of practitioners, came the Dutch, gaining 14 mentions for both past and current work. In the latter, the new thrust in systematic research was much appreciated by Britons well aware that their own Establishment had until recently had all the vigour of a beached whale and so had administered no remedy to abate the taxonomic diarrhoea of the twitcher constituency. North America, in spite of its greatest mass of professional ornithologists and birders, came next in sixth position, achieving 12 mentions for past work and ten for current contributions. Australia and New Zealand led the tail of other respected ornithological nations which also included Israel and Spain. None of this quartet, however, were noted more than four times for either past or current work. Given the historical perspective sketched earlier in this chapter, the absence of any mention of French and German birdwatching and ornithology is astonishing. Both had been former partners in progress and, as the Editors of *BWP* found at cost of many hundreds of hours, the ornithological productivity of former East German observers was and remains huge. Undoubtedly, my sample of British opinion on other European contributions is biased by many comparisons made of the modern obsession with identification and rarities but overall it clearly demonstrates the great respect in which Britons now hold their fellow enthusiasts around the world. All British claims to any holding of pole positions have gone.

The reverse question immediately arises. How is the current British effort in birdwatching and field ornithology seen from abroad? I sent a questionnaire to 12 European authorities, all within the regular universe of *British Birds* correspondents. Regrettably, nine fought shy of response and although I was eventually able to get some commentary from seven countries, what follows is most influenced by the Swedish and Dutch views of our behaviour and performance. Our past and recent traditions remain highly respected. Undoubtedly the current European observers that took up birdwatching in the 1940s and 1950s accepted the Witherby *Handbook*

The complete internationality of modern birding is exemplified by the presence of a Spanish Sparrow on the cover of a Finnish journal.

"Until around 10 years ago, I feel we were regarded very highly throughout the world for our expertise and thoroughness. I still feel we do a lot for conservation globally but I am ashamed at the trivial way we seem to approach birding today. The rest of the world is starting to laugh at us. We seem to have lost the plot – everything revolves round supposed *rarities. I don't* think it will change – it's too late. Fast foods – fast birding." Frank Moffatt, *in litt.* (2000)

The latest cover of Lars Svennson's meticulous guide to the 'in the hand' identification of European passerines. A must for would-be experts.

and its continuum in *British Birds* as leading voices in the international creed of field ornithology. The sequential work of Witherby, Ticehurst, Tucker, Hollom and Ferguson-Lees attracted close readership and they and their associates were seen to set new standards in detailed studies and well-edited results. Our current vision and its products are, however, less universally approved. While the large scale and force of our ability to drive conservation and other ornithological projects and publications is still fully acknowledged, there is a feeling that our lack of linguistic facility (or sheer idleness with foreign tongues) has allowed what one correspondent diagnosed as 'Anglocentrism' to restrict our appreciation of the international diversity of ornithology. The main charge put by our colleagues is the large number of original perceptions in the European literature apparently missed by British readers, for example the brilliant renderings of voice by Erik Rosenberg in *Faglari Svevige* (1953) and the new and refreshing logic of Dutch taxonomic thought. Against such apparent neglect can be put the continuing British initiatives in field studies, particularly in North America and the Antipodes, and the British supply of much improved identification literature, beginning in India and South-east Asia and tackling recently East and West Africa. These may have been missed by foreign eyes.

Nevertheless, one particular set of Nordic observations deserve closer attention, coming as they do from Lars Svensson of Stockholm, well known to and respected by British ringers (and observers) for 30 years. His small but meticulous, classic 'in the hand' *Identification Guide to European Passerines* (1970-92) is called 'Svensson' in almost as warm a tone as the original *Field Guide* is named 'Peterson'. Lars is a cultured man with an enviably wide and intense experience of the Palearctic region in both field and museum. He respects the Witherby tradition and its legacy completely but points out the many parallel developments in Swedish ornithological skill, knowledge and literature that also began in the 1920s and similarly flourished in the 1950s and 1960s. In terms of textual quality, Lars *still* sees Erik Rosenberg's book as *the* best bird book written in Europe. He also points out that as a populist of conservation, Per Olaf Swanberg was active as early as Max Nicholson. Gustaf Rudebeck and Carl Edelstam made huge contributions to the development of the network of European bird observatories, assisting the foundation of Falsterbo, Ottenby and Capri, and also to the identification of raptors and the study of visible passage. They and their successors have seen to it that the full range of activities in modern purposeful birdwatching exists in Sweden. Although able to draw on a national population only a fifth as big as that of Britain, the Sveriges Ornitologiska Forening has a membership not far short of that of the BTO and even the number of Swedish twitchers approaches that of their British siblings. On a lighter note, the magazine of the latter is aptly entitled *Roadrunner*, its proceeds nevertheless contributing directly to conservation and research.

The excellence of Swedish and other Baltic birdwatching and ornithology might be thought to have insured its sure inheritance by younger observers but Soren Svensson and Lars Svensson (for Sweden) and Lasse Laine (for Finland) confirm that worryingly, as in some British eyes, the ranks of trained purposeful observers are ageing and thinning. The numbers of young recruits are steadily reducing in the face of the late 20th century's alternative seductions in computers, cellular phones, heavy rock music, roller skates and graffiti (*Ornis Svecia* 7:175-9). No change there, then, but such turning away from natural and scientific recreation is not universal. From Holland, Arnoud van den Berg makes the telling point that where ornithology organisations are relatively youthful – both SOVON (a BTO-like body) and the Dutch Birding Association have yet to celebrate 25th anniversaries – they continue to attract young recruits and so ensure their continuous rejuvenation. Is this why *Dutch Birding* is such an urgent and convincing journal? Good examples are always followed, witness the youthful and vibrant audience that I faced at the Dutch Birding Day in February 2002. The globalisation of the birdwatching and ornithological community is nowhere more evident than at the British Birdwatching Fair. In August 2001, over 30 countries were represented by live participants waving all sorts of banners and over the three days, I also heard in conversation news of ornithological doings in 12 other nations. With the internet running through every second of each day, the multilogue of peoples brought in touch by birds must produce a communication facility of enormous potential benefit, for those with hours free for mouse movements and cyberbirding.

My failure to attract European correspondents means that I cannot close this chapter with any real measure of the wider community of western Eurasian birdwatchers. In re-united Germany, Holland and Finland, a total population of 81 million people includes about 160,000 with an interest level equivalent to British bird club membership. In Holland, about one in ten of such people, or some 5,000, are described as 'hard core observers'; 500 of them carry pagers. In Finland, the last corps numbers 1,000, this figure suggesting that the Finns are *pro rata* the most rarity-conscious birdwatchers in Western Europe. Adding British information to the above, I estimate that the current incidence of birdwatching as a prime hobby in Western Europe runs at no more than 2% of its population. In Eastern Europe, recreational interest in birds appears to be much lower or less visibly expressed. Within Russia's 115 million inhabitants, 1,000 professional ornithologists have at most only 2,000 amateur colleagues. In Lithuania, there are only 300 birdwatchers in a nation of 3.7 million souls. Conversely for Israel with 6 million people, a total of 10,000 birdwatchers is reported. This figure includes, however, hundreds of bird tourists and only 200 'native experts'.

Elsewhere, the most numerous birders are undoubtedly those in the United States. They are also the most studied, with both Federal agencies and the American Birding Association undertaking research programmes into their behaviour. In less than a century, the hobby of birdwatching has mushroomed from a small field dominated by 'Yankee gentlemen' and a few clubs in north-eastern cities to its current status as the fifteenth most popular outdoor activity of 285 million Americans. This means that at least 70 million (33% of the over 16 year olds) go out birding at least once a year. Enthusiasts, so named because they go out on at least 50 days, number 19.7 million (9.3% of the over 16 year olds). How many of these would match the skill and productivity of purposeful European birdwatchers is not deducable from the research but certainly the sexual profile of the American

Dutch Birding is not all Dutch; half its text is in English. The second issue in 2000 contained Magnus Robb's accomplished analysis of Crossbill voices. A CD accompanied the text and sonagrams and demonstrated that there could be six 'choirs' within a supposedly single species. Staggering ornithology from a musicologist.

"*Professionally, it has to be the USA. They have the money to do good science, and taxonomy and evolutionary biology are not endangered species like they are in the UK.*" David Parkin, *in litt.* (2000)

The question sought those countries to which the respondent looked for new knowledge.

enthusiasts differs strikingly from their European counterparts. Mature ladies make up 62% of keen birders. The major growth factors in the advance of birdwatching in the USA include the near 70-year availability of field guides (two and a half decades more than in Europe), the widening of binocular ownership after the Second World War, the broad achievement of 'high school' education and the demand for leisure pursuits before and after child-rearing. In the USA, there was a marked surge in birding in the 1980s and 1990s and as yet there is no sign of any disaffection with the hobby. Many European birdwatchers still return from North America full of envy for its wide spaces and their still-teeming birds. Perhaps for once, big is beautiful.

A Cambridge Bird Club party explored north Norway in July 1957. The members got a free house and local 'gen' from Herr Schaanning of the Pasvik River. JC.

The Flamborough Ornithological Group visit Oman. Five-star hospitality was provided by (on right) the famous bird photographers, Hanne and Jens Eriksen. The latter is a Danish professor who co-runs the database of modern Omani ornithology. Their guests here are (from left) Andrew Lassey, Brian Hill and Andrew Grieve. DIMW.

Andrew Grieve and Andrew Lassey extract an Isabelline Shrike from a mist-net set in the mangroves at Shinas on 22nd January 2002. DIMW.

Another beautiful camp-site in the acacia woodland of Mussandam on 23rd January 2002. DIMW.

Happiness is a camp fire at Tiwi on 19th January 2002. The fuel is fiercely burning acacia, gathered from the ground. From right, Andrew Grieve, Brian Hill and Andrew Lassey. DIMW.

Stray Feathers

Smoke and perhaps fire – the habit of bird-inspired writing – some of its recent exponents – the pleasure of international correspondence – examples of a half-century of private ornithological literature – reverie, the antidote to disenchantment – an answered prayer

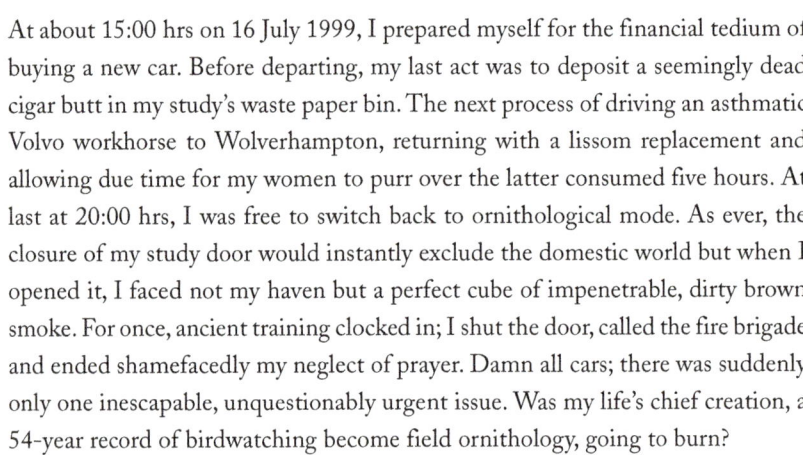

Humorous writing about birds and birdwatchers is too rare. Bill Oddie and Anthony McGeehan provide the chief exceptions to this rule. Shown here is the latter's first byline and a typical photograph and caption.

Of several claims for the Short-toed Treecreeper for Epping Forest, Essex, only one – a singing bird on 26 May 1975 – has made the statute book. Where the coastal waifs end up is a mystery.

At about 15:00 hrs on 16 July 1999, I prepared myself for the financial tedium of buying a new car. Before departing, my last act was to deposit a seemingly dead cigar butt in my study's waste paper bin. The next process of driving an asthmatic Volvo workhorse to Wolverhampton, returning with a lissom replacement and allowing due time for my women to purr over the latter consumed five hours. At last at 20:00 hrs, I was free to switch back to ornithological mode. As ever, the closure of my study door would instantly exclude the domestic world but when I opened it, I faced not my haven but a perfect cube of impenetrable, dirty brown smoke. For once, ancient training clocked in; I shut the door, called the fire brigade and ended shamefacedly my neglect of prayer. Damn all cars; there was suddenly only one inescapable, unquestionably urgent issue. Was my life's chief creation, a 54-year record of birdwatching become field ornithology, going to burn?

To my generation, the habit of writing down nearly every place searched and most birds seen came early, with in my case day lists kept from 1943 and count registers, site maps, outbursts of purple prose and abundant field notes and sketches all fully fledged by 1951. Unaware in those years that similar records had in fact been kept from at least the 17th century, we took our tutorials in ornithological letters from their more obvious modern sources, such as the daily logs in school ornithological societies, the narrative delights and jokes in the diaried history of the Isle of May (where Maury Meiklejohn never let bad weather pass without penning some delicious verses), the full logic of the Fair Isle recording discipline and, above all, correspondence with birdwatching betters. To the last fund, my own greatest contributors were Ken Williamson and Dougal Andrew. As my chief teenage mentor, the latter quickly took pole and most lasting position. His model diaries were and are amazing documents, written in full prose, meticulously typed with every bird name underlined in red and serially bound in vellum. His letters, still coming after 50 years, spill out news and humour in equal measure.

Today it seems such composed literary traffic between birdwatchers is less common. All is email and fax. Against the general aseptic trend in reportage, only the readers of *Birdwatch* and currently *Dutch Birding* and a few close friends have had the privilege of reading really engaging birdwatching essays. These were the topical and often riotous columns of Donegal and Ulster's champion observer, Anthony McGeehan. Part of Anthony's literary inspiration is the Lake Woebegon saga of Garrison Keillor, and his stimulation by such narrative gems shows that wide cultural experience enriches all interests, especially in the case of the most compulsive questor after birds, with whom I have shared precious time.

Yet there are great birdwatchers who have contributed hugely to the hobby and science without filling shelves and tea-boxes with a record of their personal experience. The late Chris Mead was one, having been for at least the last three decades a major fount of ornithological news, particularly in the recent annals of ringing and as marshaller of the BTO's recent proactive press campaign. On radio, he performed regularly as a veritable 'Moses' of birding. He should not have been "gobsmacked" (to use his own word), when the BOU gave him its chief medal for his long apostleship, but I was so, when he admitted to not keeping a bird diary. Presumably after his output of ringing schedules and recovery lists, time just ran out and he did have a Breckland garden full of feeders and birds as daily live entertainment. Each to his own reverie.

It is widely observed, however, that ornithological audiences always gobble up stories about the great characters in birdwatching, their exploits and utterances. My store of such tales resides in 50 annual logs and scores of now mouldy, even mouse-nibbled files in damp tea-boxes humped through 11 house changes. They will eventually go to a tip but every time that I attempt to begin that inevitable last consignment, I soon chance upon a letter from a fellow enthusiast written from every continent but South America. Immediately the clock spins backwards as the context of the original exchange goes live in my mind. There is no more delightful excuse to cease a tidy-up than the recovery of past debates. Here lies Horace Alexander's crusade to persuade some stuffy North American zealots that his Palearctic times were relevant to the identification of Black-throated Divers in winter plumage; there is the jumble of the many starts and stops that I shared with Peter Grant, Keith Vinicombe and others in the tortuous pursuit of trustworthy stint and treecreeper characters. All looks a bit shoddy in the current era of Lars Jonsson, whose portrayals of difficult species are the best ever, but we tried.

Intriguingly, however, much more ancient letters throbbed with equally urgent life. For example, a full century ago, Heinrich Gatke of Heligoland wished John Cordeaux of the Humber District a flock of Pallas's Sandgrouse for Christmas, while Lord Lilford discussed the niceties of Upcher's and other rare warblers with European collectors. So it is that I cannot imagine a birdwatching life without access to teasing and delightful mementoes of individual perceptions and comments. Regrettably these are not the stuff of caveat-ridden gurus or committees. There follow therefore just some of the 'stray feathers' that I have found in books or my own letterbox. Whether they will match the lost records of Allan Octavian Hume, I cannot tell, but at least printed here, they will take longer to reach the tip.

In the years when *my* rarities were accepted, I would occasionally chance upon a bird that left me, the Witherby *Handbook* and even Phil Hollom's updates floundering. My final oracle in those cases was usually Derek Goodwin of the 'Bird Room' who would always respond with something helpful and might even include some verses in celebration of the bird or event. In 1974, a trickle of Yellow Wagtails through St. Agnes in late April was totally upstaged by a truly wondrous sibling on 3rd May, brought in by a SE Force 8 conveying in turn fog banks and

"I ought to have acknowledged your last letter about Black-throated Divers long ago, but I forwarded it to the man who deals with local records, and he has not cheeped a chirp. I kept hoping I should hear something, but I suppose he decided that as he had no defence and was unwilling to admit it, he had better say nothing. I hoped that I had written the kind of letter that would respect his point of view, and not offend; but apparently not. He told me he had been in correspondence with experts in the west, where, of course, all three species commonly occur together. Now, a study of American Birds shows that, on the Christmas count, all three species are constantly being identified, and the identifications accepted, by numerous west coast observers. Well, may be some day some westerner will identify one off these coasts, and then all the world will accept it." Horace Alexander, *in litt.* (1976)

The culmination of a nine-month effort by Horace Alexander to convince an American editor that not only Pacific Coast observers but also a Briton could identify Black-throated Divers in winter plumage. At 87, Horace could still feel impatience but as ever it was always expressed in tolerant terms.

During the amazing invasions of Pallas's Sandgrouse in the late 19th century, some bred. Young birds were found in the Culbin Sands, Elgin, in June 1888 and August 1889. Here a party of seven adults settle on a sand dune.

rain. Already used to the confusing gene flow of the group, I knew that it could be a hybrid or mutant Yellow or, better still, a female Citrine Wagtail. David Hunt came just too late to see it near the Big Pool. So it became another case for Derek. His comments were encouraging; his verses were charming.

Wherever he goes
Mr Wallace of Hessle
With ornithological
Problems must wrestle;

For the birds that he sees
Are exotic and strange
And thousands of miles
From their usual range.

The voice of *S. turtur*
Is heard in our land
But it's *orientalis*
But comes to his hand

His warblers; although
They are little and brown,
Are not common species,
That others note down.

The wonderful wagtail,
That lately he spied,
Was certainly not
Grey, White, Yellow or Pied.

At rarity spotting
He's one of Fate's darlings;
Whose future is rosy.
And so are his starlings.

Inevitably, the Rarities Committee played the 'indeterminate' card but, two decades later, one of the many photographs since published of a once little-known species contained a dead ringer of an image. Better still, some moulting female Citrine Wagtails in Oman in January 2002 were identical. Patience is all?

The 'wondrous' wagtail of the Big Pool, St. Agnes, Scilly on 3rd May 1974.

As noted earlier, the mid-1970s saw a surge in the Scilly boom. In 1975, Bob Emmett, my first wife Karin and I were hosts in the Barn Cottage on Tresco to two non-birdwatching couples who had enlivened our time in Nigeria and were in need of some island bliss. The husband of one was a most endearingly alcoholic intellectual called Robert MacDonald. Prior to his Scilly visit, he had assumed that my birdwatching kit and dress as displayed on an African beach constituted a unique image. Faced with platoons of look-alikes in Scilly, he exclaimed "Jesus, there are scores of you" and, after seven days of observing us all in action, came out with a three and half stanza poem, written (he said) in an Anglo-Saxon format.

The Shadow and the Form

What does the bird know? Stiff-winged and swooping
Light over low land, lifting and looping
High over headland, strong in the storm's eye
Grey as the gull's wing, searching and stooping.

Deep in the down land, white faces waiting
Light upon lens eye, ticking and dating
Hard in the heather; the North wind unnumbered
Word-write on wind face, harsh voices grating.

What does the bird know? Thermal uplifting
Scan over sea-scape, white wave-caps shifting
Crystal the clear sky, written on wild wind
Dust-dry and wind-borne, the white paper drifting.

What does the bird know? Searching and stooping
Grey as the gull's wing, lifting and looping.

Tresco, 14 October 1975

Hardly Beowulf, you may think, but poetry in celebration of bird and watcher is not a prolific genre.

Nowadays I have little envy for the doings of rarity collectors but I can still be awestruck by the efforts and perceptions of purposeful observers and particularly the dying breed of all-round naturalists. For me, the champion of the last corps is Frank Fraser Darling whose *Natural History in the Highlands and Islands* has completed 54 years and counting of inspiring instruction. Here is an excerpt that show his infectious enthusiasm. It comes from the end to his chapter on sub-oceanic islands:

> The only thing which can dull enthusiasm for reaching the little islands must be the stiffening joints which may prevent one from getting ashore in the moment of slackening swell. Seasickness is certainly not enough to keep the naturalist on the mainland, as this poor bedevilled sufferer can testify. There are certain situations which arouse a tremendous feeling of exhilaration and physical well-being. I know of none to beat the approach in a launch to a remote and uninhabited island where the swell is whitening the foot of the cliff. Whether you are at the engine and tiller, with a kedge anchor rope running astern, or poised on the fore peak, ready to jump ashore with the forrard rope, it is all the same. You are trying to beat the elements but you are also working with them as the boat lifts and falls in the broken water at the foot of the cliffs. The island is still remote till your foot touches down, and sometimes the swell will beat you. In this way I touched Sula Sgeir with a boathook but never got ashore. James Fisher in the same month did manage to jump on to the rock, and came off again twenty minutes after with no more than a wetting. Robert Atkinson and Malcolm Stewart were luckier.
>
> The islands are a paradox: bare and remote to our eyes, they are nevertheless among the most heavily populated areas in the kingdom and contain forms not found elsewhere; on the one hand there is the falling off in the numbers of species, and on the other the immense numbers of those there are; the vast comings and goings of the creatures, and at the same time the irrevocable isolation of the tiny mammals. We tend to think of these northern islands as storm-bound and mist-wrapped, yet nowhere can there be greater brilliance of colour, the sea so blue, the grass so green, the rocks so vivid with saffron lichen. A meadow of sea pink in June contains all colours between white and deep purple, and the predominantly white plumages of most of the birds reflect the boreal intensity of the summer light.

All the island goers that I know have relished the same sort of struggle cum puzzle and delight that assaults the eye and mind in the great, still-pristine oceanic

After seven Hebridean voyages in the 'Ocean Bounty', her fortunate passengers gave their leader Dougal Andrew this cameo in remembrance of shared magic.

wilderness of north-west Scotland, into which the odd cruise but hardly a bird tour ventures.

Next comes Kenneth Williamson's evocation of first Leach's and second Storm Petrels from his wonderful description of the Faeroes and their people, *The Atlantic Islands* (1948).

> The summer night is alive with these elfin birds, shadows a shade darker than the sky flitting to and from a whirl of giddy movement. They electrify the air, dancing around like strange, nocturnal butterflies, and with the same wayward flight as big-winged, exotic butterflies. One moment a dozen are visible whilst you watch, and the next moment they are gone with the gloom. They swirl and eddy about your head, sometimes brushing past your face with a soft, ghostly touch. The night is feverish with brisk, excited movement and their clear staccato cries; for they are constantly calling, and their eerie voices are flung against the murmur of the wave-roll on the shore and the soft sweep of the wind through the long grass. It is an unreal and fascinating world which any naturalist would feel pride and joy in visiting for a while.
>
> When the last pale wash of sunset is in the sky the storm petrels come to the isle. They are the first to arrive. The path on the northern side of Rogvukollur is a good place to see them, and in parts of west Mykines they outnumber the larger kind. The old ruin on the south-east corner of the holm is another of their strongholds, and here they fly in wide circles, fast and fairly direct, and in the gloaming their quaint conversation and mysterious love-making begins. Their song rises out of the earth or comes from among the stones of the ruined walls; it is a thin, whirring song

reminiscent of the hum of a spinning-wheel. The Leach's petrels are later in their arrival, but soon after dark they are to be found fluttering in scores above the sloping brow where the pasture descends to the south side rocks, whilst one can also find them, though not in the same large numbers, on the brink of the great northern cliff.

I recently learnt from Esther Williamson that Ken wished to be remembered most for his writing. The above example shows why there should be a lasting fulfilment of such desire.

South and a smidgin west of the Faeroes lies the archipelago of St. Kilda. Since Martin Martin first described it in the 1690s, this uncompromising set of stacks and isles has evoked strong responses in human observation. This from 4th July 1956 belonged to the late John Arnott, actor, broadcaster and conservator.

A 'typical adult' and a juvenile of the St. Kilda Wren drawn and painted on 8th July 1956. Although put at risk by excessive collection, the subspecies has survived.

Wind – SE but variable with rain squalls, which marred day from c. 11:00 to 14:30 hours.

Although we had spent most of yesterday lying down, we were willing to continue doing so as soon as we had finished our meal last night until about nine or ten this morning, Messing (le mot juste) is divided between two lots of four, but this morning the menus combined to offer porridge of varying consistency, lush Macfisheries kippers and The a l'Hirta.

The hills were shrouded in cloud all morning and afternoon, and occasional showers of rain dampened members of the party making their first exploratory walks round the island. Swallows, Swifts and a Carrier Pigeon were unexpected visitors in the morning and we hope for the occasional caller considerably less common.

T2 hirtensis (the St. Kilda Wren) has been quick to show himself in the village and has provoked remarks about his appearance ranging from: 'There's not much difference really' (Joe) to 'Good Heavens, you'd think it was an entirely different species' (Ian). No nests have been found yet.

Twite and Oystercatcher are the other inhabitants of the deserted street. All the roofs are down except on the Manse, the Church and the Factor's House, and gaunt doors and staring windows reveal jumbles of rafters and beams sprawled over rusty iron bedsteads and other abandoned debris. Ten or fifteen years ago one might have got the feeling that this was a newly deserted island; that people had really lived here all their lives within the recent past; but now the buildings have decayed to such an extent that it is difficult to visualise this, and the houses are merely curious timeless relics, in the same way as broken down medieval castles or superannuated Jacobean country houses.

The smell of death also pervades the old church, where the

remains of a dead lamb sprawl over the pulpit steps and the old wooden pews lie mouldering. In the pulpit there is a book of sermons, and on the floor amongst the filth a tattered hymn-sheet containing the words 'Lead us, Heavenly Father, lead us, O'er the world's tempestuous sea' – a living image for these people who lived here in three vast dimensions under God. On the lectern and round the walls are scratched and written the names of sacrilegious visitors, one with the apparently necessary addition 'Priez pour moi'.

In the afternoon the M.C.I.C. (Mouse-Catcher in Chief), Joe caught a Genuine Bedouin St. Kilda Field Mouse, which was immediately incarcerated in a biscuit tin and sketched by Ian in the few minutes when it was not trying to escape.

The possibility of getting over to Dun was examined in the evening by three of the party, who came to the unanimous decision that we should have brought a dinghy, rubber or otherwise. It might be possible to swim across the narrow gap at low tide and scale the 100 foot cliff, but it will not be easy.

True to our mutual firm decisions to be early to bed tonight, we are still talking and drinking cocoa at 1 am while planning ringing expeditions for the morrow. Ian went off on a wild goose chase after a Whimbrel, and will probably stick to *Troglodytes* now. Readers familiar with the St. Kilda poets will recall the lines:-

> There was a young man in a cleit
> Who said he sat in the weit
> "I've been watching this Wren
> Agen and agen;
> By tomorrow I'll catch it, I beit."

The partial desecration of Hirta by a missile-tracking station was a function of the Cold War. The later rewinding of its human past by the reconstruction of several cottages seems much more bizarre. The cycle of a unique seabird economy, over-strict religion, male fatigue and the resultant evacuation in September 1930 was complete. Its purely cosmetic retrieval is a sad interruption of its proper return to full wilderness. Happily its still-teeming birds are spared the apparatus for such folly.

The life of the Inland Observation Point Scheme, reconstituted as the Daily Bird Count (DBC), was brief, lasting from 1st March 1962 to a late date in 1965 that I cannot trace. It was nevertheless an inspired attempt to map and time all bird movements within Britain, marshalled first by Humphrey Dobinson and second by Peter Davis and five other BTO stalwarts working in their free time. Its keen participants could have been numbered in hundreds but its development was prevented by chronic underfunding at a time when the reversal of the threat posted by toxic agrichemicals needed every last penny. Yet one observer's efforts were exceptionally fortunate.

HOW TO MAKE A PROFIT ON YOUR DBC

Daily Bird Counter 924 awoke on the morning of the Grand National, 27th March, and bid his DBC. The only bird of note was a Jay, the first he had had during the month. Luckily, he took this as the augury it was and backed Jay Trump in the National. It won, of course!

Forty years ago, London's skyline from Primrose Hill was the backdrop to remarkable autumn pageants of visible passage. Here a Bullfinch (bottom right), Chaffinches (bottom left), Fieldfares (top left), Starlings and Wood Pigeons move west into the wind.

One mystery that the IOP-DBC Scheme might have elucidated is the late autumn overland passage of Wood Pigeons, a classic feature of London skies in the early 1960s (and occasionally since). Yet it was seemingly denied by the lack of observatory sightings, as relayed to me by Peter Davis.

> I cannot throw any light on this passage at all. Is it conceivable that this was the morning movement from a roost? Though why pigeons should roost east of London and feed west of London I do not know. I have heard no comparable autumn movements anywhere, this year or any year.
>
> P.S. The early timing of movement suggests they could not have travelled very far.

However, the IOP motivation dies hard. The watchers at Redmires Reservoir near Sheffield, marshalled by Keith Clarkson and Steve Hunter and counting through the 1990s, found that to their first million registrations of diurnal migrants reaching the eastern Pennine scarp, Wood Pigeons contributed 503,000. Game, set and match to dawn risers? Unfortunately not, if the following comment from David Lack remains true.

> 26th September, 1962
> The comparisons I have made over several years now, between visual and radar data, indicate that nearly all the big migrations are above

visual range, by which I mean ordinary binoculars. This is because in the autumn they occur primarily with following winds, and normally they are seen in autumn only on the relatively infrequent occasions when there is a light opposed wind in the area of origin and a stronger opposed wind where the observer is situated. However, the Americans have recently discovered that if you point a twenty times prismatic telescope vertically upwards, and lie comfortably on your back, you can in fact see a great deal of what happens at higher altitudes. I would suggest this technique to you. The critical points are:-

1. the telescope should be securely clamped and have a simple focus;
2. the observer should lie comfortably on his back;
3. the eye not in use should have a black patch over it, because it is impossible to hold it shut otherwise;
4. there should be a separate recorder to take down what the first observer sees, and the two should change places at whatever interval proves most suitable.

I hope that these suggestions will prove valuable to you. They were reinforced because I have just been to Minsmere for several days with northerly winds, when one knew from radar that birds were pouring south over the land by day, and not one was visible until the wind went round to the south!

The above trenchant advice was sent to Tony Gibbs, one of the stalwarts of London birdwatching from the 1950s and co-organiser of the London Migration Watches conducted early in the 1960s. His comment on a brief trial of the technique follows:-

21 October, 1962
I have tried this using binoculars but all I have seen is mist – and a few birds out of the corners of my eyes!

Faced with Olympian comments from David Lack, one of the leading students of bird migration as displayed fully by radar, much of the enthusiasm for visible passages and indeed the whole observatory system was lost. David Lack could, however, write with a much lighter touch. Here is his poetic criticism of obsessive life-listing.

Bird-men, raise the exultant strain.
Ornithology is slain,
Photography, ringing have had their day,
And mere bird-loving has passed away.
None of these triflings will be missed.
For the one true goal is a long life-list.

Rise with the dawn,
For the rare birds fly.
Floreant orn-
ithoscopi.

David Lack's *Enjoying Ornithology* (1965) remains another excellent timepiece of the recent cusp of birdwatching and field ornithology.

Next comes a song, one of at least two sung by members of the Cambridge Bird Club to while away the slow bus journeys undertaken on days of field meetings. For its survival, we must thank David Wilson, island goer and ornithological bibliomaniac.

> I promised to let you have the words of the Cambridge Bird Club song. It was composed on a field excursion where groups were left at various places along the north Norfolk coast. Twelve went to Scolt and eleven of them got cut off by the tide. Chestney senior rescued them in his boat.

> 12 for the Bird Club Members
> 11 for the 11 who nearly went to heaven
> 10 for the terns and Twites and tits
> 9 for the Knot that were Knot or not
> 8 for the April migrants
> 7 for the Severn Wildfowl Trust
> 6 for the six Barred Warblers
> 5 for the Whimbrel on the shore
> 4 for the Handbook authors
> 3 for the di-I-I-I ivers
> 2 for the Buffel-headed Duck clothed all in buffalo
> 1 for the Solitiary Sandpiper which evermore shall be so.

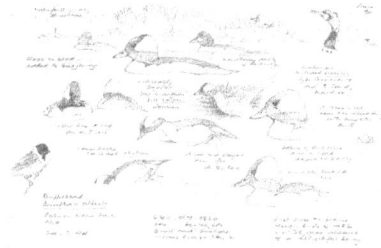

Today the Buffle-headed Duck is called simply Bufflehead. It remains a magic being. This drake graced the Colwick Water Park, Nottinghamshire, from 17th to 26th March 1994.

Chris Mylne, who was probably President (of the Cambridge Bird Club) at the time and chief composer, has checked this and made one small correction. I honestly cannot think of anything else I learnt during my three years at University.

Chris has reminded me of another CBC song but there is only one line we can both remember. To the tune of Quartermaster's Stores:

> There were Scaup, Scaup, psychoanalysing Dr. Thorpe.

> Maybe if I repeated this over and over for long enough, another line might be recalled.

Such was the recreative invention of the mainly undergraduate club members in a world of many fewer other distractions – grey TV screens, 78 rpm shelac records and unheated buses were three. Like David, I cannot recall the second song but for its best line above.

Escapes from danger are the better perils of war. There follows Norman Orr's tale of such a release coupled with his loyalty to a precious book.

"It must be difficult for young birdwatchers today to envisage the pre-1945 era when textbooks on birds were sparse indeed. For example, when I spent three wartime years flying light bombers, mainly in support of the 8th Army in Egypt, Libya and Tunisia, the only reference book on the birds of this region I could find was the massive 2-volume "Nicholl's Birds of Egypt" by Colonel Meinertzhagen. It weighs nearly a stone!

Fortunately my late wireless operator/air gunner was a tough but kindly Canadian who, inexplicably, insisted on carrying them around for me, in a kitbag! At one stage [at Tobruch in January 1942], I came very close to having to surrender my Meinertzhagen to Rommel. The incident occurred on 31st January 1942, during our retreat from Rommel's advance into Libya. My squadron (No. 55) of Blenheim aircraft which was then based at el Gubbi, Tobruch, had been ordered to pack up immediately and fly back to the Suez Canal Zone, leaving all our personal belongings. It was this last order that I was unwilling to accept at least in regard to my two volumes of Meinertzhagen! Unfortunately, just as I had made my way to my tent, a vicious sandstorm blew up making it impossible for me to find my way back to our dispersed aircraft. We had more aircrews than aircraft on the squadron; so everyone piled into any available aeroplane without a personnel check being carried out. So my absence was not noticed, my own crew thinking I must be in another aircraft!

We could take-off quite well in a sandstorm (landing was another matter!). I counted, by the noise, all the squadron's aircraft taking off, leaving me on my own in the desert. Needless to say, I was more than a little depressed! I found the one-and-only desert road under my feet; so I sat down on the side of it awaiting events. The sound of the enemy's artillery was not uncomfortably close; so all I could do was to try to prepare myself for being shot or captured. Eventually, I heard a vehicle coming slowly along the road from the west. I didn't know, of course, whether it was one of Rommel's panzers or one of ours, but I started to wave frantically.

I could scarcely believe my good luck when it turned out to be a New Zealand lorry on the retreat. The corporal driver told me to jump on the back, which I did with alacrity, still clinging onto the kit-bag with my 'Birds of Egypt' in it. A convoy of retreating vehicles gradually built up, all heading for Egypt. We were straffed continually by the enemy aircraft and suffered casualties, but I survived, the corporal insisting on sharing his meagre rations with me. (I have to confess that I didn't do too much birdwatching on that trip!)

When I eventually rejoined my squadron on the banks of the Suez Canal I found that I had been reported missing, believed POW. But I had my Meinertzhagen, and still have it today. Western Desert sands fall out of it occasionally!

In my opinion the work is a masterpiece which has never been bettered. Beautifully and economically written with almost no important information omitted, yet without superfluous verbiage."

The massive book in Norman Orr's war story was indeed Richard Meinertzhagen's *Birds of Egypt* (1930), effectively his completion of a task begun by his close friend Michael Nicholl. On occasion and contrary to Norman's description of his writing, Meinertzhagen could exult in birds and their settings. His diary of 26th July 1914, written in the mountains of Baluchistan, includes this tryst with a Lammergeier.

D.M. Henry's painted complement to Richard Meinertzhagen's exultation of a Lammergeier (from *Birds of Arabia* 1954).

The finest view I ever had of a lammergeier occurred today. I came on him but a few feet away silhouetted against a gold-red sunset, magnificent against a horizon stretching for miles and miles into golden infinity. He was quite unconscious of my presence. He sat on a rocky pinnacle facing the setting sun, wings slightly drooping and half-stretched, head turned up towards heaven. Was this the phoenix of the ancients, Pliny's bird of brilliant gold plumage around the neck, the throat adorned with a crest and the head with a tuft of feathers? Was this lammergeier conscious of his sacred relationship with the sun? The phoenix of the ancients presaged peace everywhere in the land. What I saw this evening seemed to foretell war, a long bloody war. It was the finest, most beautiful and yet most terrible, the most romantic view of any bird I have seen at any time anywhere.

Sceptics of Meinertzhagen will suspect some gilding of the lily in the above but I for one agree that for raptorial romance, the Bearded Vulture or Lammergeier has no equal.

My last 'stray feather' was posted from Britain's most exposed community on Foula in December 1954 by Christopher Mylne. He had gone to the western outlier of Shetland as 'Teacher-Missionary-Birdwatcher', quickly learnt the additional roles of 'Gardener-Housekeeper-Bachelor' and saw on the edge of the world life where it is totally real. His Christmas letter was the most transfixing that I have ever received. Here is a section of it.

The assortment of jobs to be done can only be described in detail and cannot be categorised except under the heading of 'life'.

All the shopping has to be done by post, all the transport by wheelbarrow, especially peat, all the work by Joe soap – it's just life. There are boots to be oiled and a sermon prepared – remember to feed the dog and that lavatory cistern needs cleaning this weekend – a brood of young Wheatears to be ringed – "Peter, will you play the organ at tonight's service, please" – must mend that window before the next gale – "What's that bird on the peat stack, teacher?" – 5 eggs a day from 5 hens; good going – those arithmetic books to be corrected – must visit old Betty this evening; and where are those books for the North end? – 24 birds in one drive of the trap, mostly Starlings – "Anther two dozen 2.5d stamps, please Harry" – a row of carrots to be dug – "The Women's Guild will meet in the Manse on Tuesday at 8 pm as usual" – will it be a mail day tomorrow? – "There are gale warnings for the sea areas Faeroes, Fair Isle" .. must mend those socks before Sunday – now who might help me sweep my kitchen chimney? – dead Merlin to be skinned – the bread should be about ready now – "Dear Sir, With reference to your letter about the new tractor for the isle of Foula" ... - just enough apples for the children's Hallowe'en party – a neighbour's wireless to be repaired – note: paraffin and methylated spirit from the shop tomorrow – better get down to the boat with my box for Walls – definitely too big for a Garden Warbler; must be an immature Barred – a puncture in my bicycle tyre; these roads! – "Coming out fishing tonight?" – a leg of mutton to be cooked; it'll make good soup for the weekend – time to order a new lot of library books – "the Wrens of Foula appear to belong to the Shetland subspecies" – what on earth can I give the children for handwork tomorrow? - "The funeral will be on Tuesday, Mr. Mylne, at the usual time' – several Blackcaps in the S.E. wind and a number of ectoparasites taken off a Blackbird today – must get the school record of work up to date this week – this East wind always floods the kitchen floor – "You must see my new batch of colour film; some good shots of young Bonxies" – a lovely display of the 'pretty dancers' in the North sky tonight – "Hymn number 165 in the Sankey book: 'Let us with a gladsome mind Praise the Lord for He is Kind'.

Yes, not a bad idea; it's a good life after all and Christmas comes but once a year.

At any distance a young Barred Warbler looks amorphous. Close to, its dull wingbars are diagnostic.

After that excerpt of Chris's remarkable evocation of the place of birdwatching in an unusual life, all that I can do is to add the outcome of the smoke-filled evening of 16th July 1999.

One red engine and six large calm firemen came in nine minutes. After 20 more, a visored head poked out of my study window and announced that my chief solace had been wholly soiled but not more than marginally scorched. Several experiments were needed to prove that the collision of a cigar's last spark with a tissue edge had been the cause. Do not smoke prior to a nerve-wracking exchange of vehicles, if ever, was the simple lesson. The effect of prayer, particularly in the mode of ornithological contrition, remained a mystery but I no longer bet against it. God or Superforce had been kind, indeed.

Recent Observations at Lowland Gatherings

Doubts about this book – new tribe disowned – a change of writing tack – a simplistic exchange – the British Birdwatching Fair 2001 – a magic fall – a message from David Glue – birds beset by agricultural crisis – my Swallows – the BTO Conference 2001 – Heligoland strikes again – a million birds at Redmires – two wobbles – better a birdwatcher than a bird

"Ultimately birding is a hobby and it is up to individuals to follow it as they wish. The diversity of birdwatching, from the old lady feeding "her" robin to the most manic twitcher to the most dedicated scientific researcher is part of its joy. There's probably a little bit of each person and others too, in all of us (I know there is in me)." Rupert Higgins, *in litt.* (2000)

Looking back on what I have written so far, I am uneasy. My main aim was to convey to you some sense of the magic carpet ride that birdwatching has been for my peer group and near generations. Why then have I enjoyed the partial reprise of its beginnings far more than the review of its more recent and still burgeoning estate? I could be suffering from the angst of mere age. Certainly I find the bird blues, induced by the distinctly red figures in the conservation accounts, throbbing insistently in at least one ear. What really makes me itch, however, is the growing thought that my anthology cum commentary could be an epitaph to a past fulfilment and not a conveyance to a more widely shared, ever greater enjoyment of birds.

I have coveted the role and title of purposeful birdwatcher for nearly half a century. In his fascinating *Birders: Tales of a Tribe* (2001), Mark Cocker opined that such a 'veteran warrior', or 'trained soldier' as I have named him or her, has long since been subsumed into the tribal nation of birders. Sorry but I did not attend any such joining-up parade. Furthermore, the full semantics of the modern nomenclature are debatable. For a start, Shakespeare's birders had a lethal effect on their avian subjects. I want us to retain the grace note of responsibility in the word 'watch' and so avoid the dumming down of the amateur ranks of ornithologists. I also regret the diminution of my own ornithological ancestors and birdwatching peers by yet another retrospective review of their true identities, while to denigrate the millions who pay for conservation as 'dudes' and 'Robin-strokers' – as recently done again by Chris Packham on Radio 4 – is an associated act of disgraceful, inverted snobbery. To demonstrate the reasons for my concern on the above and other current issues, I give up in this chapter my failing attempts at impartial commentary and make some direct personal observations, mainly on the events of three recent lowland gatherings of people interested in birds.

Scene I: The BTO/Birding World Conference of 1994

Our shoulders rubbing in the morning coffee queue at Swanwick, Tony Marr (soon to be chairman of the BOU Records Committee) and I fall into conversation. He has stayed in the official mainstream of British birdwatching for two decades longer than I and always has news to offer. We exchange observations on the unusual but welcome, youthful atmosphere of the joint conference. Puzzled by the recent competitive, repetitive pursuit of rarities and its absorption of so much energy, I ask Tony, "Why do young birders do nothing else?" "Because they think

It may not suit a birder's mantelpiece but it is just this sort of image that softens hearts and opens purses. Long may it and others do so.

that we've done it all", he smartly enjoins. Both question and answer are patently simplistic but the exchange leaves me with a nagging doubt about the current values in some sectors of the birdwatching or birding universe. Of course, people are free to do absolutely what they want but not to investigate (or be shown) the total potential for personal recreation in the most accessible field of all natural history is, to me, such a waste. How could I find more cheering thoughts and a fresh belief that the magic carpet has more mileage?

Scene II: The British Birdwatching Fair of 2001

Five years on but still regularly haunted by Tony Marr's observation, I decide to re-double my search for birdwatching cheer at the amazing fair held annually at Rutland Water. It is the brain- and brawn-child of the Wildlife Trusts' Tim Appleton and RSPB's Martin Davies, both conservators now doubling as ornithological impresarios, and their main sponsor, Bruce Hanson of In Focus. The year's event is their 13th offering in a series begun in 1988.

Normally I whip in, give my talk and whip out, but this time I peregrinate through the maze of two vast car parks and 24 small to gigantic marquees for all three days. I survey the product offers provided by 20 sponsors and 230 other exhibitors and sample the continuous entertainment and enlightenment streaming out from 78 talks, four panel games, a workshop, an art gallery, a photograph gallery, a playlet, a football tournament, a ten-book signing and an awards ceremony, all these having been led off by a celebrity lecture on the opening night. Pile on all this arranged and, even better, unexpected meetings with old friends, live Cuban music, all day food and drink – for me, shandy galore throughout the customary three baking days – and you have not a fair, rather *the* bird-induced extravaganza of Planet Earth. In its early years as just noted, I had been a reluctant attendee, dutifully talking for *Birdwatching* in return to that engaging magazine's loyalty to me as a columnist, but particularly as I heard in my audience's reactions an increasing playback of the eternal avian and human delights in ornithological recreation, I became as great a fan of the fair as its creators.

In 2001, about 15,000 people more or less intrigued by birds came to enjoy them and themselves. Within the crowds weave every conceivable type of tribal member, from sleep-starved, fidgetty twitcher nervously eyeing his pager to the archetypal RSPB family unit. Even sage elders as venerable as James Ferguson-Lees appear, still atlasing at 72, and the all-time record climber of the conservation establishment, Baroness Barbara Young of Old Scone, puts in her annual smiling appearance. She is now chieftainess of the Environment Agency but still up for panel games. Everywhere, all hours, birds and translations of them by human inventions – from stunning digital photographs to deliciously musty, old tomes – are rampantly consumed. Most importantly of all, a new record fund of £135,000 is raised to help keep safe the wilderness of eastern Cuba, virtually the last unspoiled habitat in the entire Caribbean, and the smallest bird in the whole world, the Bee Hummingbird, which lives nowhere else.

There are a few misanthropes who still foreswear such excess of indulgence but I bet even they could not resist the crack, as some personalities are made yet

"I must confess to being extremely disillusioned with the hobby itself. Britain must be the Number 1 country in the world in terms of the number of birders per head of population and for the amount of knowledge that is held by those people. One only has to look at the sheer number of bird books being published in this country, relating to all areas of the world, to appreciate this. In fact, I would go as far as saying that, in years to come, this period will be looked back on as some kind of 'golden age' of ornithology. What totally gets my goat though is that the whole scene is so disorganised and it seems to have been taken over by people who hate each other! The nastiness and unpleasantness is phenomenal. The funny thing is, most of the people I go birding with and who I know well here in Bristol are really pleasant, nice, ordinary blokes who enjoy their hobby (at whatever level), yet the people in control seem to be power-mad lunatics! I have often wondered if it's about time that the ordinary birdwatchers regained control of their own hobby – maybe we should have an organisation like that in America which is run by birdwatchers for birdwatchers, but I suppose that would only introduce yet another organisation that would eventually go the same way as all the others. I've got to the point where I think to myself, I'll go birding, enjoy myself, write the odd article and sod the lot of them! I think a lot of other top birders have gone the same way." Keith Vinicombe, *in litt.* (25 January 2001)

The heartfelt frustration of a birder/bird-watcher/field ornithologist/ thoughtful author at the beginning of the 21st century.

more immortal and others die yet another death by birders' slag. Endless stories are told or embellished by repeat. Now more than at any national conference, Rutland Water presents the whole kaleidoscope of British, nay international birdwatching and birding. It really swings. For birds and their refuges, the cumulative dividend has now reached a triumphant £750,000.

Evolution is also apparent, for BBWF has become subject to taxonomic radiation. A widening mushroom of regional fairs has popped up and all of them are also successful.

Scene III: Byrkley Park and Crossplains, East Staffordshire, late August 2001
Only one week after BBWF, I witness a small miracle of bird movement that may relate to the recoveries in some bird fortunes registered by the BTO from 1999. It is my first sizeable inland fall of passerine night migrants since those that Rolfe Green and I shared with glee in Regent's Park, London, back in the 1960s. There the daily migrant counts would regularly present a score or more, once even 94 newly arrived birds that would leap after food in the willows, oaks and hawthorns near the lake and canal. Horrendously, a comparison made with similar counts of the 1990s has shown a decline in the park's autumn migrants of nearly 40% over three decades. Privately, I had given up all hope of seeing another merry mob of summer visitors departing through the late August window demonstrated so well by the radar studies of Lack and Eastwood in the 1960s.

Yet on 27th and 28th August, only 300 yards of sallows and thorn bushes by two fish ponds and 200 yards of runway and stubble interface attract no fewer than 50 migrant individuals of 13 passerine species. At the ponds, I twice have four species feeding together in the same binocular view. Once again against the recent odds, I revel in a classic 'withdrawal fall' which features not drifted migrants cast fortuitously on the east coast but the totally fit progeny of British parents. The youngsters, flying south in fine weather with little cloud and a light northerly tail-breeze (the antitheses of North Sea hazards), had spotted two good 'service areas' and had immediately 'parked'. I watch them 'fuel up' with utter delight. By the 29th, all are gone on for Iberia and Africa.

Totally re-enthused, I write one of my occasional letters to David Glue, the BTO's tetraplegiac but wonderfully vital research biologist, reporting some late breeding season news and wittering on about the nostalgic fall. Back by return on the 31st comes a typical David missive. It begins with a playback of other favourable breeding bird news but then switches into regret for the collapse of migration studies in Britain. He then makes the following trenchant observation on the current crop of birders, alias incomplete birdwatchers.

Forced recently into the shade of two local reserve hides, I have listened quietly but with great sadness to some remarkable conversations. Why with so many RSPB members taught to 'proclaim' what government should do, are there so relatively few BTO members actually taking part in projects? Why has the modern identification literature failed to pass on a knowledge of bird voices?

Why is it all day lists but seemingly no diaries? Why can't people sit and watch for more than a few minutes? Where did the blinkered interest in the scarce and rare come from, and why are they so keen to score off each other? Surely the UK deserves a better informed and more able workforce of active birdwatchers … Or am I just getting old? Better stop; my blood pressure is rising but some European countries are starting to knock spots off us in the hobby and science that we led with such flourish.

A White Wagtail looks for blown insects along a runway where leisure flying causes constant disturbance of what was once a 'deer running' in primeval oak forest. The place is Tatenhill Airport, Needwood Forest, Staffordshire.

Not entirely put down, I head back to the patch to see if the day's vicious showers off a gusty NNW wind have brought anything down. This time I concentrate on the airfield at Crossplains. Large tractors had been scuffling the day before; there could be fresh plough, always good for gulls. Using the maize belts for cover, I trespass to the runways and scan around. Three planes doing 'circuits and bumps', only 210 irritated gulls and an early White Wagtail, drifted west to east for once? The big puzzle was, however, the colour of the till, yellowish – not blackish-brown. And is that a wide seed-drill attached to one of the heffalump-sized tractors? My red light comes on. With my autumn stubbles gone before September, am I also facing, as would my soon-expected plovers and larks, yet another change in agricultural tactic? The tractor drivers stop work and make to leave. Unauthorised legging across two runways does not suffice to catch them but my quandary ends with the appearance of two black labradors and a Range Rover complete with farmer.

The crux of our conversation is stark. Yet another major farming business in my study area has become loss-making. With the new capital cost of agrimechanics now beyond it, the family have had perforce to move to the last refuge of profit, share-farming with specialist contractors. So it is that in just 48 hours, 80 acres of cereal stubble has become 80 acres of just scuffed and cut surface thatch drilled with rape seed. "Not good for the birds", says Christopher Hall. "You said it," I rejoin but without any tone of blame. I trail back to my car, musing yet again on the current farming agony. At BBWF, I had heard from Chris Knights, champion grower of carrots and Stone Curlews, that even his excellent enterprise had become unprofitable. Around my own home, an aggregated herd of 350 milk cows are not paying for their cud and now the biggest arable enterprise in my patch is in negative cash flow. The entire food making and marketing equation is awry and more than tinkering with the overall £1.6 billion subsidy is needed to put it right and allow also a new strategy of countryside care. In the meantime, we must keep counting, proclaiming the rights of other beings and praying. 'We' means birdwatchers and other naturalists with purpose and as many birders as we can get on board our version of 'Greenpeace Warrior'.

No one has done more to restore the fortunes of the Stone Curlew, also known as Norfolk Plover, than Chris Knights. On his farm, ploughing tractors stop and their drivers replace nests.

As I reach my car, a Grey Partridge with just two flying chicks gets up from one of the few relict weedy patches and a tiercel Hobby appears in hot pursuit of a tit or warbler. The falcon ducks under the sprite, drives it up from the nearby canopy and has it in its talons in a trice. Normally I would have punched the air with the pleasure of a double tryst with scarce gamebird and favourite falcon but not this time.

Scene IV: Mount Pleasant Farm, Anslow, 1st September 2001

Although I kid wife and daughter that we have not had the last of summer, the birds and I sense the real chill in the north-wester that stirs the morning. My pair of Swallows still sit lovingly together on the wires but their eight young from two broods are long gone. The adults will soon go too, reproduction achieved and blessed by me for the long haul to South Africa. They did not have to listen to today's *Today* programme on Radio 4, full of woe for the general plight of their species in Britain. Fewer beasts, fewer flies, fewer insectivores: another adverse chain reaction threatens biodiversity, no doubt worsened by the sickening culls that have been our only answer to the scourge of foot and mouth disease.

Scene V: The BTO Conference of 2001

This time round, the BTO's annual jamboree combines what were the former Birdwatcher's and Ringers' Conferences. The theme is migration. I buy a day-ticket and bowl up to find a record 470 other birdwatchers and professional ornithologists already there. The Swanwick Conference Centre heaves with us as we feast on 34 often parallel lectures and 20 sideshows. For the first time since Oxford in the 1960s, I feel the full breadth and power of purposeful birdwatching and *mirabile dictu* our average age has fallen by several years.

Richard Porter's opening address on raptor migration across Europe, the Middle East and North Africa is Olympian. There are echoes of Sir Henry Rider Haggard's immortal *She* in the misty shots of montane wilderness and wheeling birds, particularly when he points out that one major sea crossing of Asian broadwings remains undiscovered. I am reminded that for five decades we have been staring up at raptors. It is rather heart-warming that they have not yet yielded up all their mysteries. Who will discover the missing passage? Does it await us low in the Persian Gulf? Dodgy place but what a challenge!

On many of us, two really inspiring performances raise neck hair. The first is Franz Barlein's Witherby Lecture on where the study of bird migration should now lead. When he displays some observations on migrant Wheatears from Heligoland that were founded in exceptional visual acuity, I am lost in admiration. ("How did the Germans lose the war?", I ask my row of the spellbound audience). It is difficult not to feel shame over the lack of similar research initiatives at the British bird observatories. Peter Howlett reminds me that the last's seeming fossilisation has stemmed partly from thin finance and then confirms that at long last some indexing of observatory counts has begun. The fact remains that there never is any excuse for a lack of imagination. We could do with reviving the Cordeaux–Gatke axis of Anglo-German enthusiasm. The second uplift comes from Keith Clarkson who describes with unstopped relish a decade's count of a million migrants at and over the trio of reservoirs at Redmires, west of Sheffield. I knew of a similar watch at Strines Gap but until Keith puts up the tally of uncommon migrants, I have no idea that yet another inland watchpoint is producing Lapland Buntings and Richard's Pipits, both supposedly coast-bound semi-rarities, and restless Bullfinches, 93 of which have denied their supposedly sedentary nature. Actually the ancestry of such observations is at least 50 years old. I instantly recall David Lack's early Cornish discoveries, recently so ably reprised by Matt Southam and Dave Lewis (*Patchwork* 1, 2:2-8), and all the effort put into the Trent Valley, Sussex and London migration watches of the 1950s and 1960s. For sheer cheek and breadth of cover, Bill Bourne's amazing watch of October 1950 which sampled Skylark movements across East Anglia into Wales may remain unequalled but the Redmires tally was an exhilarating testimony to what keen amateurs can still deliver – and how the early-rising ones can still have such fun!

Still at the conference, there is some bad news. Talk in the corridors indicates that problems with fine line printing on its maps will delay the BTO's much-touted but still tantalising *Migration Atlas* for at least nine months. (It arrived triumphantly in December 2002.) The West Midland Bird Club's laudable initiative to revive the spirits of local and county societies remains without a firm date or venue. Yet on balance, the atmosphere of Swanwick 2001 is one of increasingly confident field ornithology created by the practised mix of professional and amateur effort. The Oxford initiatives of the 1930s are alive and well. Clearly, *contra* Tony Marr, we are far from finishing our tasks.

At first sight, the two major lowland gatherings in 2001 of British birders, bird-watchers and ornithologists display very different field characters. Rutland Water is big, buoyant, almost brash; Swanwick has been small, sometimes over measured, and rarely loud. The latter's 2001 overflow of enthusiasm shows, however, that the BTO's current staff have relearnt the knack of delivering enjoyment as well as statistics. A full recreation in birdwatching is crucial to its continued broad support and there would be no harm in risking another tryst with the international birding, even twitching universe, as so successfully achieved with the help of *Birding World* in 1994.

As for the few thousand souls that make up the cumulative peak viewing attendances at some individual rarities, the associated penny finally drops. With 5.5 million Britons interested in birds and well over 1 million paying for their conservation and a slowly growing 10,000 showing at least recording purpose, the obsessives hardly form a lobby, let alone a constituency, and after all they are absolutely free to do as they wish. Otherwise, the balance of my observations allows me to feel confident about the ornithological futures – but not the avian fates.

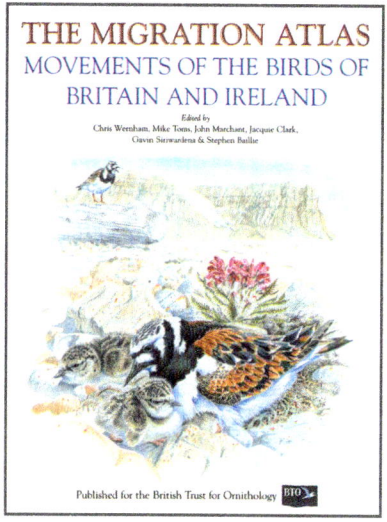

Keith Brockie's very fine paintings adorn both front and back covers of *The Migration Atlas*. The choice of Turnstone as their subject is apt; it is declining in Britain and Ireland, raising concern about breeding populations as far away as the Ellesmere Islands, Canada.

"There is absolutely no law that people have to spend their spare time doing something 'scientific', or even 'constructive', and it seems to be all too common that people within the institution of British birdwatching take up some kind of moral high ground, and put down those that simply enjoy their birding in a relaxed way. Football fans don't need to understand the trajectory of the ball through the air to enjoy their hobby. For better or for worse, ornithological science is now intellectually much more advanced, taking it outwith the scope of the casual birder. Those that have the interest, knowledge and skill to make a serious contribution will continue to do so, and this will not be affected by the simultaneous development of birding as a leisure pursuit." Jane Reid, *in litt.* (1999)

Sooty Shearwaters over the North Sea, an ocean-long migration away from their burrows on far south Atlantic isles and coasts.

The British Birdwatching Fair, 2001. The stand of the British Ornithologists' Union, manned by Gwen Bonham and Ian Bishop. Gwen gave long smiling service to the BTO as well in March 2004 she retired rightly clutching the Union's medal. DIMW.

Still providing both enjoyable employment in and researched product from purposeful birdwatching, the British Trust for Ornithology merits a large stand which was (until his sad death in January 2003) the niche of 'Moses', alias Chris Mead. DIMW.

At the *British Birds* stand, the new editor Roger Riddington beams a welcome to new subscribers. The 96-year-old journal is still the best.

Offering both their own wares and those of the Fair Isle Bird Observatory, Tim and Irene Loseby always have news of Britain's 'Heligoland'. DIMW.

Choosing a new telescope is easy when you can look at wild birds in a real marsh, but check out a second mortgage for 'state of the art'! DIMW.

Right: A pair of artists put their images into the mural of Threatened and Endemic Birds of Cuba. The real point of the 2001 fair was to assist their conservation by the raising of £135,000! DIMW.

Epilogue

The unseen rot - conservation challenges – future fields of amateur birdwatching –
a view from America - last musings and fervent wish – two ornithological rockets –
the return of my Swallows

I ended my review of birdwatching in the 1970s with some wince-making pontification. In retrospect, the most unjustified thought was that there were still enough birds in Britain to go round their increasingly abundant devotees. As the doyen of farmer-ornithologists Michael Shrubb once memorably shouted at Baroness Young, the bird-repellent practices in current agriculture gathered force in the mid-1970s and no act since then has reversed their effect. Actually the post-war renaissance in farming had done no more than restore the agricultural grasp of land to its 1875 position but it became subject to unprecedented politico-economic imperatives. Hence the flood of agrochemicals and the rise of horizon-blocking machines. Wildlife can adapt to change but only at its own pace; it has coped badly with the new technologies, which through increasing monoculture have reduced the threads of diverse life-support and left them less interwoven.

Blake's green and pleasant land seems still to exist – and is still gazed at lovingly by the three in four people who resort to the countryside for relief from stress – but the old, more natural richness of its fabric is tattered. Recently, other contagions of paucity have affected woodland and even the near wildernesses of northern Scotland. Consider just one example, the appalling equation of 20 million lost elm trees *contra* 31 million (and counting) cars, the latter for all their recent 'cleansing' still essentially pollutants on wheels. Not all the ill colours stem so directly from a nasty little beetle and our demand of mobility, but, given the general impoverishment of the environment, even the recovery potential in exceptional breeding success can be frustrated by yet another cold wet spring. We face the ends of many tethers.

I once asked Richard Porter, "Why did the RSPB not spot the overall decline in many common birds earlier?" He answered, "Maybe we spent too much time in (then smoke-filled) rooms worrying about our supposed special charges and the birds of prey which were in serious trouble". Happily the last tribe has done relatively well but such flagship successes for the conservation movement (or from the remaining full food chains) are no salve to the general wound and its conflicts. Another of Richard's memories of perplexed RSPB discussions came from 1970. It centred on how to keep Tree Sparrows out of nest boxes erected for tits. Today the plight of the sparrows is a main signal of how modern agriculture allows less and less living-room for field species. Enough wailing at the wall from me, however; let me end this homily with David Attenborough's definitions of the five forces of extinction. He saw these as over-harvesting, pollution, islandisation (the pocketing of populations), habitat restriction and change, and alien introductions. Are we to be guilty as charged, on all counts?

In the almost constant gale of wildlife crisis, the magic carpet of the thinking,

"A progressively grottier, less interesting place". Professor John Lawton, Chief Executive of the Natural Environment Research Council, describing the planet Earth's future state to Sir David Attenborough during the BBC series 'The State of the Planet'.

In contrast to that of the Canada Goose, the introduction of the Mandarin appears to be faultless. A tree nester, it found an unoccupied niche and after 250 years of hesitancy, British birds have recently increased and spread. At 7,000 birds, their number represents about a tenth of the wild Far East population.

No palled recreation was provided by the Ivory Gull that tried to winter at Aldeburgh, Suffolk, in 1999. Regrettably the local millennium fireworks scared it away and frustrated many birders keen to have it on their 2000 year list!

"One day I expect there will be a gadget that you point at a bird and which then instantly analyses its DNA and brings up the species name on a screen!" James Ferguson-Lees, *in litt.* (2000)

"I can envisage the day when all (British) bird-sounds will have been digitised on matchbox-sized computers coupled with highly sophisticated listening equipment, all to facilitate instant identification. But where the hell would be the fun in that?" Mike Rogers, *in litt.* (1999)

doing naturalist may often be buffeted but there is no good reason for it not to fly on. The half-millennium of truly amateur commitment to birdwatching has produced a wonderful tradition far from culmination and *jubilate Deo*, the taking of personal solace and fun from birds is neither a force of extinction nor a palled recreation. In an attempt to foresee where purposeful birdwatching might go in tandem with professional ornithology and conservation, I looked again at the comments of my 44 questionnaire respondents and 12 other correspondents aware of my book's texts.

Taken altogether, these suggested that *already* the study priorities of the purposeful birdwatcher have changed, particularly since joint conservation initiatives took hold in the late 1960s and when new diagnostic technology revolutionised systematics during the 1990s. Thus, however interesting and contentious, what Witherby called Taxonomy, Character and Allied Forms are now seen to have been taken over by a closeted tribe of near monopolists who speak in the strange, private tongue of geneticists. If avian systematics are again to attract worthwhile amateur efforts and not just chatter, the role of the field observer needs more detailed new instructions than have so far been offered. Within the associated subject of Linnaean and English Nomenclature, the rash of changes in the 1990s still causes heads to shake. Except where a new taxon is firmly positioned within its generic or familial radiation, enough is regarded as more than enough.

Intriguingly for Field Identification and (Plumage) Description, such has been the advance in optics, books and rarity access that some boredom shows. The multiple reviews of difficult groups are seen as caveat-ridden but only a small minority want a debate on whether observers should continue to squint over ever more minute details or step back and look again at the whole beings and for their own signals. Some concern is also expressed over the lack of transparent accountability in the modern system of record review, criteria setting and claimant vetting, but as ever, the question of who guards those who guard is likely to be begged. Much more sensible is the healthy recognition that in purposeful birdwatching and field ornithology, the over-touted issues of rarity fact or politic are totally irrelevant. On Voice, my circle of commentators sounds distinctly muted notes. Without the full apparatus of 'dustbin lid' reflector, high-tec recorder, CD and sonogram, the human ear seems to have had its day but, to introduce one personal thought, when were song-periods last checked? Of the studies formerly covered in Habitat, Food and Display and Posturing, there was no mention. The apparent lack of future interest in the latter two is a sad reflection of how diminished amateur interest is in behaviour, no longer the stuff of endless notes to *BB* and yet still the daily source of so much observed pleasure.

In the monitoring of Breeding, the recognition of the importance of amateur recording remains almost universal. The BTO's Nest Record Scheme is a particularly crucial database and participation confers real skills in observation and field-craft. Other recent initiatives in the measurement of both breeding and wintering communities deliver the same benefit to surveyors. Thus the Breeding Bird and Winter Farmland Bird Surveys have attracted thousands of responses and are securely set into the conservation calendar. On Distribution within the British Isles, both local and national atlasing are expected to attract fresh surges of enthusiasm. Appealing to both hunter and gatherer, atlasing has been the most widely enjoyed of the new traditions. To this work might be added, however, a modern review of the endemic British and Irish subspecies which should attract combined enthusiasm from observer and ringer. Taking tubed skins to live captures is not difficult and it would be fascinating to continue monitoring the genetic differences already noted along the western bulwark of Eurasia.

Who today would write a note to *British Birds* about a Nightingale seeing off a Cairn Terrier?

For Migrations, the enjoyment and recording of bird movements, still visible in Britain for nine months in every year, are expected to continue. Associated with this forecast was the hope that the broader resumption of migration studies would become the resting place of the rarity chasers, so helping them to build self-found lists of much more than collection value. The BTO's current Migration Watch and its masterly *Migration Atlas* may together prove to be the overdue catalysts of re-investigation into a still-miraculous phenomenon. Bill Bourne was particularly keen to see the full half-century of migration data given full phenological review, while the practicality now exists to follow the intellectually delicious connections between range changes and vagrancy theory across western Eurasia with more research and even tests in simultaneous observation. The opportunities in co-operative coastline surveys and on- and off-shore seabird sampling remain open to all. From 2002, the BTO's harvest by internet of migrant arrival dates has pointed the way to instant audience participation. The hope that the bird observatories would be refreshed and take up again their training task remains to be achieved. On Distribution Abroad, the services of the tour companies and the sharper thrusts of expeditions should be put together to strengthen the now necessarily international chains of conservation responsibility. The fledging flock of international societies and the BOU list series has opened new conduits but, to add another personal thought, perhaps a review of Birdlife International's Important Bird Areas and the increased direction of eco-tourism to them would make sense. Travel spend goes round seven times faster in the native communities than any other monies, and expressed wonder at the natural world is contagious.

Outside the agenda of the Witherby *Handbook*, the concerns of my circle of correspondents on current ornitho-politics are varied and most have been addressed in my earlier chapters. What bears repetition here are the twin thoughts that local reports should adopt a more standard presentation of data, so facilitating their intake into the national analysis of bird fortunes, and that the ornithological year should not start and end in winter. 1st March to 28th or 29th February? That debate could go on for ever. The views of the 56 correspondents given above show how the studies of the amateur birdwatcher and field ornithologist have narrowed and where they are likely to be concentrated in future. Some individual forecasts of change were more severe. Interrupting work on her PhD, Jane Reid reminisced rhapsodically on impressing male twitchers, adding a frigate bird to the West Palearctic list and working on Fair Isle as an assistant warden. Nevertheless she still felt that the years of the individual amateur observer being able to contribute significantly to ornithology were numbered. Worried by Jane's comments, I decided to solicit a last critique of British birdwatching from one of its most precocious post-war apostles. So I approached Dr. I .C. T. (Ian) Nisbet and from his impartial constituency of Massachusetts came a typically incisive reply.

Viewed from afar, the distinctive feature of Britain is its location at the northwest fringes of the European landmass, and in the absurdly warm northeast Atlantic. Hence, the truly distinctive features of British birds are related to this location – subtemperate breeding seabirds, wintering waders, ducks and landbirds that ought to be much further south, transoceanic migrants from Iceland, Greenland and Canada, migrant pelagics, eastern vagrants far outside the main stream of European migrants. Britain's distinctive habitats – Atlantic cliffs and islands, outer continental shelf, machair, turloughs, subalpine lakes and moorland, sandy heaths – are also related to its location on the maritime fringe of the land mass.

Hence, I think the interesting parts of Britain for birdwatching or field ornithology are the north and west – Scottish Islands, west Highlands, western Ireland, and extreme southwest corners of Wales and Cornwall …

In light of this, I am not very impressed by the great bulk of birdwatching or field ornithology that has been done in Great Britain since the 1960s. I share the impatience with re-creational twitching, which is fun to do once in a while, but has not produced anything significantly new for 30 years. I have been very disappointed with the output from the Bird Observatories, especially those on the south and east coasts, which ought to have been able to come up with some useful information about migration, but haven't (it took radar to demonstrate that British birds migrate SSE-NNW, unlike nearby continental birds). The most productive Bird Observatories have been those in the northwestern fringe – Fair Isle, Cape Clear, and St. Agnes – probably because these were off track of real migration. However, even these have yielded only two really new ideas about migration (Williamson's downwind drift and Rabøl's mirror-image misorientation, neither of which has held up well under tests with radar or orientation cages) …

The real, neglected opportunities for ornithological exploration in the British Isles are nearly all in the north and west – breeding seabirds, migrating seabirds, migrants from Iceland and Greenland, wintering wildfowl and waders from Iceland and Greenland, continental vagrants at the western fringe, and fringe breeders in relict habitats such as Red-necked Phalaropes, Dotterel, Black-necked Grebes, Greenshank, Corncrakes, etc.

One of the few largely unexplored areas of European ornithology is the seabirds of the continental shelf south and west of Britain. A lot has been learned by seawatching from headlands in Ireland and Cornwall in recent years, but our experience on the Atlantic coast of North America is that what you see from headlands in onshore winds bears little resemblance to what you see when you explore offshore waters by boat in normal weather. Nowadays, there are dozens of boat trips every day in the summer off the eastern USA to look at marine mammals and/or marine birds. I think that there is real opportunity to do similar trips out of such places as Dingle, Baltimore, Falmouth, Brest, Quiberon, Santander, La Coruna,

Most of the local Song Thrushes of the Outer Hebrides left in late summer to return in February, according to the Witherby *Handbook,* but nowadays some certainly remain all year and others go to Ireland, according to *BWP.* Do none reach the British mainland? The sketches are of a bird on Mingulay on 13th June 1994.

In the wet afternoon of 31st May 1997, a Red-necked Phalarope flew down to inspect the wake of a Minke Whale, off Glas Eilean Mor, North Uist.

On Rossan Head, Donegal, Wheatears of the nominate European race breed. They and others from Iceland and Greenland also appear on migration. Here among the autumn peat cuttings is at least one Greenlander (bottom left). The tracks of the last's spring migration remain a choice puzzle.

"One cause may be the lack of truly interesting and imaginative writing which can appeal to the whole birdwatching community and is not merely devoted to rarities. Who knows what opportunities might eventually arise should a sort of ornithological Messiah choose to start putting his message over the Internet, some down-to-earth Bill Oddie backed up by the right style of camerawork, unfettered by commercial editing?" Mike Rogers, *in litt.* (1999)

Lagos, or Cadiz. I know that weather and sea conditions are not as favourable in the eastern Atlantic as they are in the west, but that should add to the challenge. As in Canada and the USA, there must be lots of fishermen with idle boats who would welcome some extra income taking crazy birdwatchers or whale-watchers out for the day …

The other big unsolved problem of west European ornithology, in my mind, is the spring migration of the Greenland Wheatear. This is one of the most remarkable landbird migrations of the world, but has never been addressed by anyone except perhaps David Snow (*Ibis* 1953). Little enough is known about the autumn migration of the Greenland Wheatear, except for some early work by Ken Williamson and scattered observations of puzzling small numbers arriving in Scotland and Ireland. Departures from Greenland and Canada in autumn have never been studied. But the spring migration (against the prevailing winds and with little chance of finding Iceland for stopover) is even more remarkable, and I don't think has ever been documented in any way. We don't even know whether Wheatears depart from Morocco, Portugal, Spain, Ireland, Scotland, or Norway. Here, I think, is a big challenge for inspired exploration – one of the last.

There will be many who would accredit British birdwatching and field ornithology with much wider achievements than those on which Ian Nisbet advises greater concentration. His thoughts are nevertheless salutary in their indication of how adventurous amateur observers could still make major contributions to avian science along and within the distinctive bulwark to Eurasia that their British and Irish homelands form. Ian may yet be impressed by the recent successful exploration of Barra in the autumn of 2002 (*Birding Scotland*, 5 (4), 2002) and the budding (at last) promise of pelagic trips off Lyonnesse. In the former, a team of

Scottish birders found again an astonishingly high incidence of so-called rarities. In the latter, a small rush of southern ocean species now includes the seemingly impossible Brown Skua from the Antarctic!

Moving back once more to the general themes discussed by my correspondents, I must note again their almost universal witness to the more aesthetic and emotional benefits of birdwatching and their full expression in the pre-1975 literature. This suggests that the quality of modern interpretation is less good. Is this why the television audience in armchairs remains double the number of people who actually go out to be graced by an actual brush with birds? Perhaps the new multi-media, once fully interactive, will add impelled experience to static dreaming.

The sheer grace of flying beings remains a motivation for amazement and pleasure. Here some Common Terns inspect the Cramond Estuary, Mid-Lothian, for fish.

I often think that I belong to the most blessed and accursed of British birdwatching generations. For over 60 years sustained as much by birds as any other beings, I find it hugely frustrating that they are increasingly at risk at a time when, work over and pensions more or less secured, I had thought to revel in them most. Birds will still see to my lack of idleness but I was not prepared to worry about their fates every day. Nevertheless, I wish fervently that this book will have shown you how much the amateur, through delight, purpose and generosity, has already given, and can still offer to birdwatching, field ornithology and conservation.

On my first visit to Fair Isle in September 1951, I suffered from the schizophrenia still likely to grip any hunting and gathering birdwatcher at a magic place for birds. In the aftermath of a migrant fall, I dragged the isle's then Laird, George Waterson, half a breathless mile to look at an Aquatic Warbler. He took one look at it and pronounced it a young Sedge, adding with a stern smile "They're not all rare, laddie". A few days later, the prevailing zephyr had removed most migrants and I was put to the extraction of Starlings from a battery of Potter traps (small wire cages with trip-fall doors) set on a maggot-heaving bank of seaweed. I got carried away with speeding the unions of birds with rings. Sensing my incomplete examinations, Ken Williamson opened the laboratory window and firmly enjoined, "You're forgetting to weigh the re-traps; you must measure their weight gains". In the heat of going for a record score of captures, I had forgotten one of the two chief scientific purposes for making them.

Fifty years on, I would not be duped by a young Sedge Warbler but I still flinch from every last piece of science. Like most of my like-minded friends, birdwatching has been for me a hobby, a recreation, even indeed a freedom, not an employment and until recently not a bureaucratic chore. If in this book I have tapped the ankles of birdwatching's modern establishment and criticised some of its manners, it is only because I feel very strongly that such mores should not cloud the wonderful estate of experience and reverie that purposeful birdwatching bestows. At its simplest and best, this arises from the daily chance to meet with the other most visible beings of a shared planet and to be posed another question by them. Nowadays the public responsibility to provide some common answers in the cause of conservation is over-riding but as my peers' performances attest, it is the precious joys, even triumphs of personal perception that lead to lifelong private enthusiasm.

At 13:57 hrs on 8th April 2002, a Swallow landed on the same inches of wire over my flower garden that last year's cock bird favoured as his central song-post. Barging through the house, I exclaimed to my other Swallow (my second wife Wendy's pet name), "He's back. The cock Swallow, singing now, on the wire". With a degree of scepticism that would have done credit to the Rarities Committee, she countered, "How do you know? It could be his son". Bins out, long stare. Curses, Moriarty, the tail streamers do look a bit short for our veteran male and is the song less vehement? May be it is a son and not old Dad but never mind, one of the ten that went from our Staffordshire home to South Africa has made it back. Fingers crossed for the ninth passing-on of genes under our watch…

So what happened to the old boy? He took his time, appearing at 08:10 hrs on the 19th. His tail streamers were impossibly long and I could swear that as I peered at him, he winked back. Two hours later, he started to sing … At 11:30 hrs, a hen appeared. Dawn on the 20th lit up a pair, visible from bed, and soon bonded, they were pair-flying and inspecting our three nest sites. By the 22nd, the odds are on the rafters of the store … And by the end of the summer, not one but two pairs had got off four broods and at least 16 young.

No Swallow nor any other bird disappoints.

The pair of Swallows that enticed me into buying Mount Pleasant Farm, Anslow, Staffordshire in May 1994. Their genes flow on … so far.

Envoy

For my life-long enthusiasm for wild places and their birds, I owe a whole team of inspiring mentors who somehow found me worth their time.

The first was Dad with whom I enjoyed "an exceptionally close bond", as Kenneth Williamson noted, whether our time together was spent by the sea or up the hills. His feet always impatient for the next summit, he marched me into Scottish wilderness. As my need for bird knowledge grew, it was also he who supplied the crucial books and early tutors.

Among letters from the last, the one that I treasure most came from the "Good Ladies" of Scottish ornithology. It gave thanks for my school's records, news of birds and a great Highland birdman and wishes for good hunting.

William Jackson Wallace 1906-58 on the South Haven seawall, Fair Isle. September 1951. Clan motto: "For Freedom".

> Balcruvie
> Upper Largo
>
> Dear Ian,
>   You have sent us a most interesting series of observations on the Waxwings & we are most grateful. Thank you very much for writing so fully, it is such a help to get data like yours. We have had a very good response to our request for information; we have notes from all parts of Scotland from Shetland to the Nith. There are none from any of the Hebrides except from Skye where Mr Seton Gordon tells us Waxwings have been seen. In 1946 we got lots of notes of flocks of hundreds, this time we have only one flock of 100 + one or two of 60
>
> but lots of notes of little parties of 3-10 or so. So you see your numbers are the largest we have. So far we have not seen any here & have very few records from Fife at all, but we saw Northern Gt Spotted Woodpeckers at the East Neuk & have had them (or the British) in the garden here off & on all autumn. Thank you all again very much & with good wishes for a happy Christmas & good hunting in 1950.
>   Yours sincerely,
>   Leonora Jeffrey Rintoul
>   &
>   Evelyn V. Baxter

If only committees would write as they did!

Selected References and Better Reads

The classic culmination of a good bird book is a detailed bibliography. I did not, however, set out to write a tome. So while the following list does contain some expanded references, it is mainly a selection of the bird and other books that have illuminated my hobby over six decades and should also light your way to fulfilling and hopefully purposeful birdwatching. For other suggestions and good advice, join the nearest Members' Group of the RSPB or your county bird club. Your town library should have the latter's reports and if not, *The Birdwatchers' Yearbook* will have the missing clue!

General Natural History

Durrell, G. (1983). *The Amateur Naturalist*. Random House Inc., London. An invitation from a charming interpreter to "greet the natural world with curiosity and delight" and an antidote to the sometime tunnel vision in birding.

Fraser Darling, F. (1947). *Natural History in the Highlands and Islands*. Collins, London. Young farmer, educated biologist, sensitive observer and humanist, Fraser Darling never wrote a bogus word; instead, he gave us in just one book "the *real* essence of Scotland's land and sea". It remains my favourite inspiration.

Phillips, A. (1963). *The BBC Book of the Countryside*. BBC, London. An anthology of the seductive narrative pieces from the much missed *Countryside* programme.

White, Gilbert (1788, annotated by Richard Kearton in 1902). *The Natural History of Selborne*. Cassell, London. All of the 'Hussar parson's' letters and 123 evocative photographs of Cherry and Richard Kearton; still a treasure trove.

General Ornithology

Attenborough, Sir David (1998). *The Life of Birds*. BBC. The famous TV series reprised in write and glamorous photographs.

Fisher, J. and Peterson, R. T. (1964). *The World of Birds: A Comprehensive Guide to General Ornithology*. The grandest product of a famous trans-Atlantic partnership and still a fount of broad knowledge and fascinating detail.

Birdwatching Primers

Cromack, H. A. and D. (2002). *The Birdwatcher's Yearbook and Diary 2003*. Buckingham Press, Peterborough. The 23rd and much improved issue of the birding Baedeker gives astonishing service to beginner and expert alike.

Hume, R. A. (1992). *Discovering Birds*. RSPB/Duncan Petersen, London. It leads you "through any kind of countryside, to all the interesting bird species", using pointers on 100 landscape photographs.

Moss, Stephen (2002). *How to Birdwatch: A birdwatcher's Guide*. New Holland, London. A new "beginner's book" full of practical advice and personal touches.

Wallace, D. I. M. (1979). *Discover Birds*. Whizzard Press/André Deutsch, London. Still "the best introduction to a marvellous hobby and a life without boredom", says Bill Oddie.

Identification Guides and Handbooks

Beaman, M. and Madge, S. (1998). *The Handbook of Bird Identification for Europe and the West Palearctic*. Christopher Helm, London. With helpful introductions to the systems of bird identification and all families, this précis of the characters of almost 900 species should have earned a 'first reference' position on many bookshelves.

Cottridge, D. and Vinicombe, K. (1996). *Rare Birds in Britain and Ireland: A Photographic Record*. HarperCollins, London. Beginning with an imaginative discussion of vagrancy (itself essential reading), the book catalogues all the past and current 'official' rarities that were secured on film.

Grant, P. and Mullarney, K. (1989). *The New Approach to Identification*. Privately published and earlier in *Birding World*. The booklet that took "identification much further than just naming the bird" by showing that through new optics, primed eyes could see almost as much as the museum worker.

Harris, A., Tucker, L. and Vinicombe, K. (1989). *The Macmillan Guide to Bird Identification*. Macmillan Press, London. But for her early death, Laurel Tucker would undoubtedly have taken the distaff lead in the modern development of identification but at least her concept shone in this excellent selection of illustrated papers on 'difficult' species.

Harrison, P. (1987). *Seabirds of the World: A Photographic Guide.* Christopher Helm, London. The "pocket-sized companion" to Peter's initial odyssey after seabird identities – *Seabirds: An Identification Guide* (1983) – and the first guide to offer photographs of nearly all seabirds.Hayman, P., Marchant, J. and Prater, T. (1986). *Shorebirds: An Identification Guide to the Waders of the World.* Croom Helm, London. A "landmark volume", wrote Roger Tory Peterson, knocked out by the 1800 figures in 88 astonishingly detailed plates, and still the standard by which other similar family guides are judged.

Jonsson, L. (1992). *Birds of Europe, with North Africa and the Middle East.* Christopher Helm. London. Wonderful paintings, astute texts in the only guide that sits as well on the bedside table as anywhere in the field.

Lewington, I., Alström, P., and Colston, P. (1991). *A Field Guide to the Rare Birds of Britain and Europe.* HarperCollins, London. Self-taught, Ian Lewington's facility with bird illustration is already legendary; the accompanying text can be difficult to mine but still an essential reference.

Mitchell, D. and Young, S. (2nd edition, 1999). *Photographic Handbook of the Rare Birds of Britain and Europe.* New Holland, London. A brisk and profusely illustrated exposé of 284 so-called rarities, containing useful European status notes.

Mullarney, K., Svensson, L., Zetterstrom, D. and Grant, P.J. (1999). *Collins Bird Guide.* HarperCollins, London. Fifteen years in the writing and painting, it now vies with 'Jonsson' for first place in its marketplace. A model of disciplined presentation.

Peterson, R.T. (1934). *A Field Guide to the Birds.* Houghton Mifflin, Boston. "A 'boiling down', or simplification, of (field characters) so that any bird could be readily and surely told *from all the others* at a glance or at a distance" was the aim of the first American master of bird identification; still worth study in order to understand the art of holistic perception.

Peterson, R.T., Mountfort, G. and Hollom, P.A.D. (1954). *A Field Guide to the Birds of Britain and Europe.* Collins, London. The book that brought Peterson's skills across the Atlantic and led the genre of field guides for nearly four decades. The text of the 5th edition still competes.

Sandars, E., (1927). *A Bird Book for the Pocket.* Oxford University Press, Oxford. Now a curio but still worth more than a glance for the charming 'nutshell' perceptions in the texts.

Sharrock, J. T .R. ed. (1980). *Frontiers of Bird Identification: a British Birds guide to some difficult species.* A re-presentation of 29 classic identification papers or British perceptions from 1960 onwards; a model text for future authors.

Svensson, L. (4th edition, 1992). *Identification Guide to European Passerines.* Privately published, Stockholm; available from BTO. This trusted précis of passerine diagnoses has not been far from any ringers' hands since 1970.

Wallace, D. I. M. (1983). "Undiscovered field characters", a letter to *British Birds* (76: 318-319). An analysis of 10 gaps in the litany of field characters and bird behaviour, written halfway through the preparation of the *BWP* texts.

Ornithological Dictionaries

Lockwood, W. B. (1993). *The Oxford Dictionary of British Bird Names.* Oxford University Press, Oxford. The satisfaction of "a certain curiosity about the names we give our birds, for they are often so puzzling".

Richards, A. J. (1980). *The Birdwatcher's A-Z.* David & Charles, Newton Abbot. Still helpful as a selection of birdwatching jargon and ornithological terminology.

Thomson, Sir A. Landsborough, for British Ornithologists' Union (1964). *A New Dictionary of Birds.* Nelson, London and Edinburgh. The third great dictionary, "undertaken as a service to ornithology", "a comprehensive book of reference, in a world context, for all who are interested in birds".

Habitat Descriptions

Fuller, R. J. (1982). *Bird Habitats in Britain.* T. & A.D. Poyser, Calton. The book that turned the Register of Ornithological Sites into a brisk description of British habitats and their avian communities. Donald Watson's drawings are a joy.

Fuller, R. J. (1995). *Birdlife of Woodland and Forest.* Cambridge University Press, Cambridge. "The gamut of British Woodland" and its birds is enthusiastically explored; a passport to a quiet habitat shunned by too many birdwatchers.

Nicholson, E. M. (1951). *Birds and Men: The Bird Life of British Towns, Villages, Gardens and Farmland.* Collins, London. A reprise of much of Max Nicholson's early work on "the impact of civilisation on our bird life".

Rackham, O. (1994). *The Illustrated History of the Countryside.* Weidenfeld and Nicholson, London. The "multi-layered story of the British landscape" and the changes brought about in it by Britons, told with expertise, passion and humour and illustrated with great imagination.

Simms, E. (1979). *A Natural History of Britain and Ireland*. Dent, London. Lyrical descriptions of 15 areas of Britain full of historical and personal asides.

Ornithological Atlases

Gibbons, D.W., Reid, J.B., and Chapman, R.A., for the BTO, Scottish Ornithologists' Club and IWC (1993). *The New Atlas of Breeding Birds in Britain and Ireland: 1988-1991*. T. & A.D. Poyser, Calton.

Lack, P., for the British Trust for Ornithology and Irish Wildbird Conservancy (1986). *The Atlas of Wintering Birds in Britain and Ireland*. T & A.D. Poyser, Calton.

The greatest creations of the British and Irish "hunter-gatherers" of bird distributions, with maps of course, but also species notes, population estimates, conservation comments and wonderful drawings.

National and County Avifaunas

Ballance, D. K., (2000). *Birds in Counties*. Imperial College Press, London. An ornithological bibliograpjhy for the counties of England, Wales, Scotland and the Isle of Man, filling in a 70-year span in national data reference.

Lovegrove, R., Williams, G. and Williams, I. (1994). *Birds in Wales*. T. & A.D. Poyser, London. The third and most needed of the new national avifaunas in the Poyser stable gave the Principality an authoritative statement at long last.

Mather, J. (1986). *The Birds of Yorkshire*. Croom Helm, London. The last of county bird books written in the "grand (discursive) manner" of earlier centuries. A real read.

Taylor, M., Seago, M., Allard, P. and Dorling, D. (1999). *The Birds of Norfolk*. Pica Press, Robertsbridge. The fourth great book about the birds of the brave place which has provided acute observations since the 17th century. Excellent essays precede full species accounts.

Venables, L. S. V., and Venables, U. M. (1955). *Birds and Mammals of Shetland*. Oliver and Boyd, Edinburgh. A forerunner of the modern genre of 'county books'.

Wilson, A., and Slack, R. (1976). *Rare and Scarce Birds in Yorkshire*. Privately published. The complement to John Mather's magnum opus takes over 32,000 records of almost 300 taxa and turns them into precise lists and fascinating histograms of occurrence patterns. A good book from the computer age.

Lifelong Reference Books

Bannerman, D. A. (1953-63). *The Birds of the British Isles, Vols. 1-12*. Oliver and Boyd, Edinburgh. A "marvellous feat of sustained application: erudite, expansive and thoroughly enjoyable to read", the perfect antidote to the read-bites of today. Stroll through his essays and wonder at the doings of pioneer ornithologists.

Cramp, S., *et al.* (1977-1994). *Handbook of the Birds of Europe, the Middle East and North Africa: The Birds of the Western Palearctic, Vols. 1-9*. Oxford University Press, Oxford. "Monstrous", in the opinion of Andy Richford, but *BWP* is well worth mining in a library if my earlier chapter on it does not tempt you to buy it.

Snow, D. W., and Perrins, C. M. (1998). *The Birds of the Western Palearctic: Concise Edition*. Oxford University Press. The two-volume version of *BWP*, with superb new maps and many new plates, is the best "short" handbook of the birds of western Eurasia.

Witherby, H. F., Jourdain, F. C. R., Ticehurst, N. F. and Tucker, B. W. (1938-41). *The Handbook of British Birds*. H.F. & G. Witherby Ltd., London. A "work of real practical utility, not only to the professed ornithologist, but (also) to the beginner" 60 years ago and now.

Conservation Review

Mead, C. (2000). *The State of the Nations' Birds*. Whittet Books, Stowmarket. The 20th century fortunes and prospects of almost 250 breeding species follow a summary of BTO surveys and habitats.

Monthly Journals

Gantlett, S. (ed.) *Birding World*. The Bird Information Service, Cley next the Sea. Once called *Twitching*, this supremely topical, never late journal has matured into a thoroughly useful addition to the record of birding thrills and identification lore.

Riddington, R. (ed.). *British Birds*. BB 2000, London. Harry Witherby's creation has recently redoubled its efforts to be "the leading journal for the modern birder in the Western Palearctic". About half its pages present annual reports on fascinating subjects such as scarce migrants and rare breeding birds. The Rarities Report lacks interpretation but it is official. The rest is an excellent mix of professional and amateur papers and notes.

Monthly Magazines

Cromack, D. (ed). *Birdwatching*. Emap Active, Peterborough. Constantly refreshed, this magazine delivers more guidance to beginners and bird fans than any other; special features include bird-rich walks and the widest span of local reports in the topical press.

Mitchell, D. (ed). *Birdwatch*. Solo Publishing, London. More popular with experienced birdwatchers than *Birdwatching*, this magazine's letter columns and reviews are brisk and forthright. Its Artist of the Year awards leaven the glossy photograph diet of today.

Humour

McGeehan, A. (columnist). 'Total Birding', in *Dutch Birding*. Dutch Birding Association, Postbus 75611, 1070 AP Amsterdam. To live without Anthony's "crack" on people and brilliant observations on birds is lunacy.

Ornithological History

Fisher, J. (1966). *The Shell Bird Book*. Ebury Press and Michael Joseph, London. "Something of a rehash" of his earlier books, the author admitted, but still an amazingly broad and deep summary of the British interest in birds from their earliest fossils to those alive in 1964. The chapter on birds in literature, music and art is typical of Fisher's unconfined, sophisticated writing.

Holloway, S. (1996). *The Historical Atlas of Breeding Birds in Britain and Ireland 1875-1900*. T. & A.D. Poyser, London. The introduction and chapter on the late 19th century environment are packed with factual history and portraits of ancient scribes.

Raven, C. E. (1942). *John Ray, Naturalist, his Life and Works*. Cambridge. Prize-winning account of the first British naturalist of international importance who, although primarily a botanist, edited and improved our first major bird book begun by his pupil and friend Francis Willughby.

Tate, P. (1986). *Birds, Men and Books*. Henry Sotheran, London. A brisk readable crossing of the stepping stones of British ornithology with developed portraits of its major exponents.

Wallace, I. (1981). *Birdwatching in the Seventies*. Macmillan, London. A 'snapshot' of the decade that completed the launch of the modern bird-counting projects and the contemporaneous bird fortunes, with an early attempt to fix the numbers and describe the behaviour of birdwatchers.

Wollaston, A. F. W. (1921). *The Life of Alfred Newton*. Murray. Extensive account of the greatest figure in 19th century British ornithology, including comment on the foundation of the British Ornithologists' Union, The Great Auk and Dodo and early birdwatching and conservation initiatives.

Olympian Reviews

Burton, J. F. (1995). *Birds and Climate Change*. Christopher Helm, London. A review of the long term changes in European bird distribution related to climate fluctuations, it includes Ken Williamson's last essay on Atlantic passerine migrants.

Evans, M. I. (1994). *Important Bird Areas in the Middle East*. BirdLife International, Cambridge. Out from Girton Road, Cambridge, comes a stream of conservation-oriented inventories, of which this was the second.

H. R. H. The Prince Philip, Duke of Edinburgh, and Fisher, J. (1970). *Wildlife Crisis*. Hamish Hamilton, London. An example of how "the great and good" attempted to marshall support for the world's wildlife; Fisher's descriptions of the fates of 15 "paradises" have become even more haunting.

Mearns, B., and Mearns, R. (1998). *The Bird Collectors*. Academic Press, London. The history of the stocktaking of bird specimens for the great houses and museums of particularly the 19th and early 20th centuries is packed not just with facts and figures but also tales and human cameos.

Meinertzhagen, R., (1959). *Pirates and Predators*. Oliver and Boyd, Edinburgh. A wide-ranging "contribution" on the piratical and predatory habits of birds (and crocodiles, men, etc) from the enigmatic soldier-ornithologist.

Moreau, R.E. (1972). *The Palearctic-African Bird Migration Systems*. Academic Press, London. An awe-inspiring review of 5,000,000,000 birds moving between Eurasia and Africa, written by a modest but so enthusiastic man fighting his last illness. "No bum swan-song" indeed!

Walters, M. (2003). *A Concise History of Ornithology*. Christopher Helm, London. An international review of the 'founding figures' of the science.

Ripping Yarns

Atkinson, R. (1949). *Island Going*. Collins, London. Part-narrative, part-anthology, a precious evocation of the "remoter isles, chiefly uninhabited, off the north-west corner of Scotland", still inspiring other island-goers after half a century.

Chapman, A. (1889, republished 1990). *Bird-life of the Borders*. Spreddon Press, Stocksfield. A "classic muniment of observation and field work", vivid, accurate, loving. If only modern authors could write with such transparent enthusiasm!

Mountfort, G. (1965). *Portrait of a Desert*. Collins, London. The third of Guy's expedition tales with plenty of attention paid to scenery, Bedouins and the already rising tensions of the Middle East in 1963.

Scott, P. (1938). *Wild Chorus*. Country Life, London. Magical paintings and great tales from goose grounds between The Wash and the Caspian Sea.

Seebohm, H. (1901, republished 1976). *The Birds of Siberia*. Alan Sutton, Dursley. The excitements detailed of two Siberian traverses set the collecting of bird specimens and observations into their complete context of weather, habitat and Czarist Russia. The 20th chapter tells of a visit to Heligoland.

Williamson, K. (1948). *The Atlantic Islands*. Collins, London. A British birdwatcher was sent to guard the Faeroe Islands, took them, their people and animals to his heart and then wrote this perceptive portrait.

Bird Art and Photography

Brockie, K. (1993). *Mountain Reflections*. Mainstream Publishing, Edinburgh. A supreme example of a stalker-artist's vision and the perfect antidote to too many glossy photographs.

Fitter, R.S.R. *et al.* (c. 1950). *British Birds in Colour*. Odhams Press, London. The text is dated but for a few pounds you get 108 of John Gould's plates from *The Birds of Great Britain* (1873) in between photographs by Eric Hosking and George Yeates.

Flegg, J. and Hosking, D. (1993). *Eric Hosking's Classic Birds: 60 Years of Bird Photography*. Over 190 of the master photographer's unsurpassed black and white images of birds decorate a life story and pre-digital photographic advice.

Hammond, N. (1998). *Modern Wildlife Painting*. Pica Press, Robertsbridge. A masterly critique of the wildlife artists of the 20th century.

Scott, Phillipa (1992). *The Art of Peter Scott: Images from a Lifetime*. A fascinating display of drawings and paintings from over 70 years showing how early talent can be constantly enhanced by practice and how practice leads to perception.

Tunnicliffe, C.F. (1945). *Bird Portraiture*. The Studio, London. One of the 'how to do it' series and still an excellent primer for the would-be bird artist.

Nostalgia

Cocker, M. (2001). *Birders: Tales of a Tribe*. Jonathan Cape, London. Explains "what makes birders tick", said Bill Oddie.

Fuller, E. (1995). *The Lost Birds of Paradise*. A remarkable evocation of the heyday of plume-hunting, with asides on characters from Lord Rothschild to Errol Flynn.

Holloway, J. (1984). *Fair Isle's 'Garden' Birds*. The Shetland Times, Lerwick. A charmingly illustrated diary of the birds seen in Stackhoull Garden and others elsewhere on Britain's 'Heligoland' from 1977 to 1983.

Vaughan , R. (1998). *Seabird City*. Smith Settle, Otley. A guide to the breeding seabirds of the Flamborough Headland with an evocation of their human predators.

Singular Voices

Alexander, H. G. (1974). *Seventy Years of Birdwatching*. T. & A. D. Poyser, Berkhampstead. Fourteen chapters of personal birdwatching experiences and ornithological reporting from the model observer of the early 20th century.

Axell, H. (1992). *Of Birds and Men*. Book Guild, Lewes. The autobiography of a confident ringer cum conservationist who took a leading role in making the British change from "the world's worst bird-killers" to "the world's best bird protectors".

Barnes, S. (1992). *Flying in the Face of Nature*. Pelham, London. A "special place … full of special birds", the RSPB reserve of Minsmere held throughout 1990 the attention of one of Britain's most forthright environmental commentators. The result was a remarkable insight into the place and its birds and men.

Beer, S. (1995). *An Exaltation of Skylarks*. SMH Books, Pulborough. One hundred and seventy-three poems and pieces featuring the 'blithe spirit' or 'bird of bright infinity'!

Bourne, W. R. P. (1992). The Study of Bird Migration at Cambridge. *"75th" Anniversary Proceedings of the Cambridge Bird Club.*

Bourne, W. R .P. (1992). FitzRoy's foxes and Darwin's finches. *Archives of Natural History* 19: 27-35.

Bourne, W. R. P. (1999). Information in the Lisle letters from Calais in the early sixteenth century relating to the development of the English bird trade. *Archives of Natural History* 26: 349-368.

Three examples of the attention of Britain's most followed doctor-ornithologist – scourge to his chief delight, birds in all sorts of circumstances.

Hunt, D. (1985). *Confessions of a Scilly Birdman*. Croom Helm, London. The extrovert autobiography of a hedonist-birdwatcher who earned eventually both ornithological laurels and economic security in Scilly, only to be killed by a tiger while leading a bird tour party in India.

Jackson, K. H. (1951). *A Celtic Miscellany*. Routledge and Kegan Paul, London. For nature described with "vivid imagination and freshness of approach", the early Celtic literature is unmatched. This book contains 24 examples.

Lack, D. (1965). *Enjoying Ornithology*. Methuen, London. Twenty-two interesting essays and perhaps the first ever quiz (in 1949) from a master ornithologist who remained a birdwatcher at heart.

Monographs

Clarke, R. (1996). *Montagu's Harrier*. Arlequin Press, Chelmsford. The past, present and troubled future of a beautiful migratory raptor told by the current harrier sleuth who includes biographical details of George Montagu in the bird's early history.

Clement, P. (1995). *The Chiffchaff.* Hamlyn, London. One of the Hamlyn Species Guides which pack in much information to paperback format. Written before the latest 'splits' within the species group, it is still a model summation of a fascinating little bird.

Fisher, J. (1952). *The Fulmar*. Collins, London. Fifty years ago "probably the most complete study of an important species of a wild bird yet undertaken", drawn from 2,400 references and still considered to have set the modern standard for monographs.

Newton, I. (1972). *Finches*. Collins, London. The first of the New Naturalist reviews of a whole group of birds by their most accomplished student. A model introduction to finch biology.

Tinbergen, N. (1953). *The Herring Gull's World.* Collins, London. The mysteries of gull sociality unravelled objectively but described with Dutch charm. A model exercise in paying attention to another being.

Migration

Elkins, N. (1983). *Weather and Bird Behaviour*. T. & A.D. Poyser, Calton. A professional meteorologist's review of the phenomena of bird migration and other responses of birds to weather and climate.

Pashby, B. S. (1985). *John Cordeaux – Naturalist*. Spurn Point Observatory Committee. Due homage paid to the Lincolnshire farmer who organised the most productive migration watches ever, anywhere.

Thom, V. (1989). *Fair Isle: an Island Saga*. John Donald, Edinburgh. Valerie's completion of George and Irene Waterston's wish for a book telling the whole story of 'a lofty, precipitous rock, rising where two seas meet', tenanted by more than vagrant birds.

Wernham, C. *et. al.* (2002). *The Migration Atlas*. T. & A.D. Poyser, London. The movements of 265 birds of Britain and Ireland displayed in the light of over 500,000 ring recoveries and other disciplines of migration study and discussed in European and world contexts. An amazing distillation well worth the wait!

Index of Main Human Characters

A good index shows you the structure as well as the content of a book. This is not such a list, being biased by my particular interests and the frequent mentions of friends who assisted my study of them. For the main themes of the chapters, please see their headings.

143-4, 147, 159, 169, 220, 247-8, 254

Filby, Dick 189

Fisher, James M. 13, 15-8, 20, 24, 75, 81, 94-5, 123, 126, 150, 193-4, 204, 225, 236

Fitter, Richard A.R. 15, 54, 97, 188

Flumm, Dave 187

Fraser Darling, Frank 126, 195, 236

Frisch, J.L. 155

Foster, Stephen 179

Fuller, Robert J. 127

Gafrilov, Edward and Andrei 161

Garden, E. (Betty) 85

Gantlett, S.J.M. (Steve) 115, 248

Garner, Martin 104

Gatke, Heinrich 46-8, 223

Gesner, Conrad 224

Gibbons, David 128

Gibbs, Tony 241

Gillings, Simon 128

Gillmor, Robert 86, 118-9, 128, 206, 221

Glue, David E. 127, 130, 248

Golley, Mark 205, 219

Goodwin, R.P. (Derek) 98, 137, 233-4

Gosney, Dave 217

Gould, John 20, 39

Grant, Peter J. 94, 102-3, 127, 165, 188-9, 196, 219, 233

Green, Rolfe 248

Greensmith, Alan 169

Grey, Edward, Viscount Fallodon 75, 95, 199

Grieve, Andrew 172, 207, 230-1

Gurney, John 63

Hamilton, Frank R. 98

Hanson, Bruce 247

Harbard, Chris 213-8

Harber, D.D. 113, 136

Harris, M.P. (Mike) 195-7

Harrison, J.C. 83

Harrison, Jeffrey G. 106

Harrison, Peter 96, 194

Harrison, Tom 62, 65, 67

Hartert, Ernst 66, 68, 78-9, 116

Hartley, Rev. P.H.T. 123

Harvie-Brown, John 41, 47-8, 56, 199

Hayman, Peter 96, 119

Heinzel, Hermann 94

Hicks, Francis 165, 184, 188, 191

Hicks, Lewis and Alice 165, 181-2

Higgins, Rupert 95, 130, 209, 246

Hill, Brian 230-1

Holgersen, Holger 88-9

Hollom, P.A.D. (Phil) 7, 8, 67-8, 93, 97-99, 101, 106, 116, 122, 147-150, 171, 233

Holloway, Lawrence 175

Holloway, Simon 56

Holman, Dave J. 185

Holness, Paul 185

Hosking, Eric 143, 170

Hudson, R. (Bob) 118

Hudson, W.H. 95, 199

Hume, A.O. 39, 233

Hume, R.A. (Rob) 93, 204, 210, 213

Hunt, David 86, 184-5, 234

Huxley, Sir Julian 75, 170

Inskipp, Tim 165

Izzard, John and Sheila 132, 141-2

Jenner, Edward 46

Jennings, Michael 171

Jerdon, T.C. 39

Johns, R.J. (Ron) 133, 136, 142, 185, 188, 219

Jonsson, Lars 94, 96, 233

Jourdain, Rev. F.C.R. 64, 68, 72

Kay, John 15, 223-4

Kear, Janet 95, 140-1, 208

Kearton, Richard 155

Kermode, P. 48

King, Bernard 138, 185

Kinnear, Norman 51

Kirtland, C.A.E. (Colin) 132

Kist, Jan 222

Knebley, Rev. E.P. 48

Knights, Chris 249

Knox, Alan 111

Knystantus, Algirdas 145, 175-6

Koekkock, M.A. 92

Lack, David 62, 64, 75, 83, 86, 95, 113, 150, 240-2, 251

Lack, Elizabeth 127

Lack, Peter 127

Laine, Lasse 228

Landsborough Thomson, Sir Arthur 83, 156

Latham, John 17, 24, 38, 54

Lassey, P. Andrew 25, 90, 145, 161, 174, 178-9, 195, 230-1

Lawrence, T.E. 110-1

Lawton, John 102, 253

Leach, Elsie P. 156

Lemon, Margaretta 198-9, 210

le Sueur, Anthony 159

Linnaeus, Carolus 16, 24, 38, 224-5

Lodge, George E. 92-3, 100

Lockley, Ronald M. 157-8, 194

Longolius, Gybertus 224

Loseby, Irene and Tim 252

MacDonald, Duncan 216

Macgillivray, William 19-21, 24

Mackenzie, Peter 187-8

Marchant, John 129

Marr, B.A.E. (Tony) 246-8, 251

Martins, Rodney 171

Mayhew, H. 30

McDowell, Willie 179

McGeehan, Anthony 104, 135, 166, 209, 232

Mead, C.J. (Chris) 8, 32, 122, 129, 156, 160, 233, 252

Mearns, Barbara C. and Richard J. 111

Medhurst, H.P. (Howard) and Marianne 132-3, 141-2

Meiklejohn, M.F.M. (Maury) 80, 86, 88, 97, 232

Meinertzhagen, Annie C. 66

Meinertzhagen, Richard 19, 37, 107-12, 170, 188, 244

Merrett, Christopher 54, 224

Middendorf, A. von 46, 77

Millais, J.G. 91-2

Minton, C.D.T. (Clive) 132, 159-60, 174-5

Time to roost. Caspian, Baltic and Lesser
Black-backed Gulls leaving the winter flood-
plain of the River Dove, Staffordshire.

Printed and bound by CPI Group (UK) Ltd, Croydon, CR0 4YY

22/04/2026

02095084-0001